Handbook for the Academic Physician

Handbook for the Academic Physician

Edited by
William C. McGaghie and John J. Frey

With 37 Figures

Springer Verlag
New York Berlin Heidelberg Tokyo

William C. McGaghie, Ph.D.
Office of Research and Development
for Education in the Health
Professions
University of North Carolina
School of Medicine
Chapel Hill, North Carolina 27514
USA

John J. Frey, M.D.
Department of Family Medicine
University of North Carolina
School of Medicine
Chapel Hill, North Carolina 27514
USA

Library of Congress Cataloging-in-Publication Data
Main entry under title:
Handbook for the academic physician.
Includes bibliographies and index.
1. Medicine—Study and teaching. 2. Medical teaching personnel. 3. Medicine, Clinical—Research. 4. Communication in medicine. 5. Medical ethics. I. McGaghie, William C. II. Frey, John J. [DNLM: 1. Education, Medical. 2. Ethics, Medical. 3. Interprofessional Relations. 4. Physicians. 5. Research. W 21 H236]
R834.H33 1985 610'.7'11 85-14726

Typeset by University Graphics, Inc., Atlantic Highlands, New Jersey.
Printed and bound by Halliday Lithograph, West Hanover, Massachusetts.
Printed in the United States of America.

9 8 7 6 5 4 3 2 1

ISBN 0-387-96178-X Springer-Verlag New York Berlin Heidelberg Tokyo
ISBN 3-540-96178-X Springer-Verlag Berlin Heidelberg New York Tokyo

Dedicated to the memory of Merrel D. Flair
Our teacher, colleague, and friend

Foreword

This book is a bold and useful tool that provides the concepts, principles, and facts needed to build and to strengthen a career in academic medicine. Developing a high level of competency in academia requires the development of skills in addition to those in one's own specialty or discipline. One needs skills for conducting research, meeting administrative responsibilities, and educating students and colleagues. These skills are *not* bells and whistles. They are the elements of academic life that make the position truly academic. This book provides the critical information needed to succeed in that world.

Until now many academicians have learned about elements of their job outside their individual discipline by experience and through the observation of role models and mentors. In the complex, highly competitive, rapidly changing world of academic medicine there is no longer time for a prolonged apprenticeship. The institution is endangered when individuals are selected for critical posts based upon skills in areas that may not be central to the principal responsibilities of the new position. How often one hears:

"He is a great scientist but he runs his department with a shoe box mentality."
"She is a fantastic clinician, but she runs a committee as if she knows everything. I hate working with her."
"How can a full professor be such a lousy teacher?"

All of the above are symptoms of the need for special skills.

There is a science and literature about academic administration, medical education, and the process of conducting research. There are known concepts and principles in these areas that can yield outcomes of higher quality in a more efficient manner. The authors and editors of this text have distilled this information into a series of chapters that makes it possible for readers to develop their abilities as educators, researchers, and medical administrators.

Frederic David Burg, M.D.
University of Pennsylvania

Preface

This is a book for academic physicians, doctors who not only see patients but who also teach, conduct research, speak and write for scholarly audiences, participate in organizational governance, and act as examples of ethical professional behavior. Academic physicians fill multiple roles and respond to many different demands. Yet despite the complexity of their work, academic physicians are usually trained only for patient care. The discrepancy between professional expectations and the preparation of academic physicians is the reason why we wrote this book. Our goal is to provide instruction and insight about academic life for physicians who are planning or are engaged in academic careers.

The *Handbook* is intended to be a practical and comprehensive source of information about shaping a successful career in academic medicine. Academic success can take several forms, which is why the book addresses five nonclinical topics of interest: (a) professional behavior, including career management and how to function effectively in a large organization; (b) medical education and evaluation; (c) clinical research; (d) professional communication, especially writing and speaking; and (e) ethics in teaching and patient care. While we do not presume that the volume is a "bible," a panacea for the professional lives of academic doctors, we do believe the book contains useful guidelines for those who seek a career consult. Some issues like career management are treated in depth while others including statistics, clinical teaching, and computers are introduced briefly. However, throughout the book there are many ideas about how readers having special interests can deepen their knowledge of particular topics. We have been deliberately (and appropriately, we believe) liberal with citations and recommended readings, not to point out shortcomings in the *Handbook,* just its practical limits.

The book originates from a faculty fellowship program we have administered together for the past six years. The faculty fellowship aims to equip physicians who are new to the academic role—in our experience, family physicians—with the basic knowledge and skill needed for professional success. The five part fellowship curriculum is mirrored by the five parts of the *Handbook,* which underscore the point that academic physicians have multiple responsibilities in addition to their clinical work.

As administrators of the faculty fellowship program, even more as teachers of its clientele, we quickly grew aware of the need for teaching materials appropriate for our learners and our goals. Journal articles, several textbook chapters, and various worksheets and exercises were available from scattered sources. There was, however, no single book or series of textbooks that we could use as primary teaching material. Consequently, the *Handbook* was started to fill the immediate need for educational material for the faculty fellowship program. As the work proceeded, and as we discussed the project with colleagues in a variety of medical disciplines and settings, it became clear that there was a generic need for the volume throughout academic medicine. Thus while the book was originally conceived to meet the professional needs of academic family physicians, we believe its contents will be equally useful for physicians in such disciplines as general internal medicine, pediatrics, obstetrics and gynecology, psychiatry, and surgery. Subspecialty clinicians and physicians who work in administrative capacities will also find that parts of this volume address their responsibilities.

Work on a major project usually involves striking a balance between professional expectations and personal priorities. Writing and editing this *Handbook* has been no exception.

Professionally, we wish to acknowledge several organizations and a number of colleagues for their influence on the preparation of this volume. The Bureau of Health Professions of the Health Resources and Services Administration, U.S. Department of Health and Human Services, generously supports the faculty fellowship program whose curriculum is reflected in these pages. Grant No. 2-D15-PE54008 makes it possible for us to present that program. For permission to reprint copyrighted material we are indebted to the Association of American Medical Colleges; Gene W. Dalton; Marcel Dekker Inc.; NTL Institute; Planned Parenthood of Orange County, North Carolina; Prentice-Hall, Inc.; and Viking Penguin Inc. We are grateful to William D. Droegemueller, John D. Engel, Andrew Greganti, William N. P. Herbert, and Lawrence LaPalio for their critical reactions to early chapter drafts. Special thanks are due to Nancy Bruce and Linda Frank for bibliographic assistance, especially literature searching. To Millie Grace for clerical help and good cheer, hats off! Marty Hawks, in turn, receives a low bow for processing our words with skill and style.

Our spouses and children should be acknowledged because their charity gave us the quiet time needed to work on the book, or to procrastinate whenever we could. Pamela Wall McGaghie, Michael, and Kathleen were helpful in one camp while Julianne Hron Frey and Benjamin did the job across town. Thanks much, we needed a conscience.

Finally, like all honest authors and editors, we wish to express a disclaimer. To the best of our knowledge the material contained in the *Handbook* is fresh and new, at least in its current form. Factual and stylistic errors are inadvertent. While we accept responsibility for flaws, we ask readers to call our attention to them so that warts can be removed in subsequent editions.

William C. McGaghie
John J. Frey
Chapel Hill, North Carolina

Contents

Foreword
Frederic David Burg vii

Preface ix

Contributors xvii

I PROFESSIONAL DEVELOPMENT

Introduction 1

1 Professional Settings for the Academic Physician
David Dill 3

2 Managing the Role of the Academic Physician
David Dill and John Aluise 11

3 Advancement and Promotion: Managing the Individual Career
Stephen Bogdewic 22

4 Managing Academic Committees
David Dill 37

5 Academic Medical Organizations
John Aluise 54

Suggested Reading 69

II MEDICAL EDUCATION

Introduction 75

6 An Experience in Curriculum Development
Charles P. Friedman and Richard M. Baker 76

7 Clinical Instruction
Frank T. Stritter, Richard M. Baker, and Edward J. Shahady 98

8 Evaluation of Learners
William C. McGaghie 125

9 Evaluating Educational Programs
George B. Forsythe, James C. Sadler, and Ruth de Bliek 147

Suggested Reading 170

II Clinical Research

Introduction 173

10 The Role of Research in Primary Care Medicine
Peter Curtis 175

11 A Research Case Study
Peter Curtis and Jacqueline Resnick 184

12 Planning a Research Study
Carl M. Shy and William C. McGaghie 205

13 Conducting a Research Study
William C. McGaghie 217

14 Resources for Clinical Research
Jane E. Arndt 234

15 Research Data Management
Paul Gilchrist 251

Suggested Reading 277

IV Professional Communications

Introduction 279

16 Principles of Professional Communication
James W. Lea 281

17 Writing for Publication
James W. Lea 294

18 Techniques of Oral Presentation
James W. Lea 305

19 Visuals for Written and Oral Presentations
Madeline P. Beery 312

Suggested Reading 334

V ETHICS: TEACHING AND PATIENT CARE

Introduction 337

20 Ethical Decisions
Larry R. Churchill, Harmon L. Smith, and John J. Frey 339

21 Teaching Ethics
Harmon L. Smith, Larry R. Churchill, and John J. Frey 358

22 Case Studies in Ethics
John J. Frey, Harmon L. Smith, and Larry R. Churchill 373

Suggested Reading 387

Index 389

Contributors

Unless noted, all are affiliated with the University of North Carolina, School of Medicine at Chapel Hill.

John Aluise, Ph.D., Assistant Professor of Family Medicine

Jane E. Arndt, M.A., Research Associate in Family Medicine

Richard M. Baker, M.D., Professor of Family Medicine

Madeline P. Beery, M.Ed., Instructional Development Consultant, Office of Research and Development for Education in the Health Professions

Stephen Bogdewic, M.A., Lecturer in Family Medicine

Frederic David Burg, M.D., Professor of Pediatrics and Associate Dean for Academic Programs, University of Pennsylvania School of Medicine, Philadelphia, Pennsylvania

Larry R. Churchill, Ph.D., Associate Professor of Social and Administrative Medicine and Associate Professor of Religious Studies

Peter Curtis, M.D., Professor of Family Medicine

Ruth de Bliek, M.A., Graduate Fellow, Office of Research and Development for Education in the Health Professions

David Dill, Ph.D., Associate Professor of Education, School of Education

George B. Forsythe, Ph.D., Permanent Associate Professor of Behavioral Sciences and Leadership, United States Military Academy, West Point, New York

John J. Frey, M.D., Professor of Family Medicine

Charles P. Friedman, Ph.D., Associate Professor and Associate Director, Office of Research and Development for Education in the Health Professions

Paul Gilchrist, Ph.D., Assistant Professor of Family Medicine

James W. Lea, Ph.D., Associate Professor of Family Medicine and Director, Program for International Training in Health

William C. McGaghie, Ph.D., Associate Professor, Office of Research and Development for Education in the Health Professions

Jacqueline Resnick, B.A., Research Associate of Family Medicine

James C. Sadler, M.Ed., Research Associate, Office of Institutional Research

Edward J. Shahady, M.D., Professor and Chairman of Family Medicine

Carl M. Shy, M.D., Dr.P.H., Professor of Epidemiology, School of Public Health

Harmon L. Smith, Ph.D., Professor of Moral Theology, Duke Divinity School and Professor of Community and Family Medicine, Duke Medical School, Durham, North Carolina

Frank T. Stritter, Ph.D., Professor and Director, Office of Research and Development for Education in the Health Professions

I Professional Development

What knowledge and skills are necessary to be an effective academic physician? The answer is simple: clinical skill, ability to do research, and perhaps the ability to lecture clearly. But if the answer is so simple, why is the turnover rate in academic medicine so high? Why do so many new academic physicians become disillusioned and enter private practice? Part of the answer is the poor preparation for the role of academic physician that most doctors receive. In contrast, Ph.D. students in the arts and sciences have opportunities to develop their teaching skills as instructors and to practice their research skills in a dissertation. They often observe the life of a faculty member at close range and come to appreciate both the flexibilities of the role and its inherent ambiguities. Another part of the answer is the unique complexities of an academic medical center and its combination of pressures on an academic physician. An additional part of the answer is that physicians have choices, and can act in accord with their personal preferences.

These circumstances suggest that the conventional answer is true but insufficient. The role of the academic physician is substantially different from that of the physician in a solo private practice, group practice, subspecialty clinic, or major hospital, the roles for which the majority of physicians prepare. Certainly teaching and research are necessary skills, but also necessary is some comprehension of the academic setting and some elementary skills of career management. We have termed these the skills of Professional Development.

The organization of the Professional Development Section of this *Handbook* follows the natural career growth of an academic physician. The first chapter sets the stage by describing the characteristics of academic medical centers that influence the academic physician role and its activities. A new academic physician might become aware of these characteristics during the interview process, but their implications may not be obvious.

Chapter 2 dissects in greater detail the nature of the academic physician role and its activities. Because of multiple responsibilities and high pressure, professional stress can be a particular problem. Approaches to diagnosing and managing that stress are introduced.

The special characteristics of the academic medical center create ambiguities regarding professional responsibilities, performance evaluation, and career development. Chapter 3 suggests how to focus on personal interests, plan a career, and negotiate for the terms and conditions needed for effective performance.

Chapter 4 outlines skills appropriate to the work of academic committees. The nature of academic medical centers makes the committee an ubiquitous but necessary part of an academic physician's life. The work of these committees can be made both more efficient and effective if certain patterns and skills are learned. These skills are not the same as those needed for working on a medical team, and adjusting to the difference may take some time.

As academic physicians progress through their careers, and often early in their careers, an understanding of skills relevant to departmental organization and management becomes important. The fifth and final chapter introduces elementary skills of organizational development as they apply in academic medical departments. This is a complex subject which can be only briefly addressed here, but helps introduce the reader to a topic of increasing importance.

These are the skills of professional development. They begin with skills of psychological coping, progress through skills of negotiation and career development, address the issues of effective interaction in university governance, and close on some elementary skills of management and leadership. Taken together they are an important part of the skills necessary to be an effective academic physician.

1
Professional Settings for the Academic Physician

David Dill

The self-concept of the medical doctor contains archetypal memories of the individual professional. Although the vast majority of doctors carry out the practice of medicine as part of organized groups including private clinics, health maintenance organizations (HMOs), large group practices, and public health hospitals, the sense of independence and individuality is strong in the new medical graduate. Life in medical organizations and groups thereby presents new realities. These realities are most striking in academic medical organizations where new staff not only confront the obligations of organizational membership, but also confront the practice of new and untried skills in teaching and research.

To survive and prosper in academic medical settings requires skill in research, teaching, and clinical practice. It also requires skills in balancing professional and personal priorities, defining and negotiating one's career, and elementary skills of management. The necessity of acquiring these rarely discussed skills derives from the unique activities and structure of academic medical organizations. Therefore an overview of these settings is a necessary prelude to discussing the skills of professional development.

Academic Medical Settings

An academic medical setting is a multifaceted entity (1). Simply stated, the academic physician is a professional member of a medical school, but this understates the variety that exists among medical schools. An academic medical center (AMC), for example, is a complex organization made up of a medical school, one or more teaching hospitals, and affiliated research institutes. The traditional form of academic medicine is for the clinical faculty to undertake patient care, research, and teaching in the university-owned hospital. This model is changing. The medical center hospital may no longer be owned by the university. Or, as in "community-based medical schools," there may not even be a principal teaching hospital affiliation, but rather multiple affiliations with hospitals and other facilities. Even the more traditional medical schools in academic medical centers have developed affiliations beyond the medical center. These include area health education centers (AHECs), hospital consortia, and HMOs. Thus, over the past 20 years medical schools have become much more complex organizations.

Regardless of its affiliations, the foundation of a medical school remains the faculty and what the faculty does. The medical school is the place where the student and the faculty relate to each other in the education process, whether it be a traditional school, community hospital, or research laboratory.The faculty in a medical school are responsible for discovering and advancing medical knowledge and teaching this knowledge to students at the undergraduate, postgraduate, and subspecialty levels. Since the student learns by observing at the bedside and in the clinic, the faculty who teach clinical medicine also care for the sick. Although the various settings of academic medicine may place different emphases on these activities, new faculty members must adjust to these multiple requirements of their role.

Characteristics of Academic Medical Settings

The behavior of an academic medical professional is not dependent solely on the abilities and needs of the individual. Instead, performance and satisfaction are outcomes of the interaction between an individual and the characteristics of the academic medical setting. Doctors who may be outstanding clinicians in private practice may be ineffective or unhappy in a medical school because they lack the skills and values required for work in an academic setting.

A recent study of academic medical centers discovered five unique characteristics of these organizations (2–4).

Complex Roles

First, physicians who work in academic medical centers have very complex professional roles. The multiple tasks of education, research, and patient care within the medical school have a long tradition. The integration of these tasks followed the educational recommendations of the Flexner Report (5), which advocated the upgrading of medical education by merging these three tasks in the same organization. Of course, other organizations have multiple activities. A successful industrial organization, for example, conducts the multiple tasks of research, production, and marketing. What distinguishes professionals in an academic medical center from their industrial counterparts (and for that matter from doctors in private practice) is that academic physicians undertake teaching, research, and patient care simultaneously, as shown in Figure 1.1.

The academic reader may view the blending of these roles in the aca-

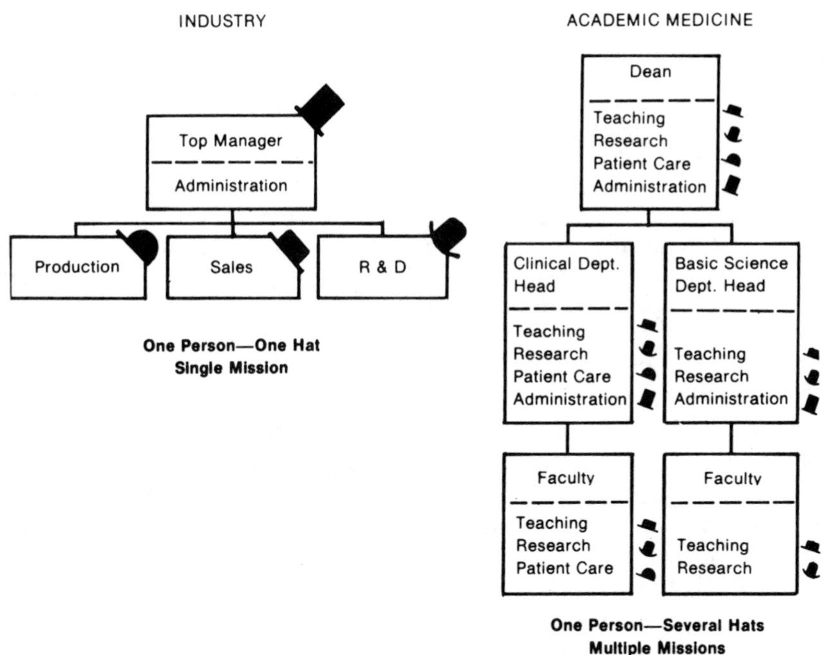

FIGURE 1.1. Contrast in role allocation between industry and academic medicine. (Reprinted with permission from *The Journal of Applied Behavioral Science,* "Three Dilemmas of Academic Medical Centers," by Marvin R. Weisbord, Paul R. Lawrence, Martin P. Charns, Volume 14, number 3, pp. 286, 290, 292–293, copyright 1978 by NTL Institute.)

demic physician's job as unremarkable. Often the same medical event will serve teaching, research, and patient care functions. But each of these tasks has its own identifiable constituencies, its own demands, supports, and constraints. Each task may involve important differences in skill requirements for effective performance. For the new academic appointee, whose skill preparation relates only to the area of patient care, the simultaneous performance of these three tasks may pose problems.

AUTONOMY

The second characteristic of academic medical centers is the tradition of professional autonomy. Medical faculty are given leeway. They exercise the right to say what should be done on behalf of patients and students, and what is necessary to get it done. Medical professors control their working conditions. The tradition of professional autonomy is supported by the Weisbord study (4). Faculty were asked to whom they were responsible for each of the tasks they performed. With the single exception of administration, over 50% of the faculty reported that they were not responsible to *anyone* for their work. A majority reported complete autonomy in research, patient care and education, and graduate degree education. Less freedom was reported in house staff and medical undergraduate education. As one dean observed:

> They'd screw the School as much as they'd screw General Motors or the Telephone Company. For them the School is just an umbrella. Doctors are very independent and they're difficult to organize. You couldn't organize doctors in the Army to all march in one direction together. They have a diversity of interests—some are interested in research and getting Nobel Prizes, others are interested in clinical health care delivery, others in teaching, and others are interested in nothing. As someone said, "There are lots of stars in orbits going their own way, some colliding, but there are no laws governing their motion." (4)

Reward System

The third characteristic is that the academic reward system is not closely connected with the task system. Whereas the task system may be influenced by various administrators including the immediate department head, the professional reward system is often uncoordinated with the administrative hierarchy. Standing committees of the faculty are responsible for determining that academic standards are established and maintained in appointments and promotion of the faculty.

The lack of connection between task performance and rewards was clearly indicated in the study of academic medical centers. In each of the centers surveyed, faculty members perceived their school's priorities as undergraduate medical education, basic science research, and primary care, in that order. When asked where they received their greatest personal satisfaction, the faculty designated patient care first, research second, and undergraduate education next to last (administration was last). The faculty members also rated financial and academic/professional rewards as deriving from separate sources: money from patient care, promotions from research. Further, Weisbord and his colleagues reported that the revenue sources of academic medical centers were changing rapidly. Funds from federal research grants were declining, and revenues from patient care were becoming increasingly more critical.

Therefore, the academic medical professional frequently works in a setting where rewards are not closely connected with task performance, where different factions have interests in different tasks, and where priorities may be in substantial and increasing conflict.

Disorganization

A fourth problem of academic medical centers, made obvious by the other characteristics already noted, is disorganization. Neither the administrative hierarchy, the work routine, nor the academic governance system has enough influence to integrate fully education, research, and patient care. Instead, integration is achieved through the loose and fluid form of a continual political process (6).

Like the Congress, medical schools have developed mechanisms for playing out this political process. The most crucial and indispensable political arenas are the intra- and interdepartmental committees. These committees are frequently interlocking, with recommendations from one level being submitted for approval of a representative committee at another level. It is through negotiation within and between these committee structures that integration is achieved among the conflicting professional groups within academic medical centers. Although committee participation is reported to be the least satisfying activity by all faculty members, participation is a "fact of life" for academic medical professionals (see Chapter 4).

DEPARTMENTS

The fifth characteristic of academic medical centers is that of departmental organization. Weisbord and his colleagues discovered that in contrast to industry, where greatest influence on policy making resides at the top of the organization, in many academic medical centers, department heads had the greatest amount of influence. In the majority of sites surveyed, department chairpersons were perceived as possessing more influence over education policy, patient care policy, and research policy than the respective medical school dean.

Although this portrait of department chairpersons as "feudal lords or oriental satraps," as one chairperson put it, suggests substantial power, it is in fact a two-edged sword (4). This is because the four characteristics of academic medical centers reviewed earlier—complex roles, professional autonomy, ambiguous reward systems, and disorganization—are issues of *departmental* management. The combination of organizational disorder and administrative power is not always manageable. As a result the administration of a department is a fourth task that may be added to the tasks of education, research, and patient care already faced by the academic medical professional. The size and complexity of medical departments sometimes require faculty members to assume part-time administrative responsibilities early in their careers. In addition, the turnover of medical faculty members sometimes results in the appointment of quite junior members as department heads. Therefore, new appointees may face an additional critical task for which they have not been trained.

These five distinguishing characteristics of academic medical centers—complex roles, professional autonomy, the ambiguity of the reward system, disorganization, and the importance of the department unit—place particular pressures and demands on medical faculty members.

Skills of Professional Development

The performance and satisfaction of academic medical professionals are influenced by the five characteristics of academic medical centers discussed above. The characteristics require resources and skills that few medical professionals have available to them or have acquired during their education. The career growth of new professionals can be enhanced by learning these skills.

First, the complexity of tasks required of academic medical professionals places substantial psychological burdens upon them. Diagnosing the ambiguity, conflict, and overload of one's role and taking steps to manage them is an elementary skill of survival and development. Of particular importance is understanding the sources and symptoms of professional stress and the means of combatting it through stress reducing techniques and time management.

A second elementary skill is what Bucher and Stelling call "role creation and negotiation" (6). As a result of the tradition of professional autonomy new medical academics rarely enter the professoriate and step into clearly defined, preexisting roles. Instead, they build their place in an organization and create the role they play. Even in the same department, a new faculty member rarely takes over the exact role of a predecessor. The research interests of the new faculty members will determine the type of laboratory space they need, and how they organize their time. New faculty members also create their own clinical role through interaction with other attending physicians.

The process of role creation requires the new professional to take control of his or her work. This is done through the skill of *open negotiation and bargaining.* This negotiation takes place with colleagues, department heads, and deans. Open negotiation and bargaining is obviously an important skill in establishing one's position on initial appointment to a medical faculty. But the characteristics of professional autonomy and ambiguity of the reward system make negotiation and bargaining of one's task responsibilities and obligations a necessary skill throughout an academic medical career.

The third essential skill of professional development is management of committees. Throughout one's academic career, even at the outset, faculty members will serve on various committees. Learning the skills of interaction that are most effective in committee settings can be important to maximizing one's influence while minimizing time expended. Further, chairing committees requires skills not normally practiced by medical doctors experienced in other task-oriented teams. Since the effective functioning of these committees is an important objective for the individual and the organization, developing skill in managing committees is also an important component of professional development.

The fourth skill is departmental organizational development. The stresses and strains of academic medical centers are most directly felt by the new academic medical professional at the level of the academic department. Even the new appointee is apt to become enmeshed in the problems of differentiating responsibilities and integrating tasks at the departmental level. The skills of department management are beyond the scope of this section and beyond the needs of most new academic medical professionals. The ability to diagnose organizational problems confronting a department and to devise strategies for strengthening these units are valuable professional development skills.

Conclusions

The conventional advice given to new academic medical professionals is that they need to add the skills of teaching and research (particularly the latter) to their established skill in clinical medicine. This advice is fundamentally true, but insufficient. Also important for longevity and satisfaction in an academic medical career are elementary skills of professional development. These skills are rarely discussed or formally taught. To the extent they are learned, though many medical faculty never learn them, they are developed through experience. Academic colleagues outside of medical school frequently develop these skills through apprenticeships and appointments in postdoctoral research positions and part-time teaching posts.

Academic physicians who come directly from residencies, or from medical practice, may not have experienced the need for professional development skills or may not have had the opportunity to practice them. The initial experience of an academic medical post can therefore be frustrating. Since medical faculty often will not tolerate occupational dissatisfaction, they may choose to leave academic medicine and enter private practice. This situation creates a degree of disappointment and "wastage" that is unnecessary. The intent of this chapter has been to describe the causes of this dissatisfaction and in the chapters following to introduce new academic medical professionals to elementary techniques for coping with these problems.

Each chapter will address one of the elementary skills: diagnosing and managing the complex role of academic physician; managing one's academic career through negotiation; effective management of academic committees; and elementary skills of departmental development.

The skills outlined cannot be learned solely from reading and thinking. They must be experienced in order to become a part of one's professional repertoire. By bringing them to the attention of the academic medical professional, and suggesting means by which they can be identified and practiced in an academic medical setting, these chapters can help physicians develop elementary skills having much professional value.

References

1. Wilson MP, McLaughlin CP. *Leadership and Management in Academic Medicine.* San Francisco: Jossey-Bass, 1984.

2. Charns MP, Lawrence PR, Weisbord MR. Organizing multiple-function professionals in academic medical centers. In *Prescriptive Models of Organizations.* PC Nystrom, WH Starbuck (Eds). Amsterdam, The Netherlands: North-Holland Publishing Company, 1977, Pp. 71–88.
3. Weisbord MR. Why organization development hasn't worked (so far) in medical centers. In *Organizational Diagnosis: A Workbook of Theory and Practice.* MR Weisbord (Ed). Reading, Massachusetts: Addison–Wesley, 1978, Pp. 168–180.
4. Weisbord MR, Lawrence PR, Charns MP. Three Dilemmas of Academic Medical Centers. *J Appl Beh Sci 14*: 284–304, 1978.
5. Flexner A. *Medical Education in the United States and Canada.* New York: Carnegie Foundation for the Advancement of Teaching, 1910.
6. Bucher R, Stelling, J. Characteristics of professional organizations. *J Health Soc Beh 10*: 3–15, 1969.

2
Managing the Role of the Academic Physician

DAVID DILL AND JOHN ALUISE

The nature of the academic medical setting places substantial pressure on the academic physician. As reviewed in the previous chapter, the pressure derives from the complexity of the academic physician's role, as well as from multiple and shifting expectations about which task is most important. Further, the need to remain "current" in teaching, research, patient care, and administration over a long medical career creates an additional burden. These conditions create a high potential for stress, frustration, and dissatisfaction among new academic physicians. This stress can be modified by a clearer understanding of the sources of pressure in the academic physician's role and by some elementary skills of coping with and managing the pressure. In the sections that follow, specific sources of stress will be outlined, approaches to diagnosing stress in a particular role will be introduced, and some ways to manage stress-producing factors will be discussed. One particularly useful skill for the academic physician is time management. Thus the application of time management techniques to the academic physician's role receives particular emphasis.

Definition of Work-Related Stress

As the previous chapter suggests, the complexity of the academic physician's role is borne by the individual. In industry any conflicts between tasks are played out as disputes between organizational units, for example, production and sales. By contrast, in academic medical centers physicians must manage individual task conflicts on their own, primarily within their own departments. Therefore, problems of ambiguity and conflict in work can become sources of psychological stress for the individual. As one hospital administrator commented:

> The role of the individual faculty member is inherently divisive, and his job has many facets. He is the key member of two large, complex enterprises and carries three different major assignments. . . . He is both a faculty member and a member of the hospital medical staff, and he is engaged in teaching, research, and patient care. Called upon to wear so many hats, he finds it simpler most of the time to

wear his own. This is often necessary because he can't get anyone to recognize him in any of his other hats. . . .*

Professional stress has been defined as a mismatch between an academic physician and his organizational role. This stress may be manifested in physiological, emotional, and behavioral symptoms. For example, new academic physicians may have a real fear of failure manifested in physical symptoms of anxiety prior to lecturing, because they have not had an opportunity to develop skills and confidence in large group teaching. Or they may grow impatient with constant pressure to conduct research and publish articles because they are strongly committed to patient care. Or they may become "workaholics," attempting to master and attain supremacy in each of the tasks of the academic physician. Or they may become increasingly frustrated, and eventually leave the medical school for private practice, because they cannot gain a clear sense of what their job involves. It is important to recognize that the sources of this stress may stem from an individual's own interests and values or from organizational sources. Developing the capacity to identify one's values and to dissect the academic physician role are preliminary means of coping with professional stress.

Sources of Professional Stress

Professional stress is produced by three work-related conditions: ambiguity, conflict, and overload. Ambiguity occurs when one is unclear about how to play the character we are assigned. For example, the new academic physician may not have adequate information about what a faculty member actually does, or those recruiting him may be vague about their expectations. As mentioned in Chapter 1, the tradition of professional autonomy and the multiple sources of authority in academic medical centers can contribute to ambiguity about what is legitimate work. The new academic physician must therefore "create" a role within the organization. But creating a role that one has not filled before, in a setting where little orientation or supervision is provided, and where there is a substantial time lag between activity and results can be disorienting. If the role is ambiguous for several years, the academic physician may become unproductive, dissatisfied, and stressed.

One way to clarify an ambiguous professional role is to ask superiors to express their expectations. For example, academic physicians can formally negotiate their responsibilities. A useful approach is to obtain, *before* accepting a new position, a written statement about the (a) goals and responsibilities of the position; (b) expected allocation of time to

---

*From Brown RE. *J Med Educ* 40(2): 126–136, 1965, with permission.

teaching, research, patient care, and administration; and (c) types of resources to be available for each of these tasks (2). The competing values of different individuals and professional groups within the academic medical center and the shifting priorities between teaching, patient care, and research suggest that without careful planning, role ambiguity can occur throughout one's career.

Career conflict is a particular problem for the academic physician. This conflict may be of two types: (a) conflict that occurs when a person is called on to play two different roles at the same time (e.g., clinician and researcher), and (b) conflict that occurs when one's personal values or goals do not match the expectations of others (e.g., the earnestly "pro-life" physician who is expected to give an objective lecture about abortion).

Conflict between roles occurs frequently in academic medicine (3). A common belief is that the tasks of teaching, research, patient care, and administration are mutually reinforcing. However, research has shown that the average faculty member is responsible for performing three of the four tasks, and that 25% of surveyed academic physicians were performing *all* four tasks. The research data do not support the view of unity between tasks. Instead, the data suggest that faculty members perceive the tasks of teaching, research, and patient care as more different from each other than the medical and scientific specialties reflected in departmental organization (e.g., pathology, pediatrics, and surgery). This insight coincides with the perception that each task is performed in a separate environment, with its own reference group, with different performance standards (3).

In short, the academic physician performs a set of very different activities that do not blend easily into a single role. As one department head observed:

> . . . I know exactly what I'm doing when I'm doing it, I'm not sure others can. The more you can separate the better you can do it. You can't do top research and top clinical care. You can't do significant research without contact with patients. On the other hand, the clinician who is constantly thinking of problems can't solve them. It's schizophrenic to pay a guy for research and to have him teach. It's time that we stopped calling teaching "care." (3)

Conflict is often redirected onto other events such as frustration with bureaucracy. For example, while working up a patient, the physician may have to search through a code book for the appropriate DRG diagnosis so that it can be properly recorded for billing. Although this extra responsibility can be shrugged off as needless paperwork, it may also be an administrative intrusion into the physician-patient relationship. At another level the antipathy many academic physicians hold toward management responsibilities is an even clearer indicator of conflict. Though the academic medical center with its multiple roles and tasks necessarily requires broad involvement of faculty members in administration, most academic physicians see administrative work as unrelated to their main efforts (4). One common tactic to deal with the frustration of managerial work is simply to ignore it. This may account for the frequent observation among academic physicians that those with administrative responsibilities fail to carry them out.

The second common source of conflict is the lack of fit between the person's interests and the professional expectations for an academic position. This conflict within a role is much more common in medical schools than in other parts of the university, because academic physicians usually have not been prepared for full-time teaching and research.

Understanding the sources of role conflict can be valuable in dealing with profession-related stress. For example, conflicts that occur when physicians try to accomplish several tasks simultaneously need to be understood and the need to combat them accepted. Many faculty members deal with conflict by "blurring" the distinctions between their tasks (5). By believing that small differences exist between tasks that are in fact quite different, the academic physician can sustain a certain psychological well-being. But this false sense of balance may come at the expense of poor job performance and increasing overload. Similarly, gaining insight into one's interests and skills and how they match the expectations of academic posts is important before one starts an academic career. Post-doctoral opportunities such as fellowships or work as a short-term visiting lecturer are effective ways to learn if an academic career should be pursued.

Third, and finally, professional stress can also be attributed to another work-related condition—overload. Overload occurs when tasks cannot be completed within a stipulated time or cannot meet quality requirements. The conditions that produce overload are evident to most academic physicians: multiple, conflicting tasks that have time, personnel, and resource constraints; little control over the demands of other people; an unpredictable workload; and tasks (e.g., research) requiring unlimited amounts of information. Managing overload is related to allocating the academic physician's most critical resource: time. Therefore, time management strategies are particularly relevant to this problem.

Throughout this section we have been outlining sources of stress for academic physicians. A currently popular term for the combination of conditions described is "professional burnout." Over time the problems of ambiguity, conflict, and overload can lead to excessive demands on an individual's energy and resources. As efficiency declines, initiative falters and interest in work diminishes. Performance can become unsatisfactory. The academic physician then exhibits the symptoms of burnout: rigidity, negativism, and inflexibility; frequent expressions of job dissatisfaction; and outbursts at authority figures. As outlined, academic physicians are particularly susceptible to burnout because of the many sources of stress they encounter. In the roles of physician-teacher-researcher, it is common for energy levels to be drained by demands that exceed one's physical and psychological capacities for work. The key to sustaining professional vitality is to learn the skills of stress diagnosis and management.

The following sections focus on methods to minimize stress coming from ambiguity, conflict, and overload. First, a technique for diagnosing professional stress is presented to help the physician determine how much stress exists and its origins. Second, suggestions for combatting ambiguity, conflict, and overload are presented. Third, the skill of time management and its practical value for treating the inevitable overload of the academic physician is explored.

TABLE 2.1. Work perception profile.

I. Indicate below the extent to which each of the statements characterizes your current position by placing the appropriate number in the Response column.

This statement describes my current position:

1	2	3	4	5
Not at all	To a little extent	To some extent	To a major extent	To a great extent

Response

____ 1. I am uncertain of other people's expectations for my role.
____ 2. I lack personnel support to share/delegate my workload.
____ 3. Advice and direction are provided me from two or more individuals.
____ 4. I have inconsistent performance standards.
____ 5. I lack resources and time to perform my tasks adequately.
____ 6. I am unable to resolve conflicting demands/expectations.
____ 7. I have large peaks and valleys in workload.
____ 8. I am unclear regarding opportunities for promotion/advancement.
____ 9. I've experienced conflicting expectations by two or more individuals.
____ 10. I am unclear regarding scope/breadth of responsibilities.
____ 11. I lack an appeals process to clarify expectations and resolve disputes.
____ 12. I am uncertain regarding the amount of work necessary.
____ 13. I have difficulty with organizing work and setting priorities.
____ 14. A recent change in leadership is directly affecting my role/responsibilities.
____ 15. My role is new to the organization.

II. Now, copy your entries for each statement in the Response column above into the space below. Then sum the three columns to determine your score for role ambiguity, role conflict, and role overload.

	1. ____	3. ____	2. ____
	8. ____	4. ____	5. ____
	10. ____	6. ____	7. ____
	14. ____	9. ____	12. ____
	15. ____	10. ____	13. ____
SUM:	________	________	________
	Role ambiguity	Role conflict	Role overload

Diagnosing Professional Stress

Since work-related stress appears to be an inevitable feature of academic medicine, it is important to determine how much stress a professional is under, and which factor—ambiguity, conflict, or overload—is most responsible.

Table 2.1 outlines a Work Perception Profile that can help suggest which of the three stress factors is most prevalent in your position. The higher the score for each factor, the more likely the factor causes stress in your professional role. Although this instrument is based on your own subjective estimates, and is therefore subject to some error, completing the profile can provide practical insight into a number of the conceptual points made throughout this chapter.

The stress rating instrument will not of course eliminate stress factors, but it may help individuals and groups identify sources of stress so that interventions can be more focused. For example, if the instrument were completed by all the faculty members in a department, discussions about common stressors could lead to useful changes in departmental policy. Stressful situations that are common to several individuals may sometimes be resolved by a group commitment, whereas an individual may feel powerless in addressing the same problem. The stress profile may also identify some individuals who require a more personalized management approach. Regardless of the group or individual findings, stress is inevitable and academic professionals must accept some responsibility for its resolution.

Managing Professional Stress

Given some insight into the nature of stress experienced by the academic physician, we can now turn to useful mechanisms for stress management.

Ambiguity

Ambiguity usually results either from a failure to rank competing activities or from working in situations that change rapidly. Academic professionals can manage ambiguity in several ways.

1. Become thoroughly informed about a new area of responsibility or a new position before it becomes part of your professional duties. It is imperative that you clearly understand the expectations that others including the department chairperson, colleagues, staff, and technical personnel have about your work. Seek information in written form whenever possible.
2. Before starting an activity or task, determine the desirable outcomes and lines of accountability. What results are expected and when?

3. Arrange a timetable for progress reports and formative reviews if the work or tasks are open-ended.
4. Since "assistant-to" positions and newly created positions are usually unclear, identify key individuals who will answer questions and clarify issues.
5. Acknowledge that in work involving education and research a time lag exists between action and visible results. Establish interim benchmarks to chart progress and boost feelings of productivity.

Conflict

Conflict occurs when two or more expectations clash. Fulfillment of one expectation makes accomplishment of the other difficult or impossible. When conflict is identified as a major stressor, several approaches may be used to solve the problem.

1. Identify the individuals or situations having conflicting expectations about your work. Confront them to establish professional goals and priorities.
2. Negotiate with superiors for tasks that will yield the greatest reward or satisfaction for you and the organization.
3. Request freedom to make career choices or decisions when conflict occurs.
4. Seek a third party to mediate interpersonal, departmental, and personal conflicts.
5. Establish an appeal or due-process procedure when conflicts cannot be resolved quickly.

Conflict, like ambiguity, should be expected in academic life. Most faculty members privately express frustration about conflicting expectations from peers or from the organization. However, professionals often lack the skill (or motivation) to confront conflict *openly* with the individuals involved. As a result, they "smooth-over" conflict and deny stress. An alternative is to develop skills in negotiation such as those outlined in the next chapter.

Overload

Professional overload occurs when one's tasks cannot be completed on time and up to quality standards. Overload often occurs when responsibilities are increased without a clear plan for their accomplishment. The most common excuse for overload is, "I couldn't say no!" Academic medical professionals are susceptible to overload because they perform different tasks, often for several superiors such as a department head or a division chief. Overload can be managed by one or more of the following tactics.

1. Keep a log of current activities.
2. Meet regularly with superiors to plan and negotiate work.
3. Obtain personnel and resource support.
4. Delegate tasks to subordinates and follow up on their work.
5. Keep an individual and organizational schedule of events and deadlines.
6. Use computerized systems, files, and other support mechanisms to organize work.

In summary, stress management does not involve *elimination* of all stressors. Instead, it involves *reduction* of stress factors according to the unique tolerance and needs of individual physicians. Stress does not have to be debilitating. It can be motivating.

Time Management

The previous discussion about academic medical work and its stressors emphasized the complexity of the role and the large volume of work that academic physicians try to accomplish. Under these conditions, knowledge of time management techniques is essential for professional growth and career satisfaction. Although time management skills have received much attention in the business management literature, physicians are rarely exposed to these ideas.

Features of academic people and their environment need to be considered when time management concepts are applied to academic medicine. Productivity results from a combination of individual traits and situational factors that vary in their influence depending on the professional event being addressed (e.g., patient care, teaching, research). Professional events also deserve attention and it helps to portray them according to six features: duration, density, location, succession, grouping, and vividness (6).

1. *Duration* is the length of time an event requires, including the preparation and aftermath. A meeting or clinical encounter may last an hour but its duration often begins well before its actual beginning and lasts well beyond adjournment. Teaching sessions often require time for preparation,whereas administrative meetings often consume time well after the session.
2. *Density* refers to the number of events occurring within a set time period. Depending on their nature, the tasks can be high density, i.e., many tasks in a short period, such as handling correspondence or telephone calls; or low density, i.e., few tasks within a long period, such as patient care, writing, or reading.
3. *Location* represents the temporal or physical position of an event.

Scheduling an event in the morning, afternoon, or evening may determine how satisfactorily it is completed. In addition, the site of the event and the physical surroundings will influence the experience. The practice site may be adequate for patient care, but it is clearly inappropriate for administrative meetings. Early mornings may be a good time to operate, but not a good time to meet with a troubled resident.

4. *Succession* means the order of occurrence of events. Succession concerns the relationships between multiple activities. A logical sequence of events can be planned on a daily, weekly, and monthly basis to prevent conflict or overload. A department head with a heavy teaching and administrative load may become inaccessible to faculty members who need advice. Creating an orderly and predictable succession between teaching and administrative office hours deliberately gives each activity the time it deserves.
5. *Grouping* is clustering of similar or related events within a given timespan. Even though professionals thrive on variety, having separate periods for doing paperwork and completing medical charts may waste time. Grouping can increase productivity.
6. *Vividness* is the degree to which an event stands out from other events. Some events are vivid because they are emotionally laden or are top priority activities. Events can also be vivid because of "whom" they involve. Trying to balance a discussion with the Dean or an irate patient with a phone call from your children will be frustrating and an inefficient use of your (and other's) time.

How can professional events be controlled to make the most efficient use of limited time? Five time management principles have been identified for use by academic physicians and other professionals.

1. *Planning.* Time is available if it is planned early. Time should be initially allotted to the most vivid tasks. Secondary or incidental responsibilities should also be planned, but given only time appropriate to their importance. Establishing an A, B, C list is recommended. This provides a priority list for yourself and others. When planning time, pay particular attention to the "actual" duration of tasks, the most appropriate location for their completion, and issues of succession and grouping.
2. *Orientation.* The more background information available about the work and environment, the more efficient the work can be performed. Meetings and informational reports in advance of major tasks can prevent unexpected misunderstandings. Advance information also reduces the need for interruptions or individual consultations by students or subordinates.
3. *Consolidation.* Contacts with people and work of the same variety can be grouped or located together to economize time and space. Working

together may also improve motivation for individuals who perform repetitive or mundane tasks. Centralization and automation also make many technical and clinical chores more efficient.

4. *Delegation.* Division of work into subtasks enables others to assist with major responsibilities. Professionals must be willing to allow others to help them. Training must precede delegation. Junior faculty, nursing personnel, research assistants, and secretaries can perform key roles in support of academic professionals. Delegation will be effective if expectations are clear and results are evaluated.
5. *Communication.* Regular meetings with key individuals to inform, give feedback, and evaluate can prevent a "crisis management" approach to academic work. Previewing the day or week and debriefing after important events may avoid misunderstanding. Written monthly and quarterly reports can provide information for planning and decision-making. One-to-one meetings are time consuming and often produce anxiety and paranoia for those not included.

Almost all techniques or suggestions about time management rely on these five basic principles. Mastering these principles can be an important factor in controlling the overload and stress of an academic medical role. A number of practical suggestions for managing professional time have been adapted from these general principles. In closing, we include these useful "pearls" from time management experts (7,8).

- Communicate goals and plans to others. Commitment to goals may occur from feeling obligated to perform since others know about your plans.
- Identify time wasters. Staff members usually know them better than you.
- Undertake the most difficult task first.
- Allow for transition time. Intervals between activities provide time to reflect, disengage, and prepare for the next set of activities.
- Keep a "C" drawer so that relatively unimportant work does not get mixed with high priority work.
- Say no by avoiding immediate responses to requests for help. Delay decisions until their consequences are considered.
- Autopsy a crisis to determine if it could have been prevented.
- Block interruptions by closing doors, keeping to a schedule, and allowing secretary/receptionist to screen calls and visitors.
- Keep meetings with individuals under 1 hour and with groups under 2 hours. Set an agenda and a time limit for each topic.
- When considering a large project employ the "swiss cheese" approach. Poke holes in it by completing small, easy to finish tasks.
- Meet regularly (but briefly) with those who work with and for you.

References

1. Brown RE. Dollars and sense in medical school-teaching hospital relationships. *J Med Educ 40* (2): 126–136, 1965 (Part 2).
2. David AK. The selection of an academic position in family medicine. *J Fam Prac 9*: 269–272, 1979.
3. Weisbord MR, Lawrence PR, Charns MP. Three dilemmas of academic medical centers. *J Appl Beh Sci 14*: 284–304, 1978.
4. Wilson MP, McLaughlin CP. *Leadership and Management in Academic Medicine.* San Francisco: Jossey-Bass, 1984.
5. Charns MP, Lawrence PR, Weisbord MR. Organizing multiple function professionals in academic medical centers. In *Prescriptive Models of Organizations.* PC Nystrom and WH Starbuck (Eds). Amsterdam, The Netherlands: North-Holland Publishing Co., 1977, Pp. 71–88.
6. Bibace R, Beattie K, Catlin R. *Psychological Aspects of the Problem of Time in Practice Management.* Department of Family Medicine, University of Massachusetts Medical School, Worcester, Massachusetts (unpublished).
7. Lakein A. *How to Get Control of Your Time and Life.* New York: Signet Books, 1974.
8. Mackenzie R. *Time Trap.* New York: McGraw-Hill, 1972.

3
Advancement and Promotion: Managing the Individual Career

STEPHEN BOGDEWIC

During the past two decades society has been moving from an industrial to an information age. Nowhere is this movement more evident than in the academic medical center. Medical schools, which were once rather simple educational institutions, have evolved into highly complex organizations. Widespread growth and increased sophistication in teaching, biomedical and behavioral research, and medical services has been the norm.

As a result, medical school departments, which form the backbone of the medical center, must respond to rapidly changing needs, technologies, and politics. These departments are often characterized by conflicting power structures, numerous and diverse programs, multiple funding sources, and goals that are often unclear or contradictory (1).

The complexity of academic medicine is an everyday reality for the departmental faculty member. Medical professors face the difficult task of balancing multiple duties including patient care, teaching, research, and administration while somehow trying to shape the direction and integrity of their own careers. How one can plan and manage a career to help ensure both advancement and satisfaction is the theme of this chapter.

Before proceeding it is important to emphasize that career advancement is the responsibility of the individual faculty member. It is of course true that both the type of academic institution (e.g., community hospital vs. tertiary care center) and certain organizational factors (e.g., funds and time available for research) have a significant influence on the academic productivity of individual faculty. Determining what these factors are, or, if you will, "how the game is played," is something that will not be done for you. It is foolish to assume that such factors as the prestige of your medical school, your scholastic ability as measured by tests, or your postgraduate training will guarantee career success. These factors are unrelated to performance in academic medicine (2,3). Instead, it is your responsibility to find a professional setting in which your goals and skills are valued. Similar to professional people, professional settings are rarely better or worse than one another—just different. Intelligent academic physicians actively seek work settings that allow them to accomplish their career goals. The point, in short, is that taking charge of one's own career is essential. Taking charge is the first step in an active approach to career advancement, an approach that defines tasks, sets goals, identifies resources, and acknowledges personal responsibility.

Assessing Career Needs

A clear definition of the term "career" is needed at the outset of this discussion. Terms such as position, job, or occupation are often used as synonyms for career. Although these words describe parts of a career, their use leaves the impression that a career decision is made only once. In this chapter, a career is defined as a sequence of positions occupied by a person during the course of a lifetime (4). This sequence is a product of an individual's values, goals, skills, sense of timing, and also political realities and circumstances. Although one cannot always influence the specific sequence of positions that define one's career, one *can* determine the congruence of any single position with one's basic career interests and aspirations.

There is no better illustration of this point than the difficult decision of selecting an appropriate location for an academic position. Factors such as family, community, and climate are clearly important. For the moment, however, consider only one factor: the type of institution. Several options exist. One might select a new community-based medical center, or a major private university with a national reputation. Yet another choice might be a regional public university with a primary commitment

TABLE 3.1. Questions about a career in academic medicine.

What has drawn you to academic medicine?
Any expectations that may be unrealistic?
Any negative factors that have influenced this decision?
What seems most appealing or glamorous to you?
What will be most difficult or challenging to you?
What are your career goals?
What current evidence is there of an academic orientation in your life?
What type of satisfactions do these activities bring?
With what professional organizations have you been involved?
What have you learned about how you work with people?
What leadership roles have you assumed?
Based on your experience, what field do you feel you could successfully teach?
What is it that most excites you in medicine?
How would you assess your strengths, priorities, and interests in the areas of patient care, teaching, research, and administration?
On what experience are you basing these insights?
How flexible are you regarding the extent of your involvement in each of these areas?

to its local area. Although the title "Assistant Professor" is the same at each of these settings, the goals, plans, and politics of the three medical institutions are probably quite different. Therefore, sound career decision-making requires academic physicians to understand their own aspirations *and* the values and goals of the institutions they are considering.

By placing the concept of a career in a life-long, developmental frame, individual career decisions become easier to comprehend. What becomes evident is that no career decision is ever made in a vacuum; career success will depend a great deal on how well one understands oneself. For example, the decision to become a physician requires extensive introspection. Certain skills, values, and beliefs have to be present before a serious commitment can be made to a medical career. A similar process of self-examination should also precede the decision to become an academic physician.

Table 3.1 provides a sample of the career questions that experienced academic physicians believe their junior colleagues should ask and answer for themselves (5,6). Because of the multiple duties of the aca-

TABLE 3.2. Factors involved in selecting an academic position.

Character of the institution
Is the institution's primary commitment toward a local, state, or national constituency?
Is this commitment compatible with your personal philosophy?
Which features of the institution match your aspirations?
Organizational features
Current organizational chart
Intradepartmental committees
Decision-making process
Stated departmental goals
Relationship of department with medical school
Job description
Responsibilities and opportunities are clearly outlined
Flexibility to accommodate needs of individual
Funding
Knowledge of funding sources (grants, clinical income, state allotments)
Expectations about responsibility for generating revenue
Stability of funding sources
Personal funding
Base salary
Merit or other salary increases
Incentive plans
Fringe benefits
Family considerations
Spouse and dependent needs
Schools
Recreation
Commuting
Promotion, tenure, and faculty development
Specific policies for promotion and tenure

demic physician, unbiased assessment of one's interest and commitment to teaching, research, and service is critical for satisfaction and advancement in an academic career.

Once the decision to enter academic medicine has been made, the next task is to secure a suitable position. As suggested earlier, factors such as community size and location, climate, and recreational and social facilities are important considerations. However, beyond these matters of personal preference there is a set of issues that should be addressed and settled before a faculty position is accepted (7). Table 3.2 presents a list of the issues. The major benefit derived from using this list during the search and interview process is that it encourages prospective academic physicians to engage in careful critical self-examination of their wants and needs.

Planning Career Objectives

All adults—physicians included—develop and change physically, intellectually, and socially during the course of their lifespan. For professionals, this course of normal adult development proceeds in-kind with the lifelong task of career development. This idea has received much attention, both among scholars and in the popular press (8–10). It is based on the notion that careers, like the adults who pursue them, proceed through a series of stages marked by different tasks and challenges. The tasks and challenges, in turn, place new demands on ever-maturing adults. Before introducing the specific tools of career management, a short explanation about the developmental nature of the academic medical career should be useful.

Table 3.3 points out three general stages of an academic career that are drawn from scholarly research and writing about the professoriate (11,12). The three stages correspond closely to the faculty ranks of assistant, associate, and full professor, respectively. The fundamental message of Table 3.3 is that the issues or considerations that will likely challenge an academic physician who is trying to establish a career are different from those that challenge more senior colleagues. So too are the approaches or solutions to career issues that one can use at different career stages. Periodic reviews of progress (discussed in detail later in this chapter) in light of an institution's promotion and tenure guidelines are critical for new faculty. Such reviews are likely to be less significant for established academics who supervise research groups, chair important committees, and edit journals.

The contents of this chapter mainly, yet not exclusively, address the career issues of academic physicians who are at the first career stage. Although associate and full professors should also profit from the chapter, their success in achieving academic tenure (in settings where tenure

TABLE 3.3. Basic stages of the academic career.

Stage	Issues or considerations	Approaches or solutions
Establishing a career	Enthusiastic and apprehensive about challenges of a new situation Role ambiguity/conflict/overload Unclear performance guidelines/expectations Desire to improve teaching and research skills Time constraints Little understanding of operations and conduct of complex academic organizations Inexperienced in negotiating for and properly using available resources	Faculty orientation program Locating a mentor Clear understanding of departmental goals including informal processes at work Determine resources allocated or available to a new faculty member Written job description Allocation of time established in writing Periodic review of progress Promotion and tenure guidelines Evaluation criteria Faculty enrichment programs, emphasis on career development Fellowships
Mid-level (integration)	Realistic awareness of institutional hierarchy and politics of academe Greater expectations from colleagues More confident of abilities Seeking recognition and advancement Concern about the future Exploration of alternative careers Seeking new incentives That which was helpful in acquiring a reputation may not be as valuable in maintaining a reputation	Serving as mentor Increase organizational identity committee involvement administrative activities exercising leadership Significant involvement in professional organizations Increased awareness of and involvement in department's external affairs
Senior (maturation)	Recognized as an expert Increased range of administrative and leadership involvement Stagnation vs. diversification Question the value of their vocation Uncertain about future challenges	Develop new area of expertise Investigate administrative positions Chair significant committees More active involvement in professional organizations Journal editorships Consulting Sabbatical leave

applies) and their professional longevity suggest a reasonable match between personal aspirations and institutional expectations.

Activity Analysis

Career management requires developing an awareness of the activities one is performing, why they are being done, and how well they are being accomplished. Professional insight will not occur by viewing one's position as a loose collection of activities. Instead, some organized way of dealing with multiple involvements must be employed. An "activity analysis" can accomplish this (Table 3.4).

The activity analysis can be thought of as an inventory of one's current professional tasks. It is both a summary and an assessment at a particular point in time. An activity analysis asks several questions: Specifically what is it that you are doing? What does it take to accomplish your work? How important and satisfying are these activities?

Completing an activity analysis involves the following steps. First, list all current involvements under the heading "key activities." Appointment books can provide a useful record of where your energy and time are being spent. Next, determine the amount of time actually required to perform these duties. Be cautious. The tendency in estimating time required is to consider what the schedule says it will take, instead of time requirements based on experience. Include the time spent in preparing for, executing, and evaluating your work.

With this breakdown of activities and time allotments completed, the

TABLE 3.4. Activity analysis.

Key activities	% Time required	Strengths	Educational needs	Relative importance		Level of satisfaction
				Organization	Self	

next step is to make a personal evaluation of your strengths and perceived educational needs for each of the activities. By being specific, this part of the exercise can help you determine how to strengthen your professional repertoire (e.g., learning how to operate a microcomputer) to accomplish the activities you have identified.

The fourth step is to estimate the relative importance of each activity separately for *yourself* and for the *division, department,* or *school* (organization). Although a simple scale ranging from 1 to 10 could be used, several options exist for assigning relative weights. For instance, a total of 100 points could be divided among the activities, creating an index of their relative importance. Regardless of the method used, this step can provide valuable insight into the "fit" between yourself and your academic organization.

The final consideration to be made in analyzing your current position is your level of satisfaction with each activity. Any scale will do. What is important is that satisfaction should be assessed in terms of what you find personally rewarding.

The activity analysis exercise is brief and straightforward. It can have many uses. First, it unravels the complexity of a professional role having multiple responsibilities. General feelings of constraint, overload, and ambiguity mentioned in earlier chapters can be framed and made objective.

Comparing the importance of each activity for the individual with its perceived importance for the organization also provides a key insight. It helps determine one's congruence with an organization. There is evidence that this type of "fit" is often lacking in academic medicine. A recent study by Gjerde and Colombo (13) found a significant discrepancy between perceptions of faculty members and chairpersons regarding academic promotion criteria.

Finally, the activity analysis helps address another question: Is this what you intended to be doing when you joined the organization? In other words, was there a design to your position, or did it simply evolve? By analyzing one's activities, areas of desired change become apparent. How to go about making those changes is the subject of the following section.

Contracting

Two basic steps have been outlined as necessary prerequisites for managing a career: (a) gaining an understanding of oneself in a career development context, and, (b) undertaking a thorough analysis of current activities. Armed with this knowledge, the next step is developing a career plan.

Since a career is developmental, change is to be expected and valued.

However, career change can be either planned or a reaction to circumstances. The exercise of conscious choice about one's career is a key responsibility of the academic physician. Career planning can be compared to an educational enterprise. For instance, one attains the career goal of becoming a physician by following a carefully designed plan called a curriculum that is prepared by a medical school faculty. The difference in planning an academic medical career is that responsibility for the plan resides with the individual physician.

The most common criticism leveled against career planning is that despite the effort, plans do not prompt action. Poor plans can't. By contrast, good plans include measures that fix responsibility for taking action. What is needed is a planning mechanism that includes a feature for keeping the plan alive.

Several models exist for designing career plans. One particularly useful tool for planning an individual career is called a Professional Development Contract (PDC). The name alone connotes an arrangement or agreement that attends to career development needs. Most faculty members maintain a rather informal contract with their department except for type of appointment (e.g., clinical vs. tenure track) and salary. These two conditions are usually specified in writing. The purpose of the PDC is to specify a course of development for the individual faculty member that will accomplish the goals of both the individual and the organization. The concept of a contract introduces an important issue. Careers cannot be managed in a vacuum. Negotiation and agreement are needed with

TABLE 3.5. Basic elements of a professional development contract.

I. Self-assessment
- Strengths and interests
- Weaknesses and current dislikes

II. Key responsibilities
- Any short-term or long-term changes
- Those you wish to continue
- Those you are interested in doing less of or eliminating
- The priority for your various involvements

III. New developments
- Specific goals you wish to accomplish
- Support/resources that will be required
- Timeline
- Priority of new developments

IV. Personal growth
- Opportunities you may need
- Skills you would like to develop
- Characteristics you would like to change
- Coaching (support/constructive feedback/counseling)

V. Evaluation of effectiveness
- Self-evaluation
- Evaluation by others

TABLE 3.6. Professional development contract: assistant professor of internal medicine.

Contract period: July 1, 1984–June 30, 1985

I. Self-assessment

A. Strengths

The area where I feel most certain of making a valuable contribution to the department is my teaching. I believe that I have excellent skills and techniques in one-on-one encounters, small groups, and comprehensive lectures. I have the ability to achieve collegial rapport with residents.

I consider my clinical skills to be strong also. I have an excellent broad base of medical knowledge, with a particular emphasis on acute care medicine. My relationship with my patients enables . . .

B. Areas for improvement

An area where I am weak is in practice management. I lack experience, especially in a high-volume practice. My business skills are limited. In general I feel poorly organized, and my lack of time management skills leads me to procrastinate.

I also recognize my limits in the area of research. I have never received formal training in research design or statistical methods. I find that my qualitative approach to research . . .

II. Key responsibilities

The activity analysis that I have completed as part of this contract outlines my various responsibilities within the department. What is most evident to me in viewing the distribution of my time is that if I am to develop my research skills, something has got to change. At present I spend only 10% of my time on research and scholarly activities. Patient care and teaching now consume . . .

III. New developments

To understand the context for the specific areas I wish to develop, it is necessary to consider my long-term goals. They are as follows:

1. Continuation of academic career, with promotion to professor within 12 years
2. Primary emphasis on graduate education with consideration of eventually becoming residency director
3. Continue emphasis on hospital medicine, with creation of a section of in-patient . . .

IV. Personal growth

I do not consider myself to be "politically" aware. Academic medicine is influenced by a wide range of internal and external factors. If I am to have a successful career in academics, I feel a need to gain a wider perspective on the role of an academic physician.

There are two ways that I believe this can happen; one is a structured involvement and the other is informal. I would like to participate in a fellowship program designed to . . .

V. Evaluation of effectiveness

I accept that career management is my responsibility. Nonetheless, if I am to be successful I believe there are two processes I must engineer into my work in the department: on-going feedback and regularly scheduled opportunities for negotiating change.

The feedback I desire is both formal (evaluation) as well as . . .

regard to the resources (e.g., clerical personnel, space, travel money) a faculty member can use to fulfill personal and departmental goals.

There is no single way to design a PDC. There are, however, several elements that should be included to ensure its usefulness (Table 3.5).

Since the purpose of the PDC is to promote development, it is useful to begin by establishing a baseline. Thus, the contract should begin with a self-assessment of strengths, interests and weaknesses. A personal assessment of one's current professional responsibilities should also be included. Completion of the Activity Analysis (Table 3.4) should provide the necessary information for this portion of the PDC.

Once the baseline is established, the remainder of the contract should clearly identify one's intended professional responsibilities and opportunities for learning and development during the coming year. The plan should include details about one's patient care, teaching, administrative, and scholarly goals; the resources that each goal will require; personal growth or learning opportunities that may be needed; and the type and source of performance feedback that will be helpful. Table 3.6 presents a sample Professional Development Contract for an assistant professor in the field of internal medicine.

The major value of a PDC is that if carefully prepared, the contract can be a deep source of insight about one's career situation and career aspirations. Department chairpersons and division chiefs should also benefit when faculty take time to develop and share PDCs because the documents can provide a basis for work negotiations with staff. Thus it is in the best interest of faculty *and* administrators to view career planning as a tool that encourages thoughtful career development, not as a bureaucratic red herring or an infringement on academic freedom. The use of a PDC or some other mechanism for career planning should encourage medical faculty to take an active approach to their career management responsibility.

Performance Appraisal

Analyzing one's professional role, determining areas for individual development, and designing a plan to meet individual career objectives will not necessarily ensure career success in academic medicine. A means to put the plan into action is necessary. A good way to begin is by developing a PDC. To ensure that this plan will be acted on, it is essential to include in the contract an agreement that one's performance will be periodically reviewed. This agreement is the starting point for the process of performance appraisal.

Performance appraisal is different from formal personnel evaluation for three reasons. First, although the purpose of personnel evaluation is judgmental and involves reaching decisions about a faculty member's

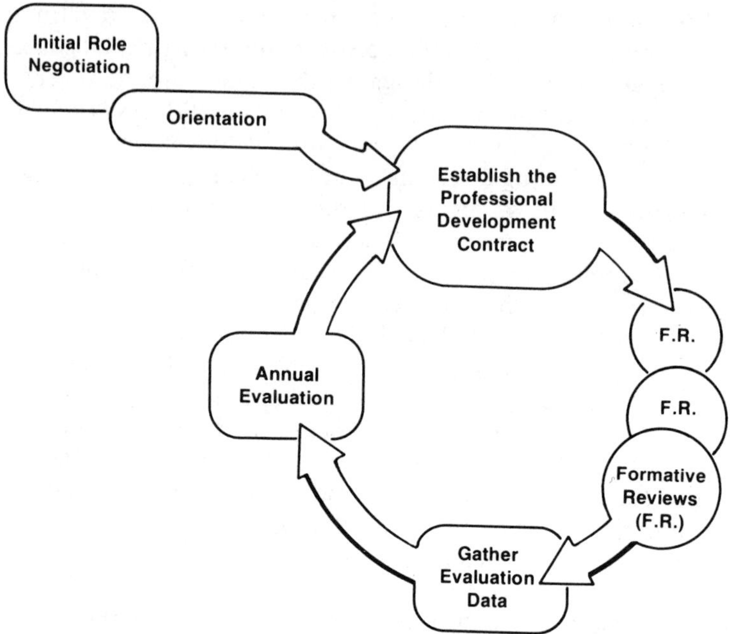

FIGURE 3.1. Performance appraisal cycle.

value or worth, the purpose of performance appraisal is to contribute to faculty productivity or to the acquisition of new skills. Second, performance appraisal differs from evaluation in terms of frequency. In business and industry, and in some medical settings, short performance appraisal sessions are often done monthly or quarterly. By contrast, formal personnel evaluation usually occurs annually. Third, performance appraisal is an ongoing, cyclic process whereas evaluation is usually done at the conclusion of a fixed period of time.

Performance appraisal and other career management ideas are not common in academic settings. Consequently, should you attempt to implement such a plan it will probably be necessary to educate the chairperson or division chief about your expectations. It is imperative that all participants in a performance appraisal system share an understanding about its purpose and mode of operation.

Figure 3.1 shows the various parts of the performance appraisal cycle. What follows is a discussion of the contribution performance appraisal can make to the development of medical faculty careers, with particular reference to a new faculty member.

Initial Role Negotiation

New members of a department can only estimate the specific features of their duties. A realistic breakdown of time and effort devoted to various activities can be determined only after time is spent on-the-job and after learning a department's formal and informal policies. Consequently, preliminary discussions about workload should acknowledge the uncertainty of early estimates.

Orientation

During initial negotiations the department head and the new faculty member should agree to meet after a fixed period of time, say, 6–8 weeks. This will allow the newcomer to become clinically, academically, and socially oriented to the setting. Such an orientation permits faculty-administrator conversation about the position on the basis of some shared experience. An Activity Analysis done according to the procedures presented earlier could conclude the orientation period.

Professional Development Contract

Once orientation is completed, a contract should be prepared that reflects the professional intentions and needs of the new faculty member and supports departmental goals. This will require articulation of departmental priorities, standards for assessment of faculty performance, and the resources available to the faculty member. For academic physicians in University settings, the contract should conform with institutional promotion and tenure guidelines. Frank negotiation is needed to produce such a contract. However, the resulting document will represent a clear set of consensual expectations about faculty productivity and departmental modes of support. The contract should also specify the expected relationship between the faculty member and the department chairperson or division chief who will be responsible for supervising and evaluating faculty performance.

Formative Reviews

Review sessions are an essential element in performance appraisal. They provide an opportunity for faculty members and supervisors to review progress toward annual goals and discuss problems or issues that need attention. Formative reviews can work only if the supervisor is clear about the dual role being played. During review sessions the supervisor must function as mentor or teacher, not as an evaluator. This same person, however, will most likely be the one who evaluates the faculty member, perhaps on an annual cycle. Separating these roles is difficult, but it is imperative. Helping the department head to keep the roles separate is in each faculty member's best interest. There is no optimal number of formative review sessions. Each faculty member should take charge of his or her own professional development by requesting formative reviews as a regular practice.

Gathering Evaluation Data

At some point, usually near the end of an annual cycle, faculty progress should be formally evaluated. Since the professional development contract includes criteria for assessing the accomplishment of objectives, one's achievements should be apparent.

Performance appraisal is a system that supports and extends faculty development. Throughout the year, the specific responsibilities and contributions of the individual faculty member are articulated and later evaluated formally. Many departments have written policies for evaluating faculty performance. These evaluations are usually related to promotion and tenure. The data used in these evaluations may or may not adequately reflect one's real accomplishments and contributions. Therefore, it is strongly recommended that each faculty member maintain a performance file. Teaching evaluations, achievements or awards, publications and presentations, as well as past professional development contracts could all be included in such a file.

Annual Evaluation

The purpose of this session is for the faculty member to receive candid feedback about progress toward stated career objectives and about accomplishment of key responsibilities. Unlike formative reviews, this is not a coaching session. An annual evaluation gives the chairperson or chief an opportunity to provide a frank, objective performance evaluation.

Issues beyond career development should be discussed if the review coincides with the department's formal evaluation scheme. The review may evaluate the total spectrum of one's involvement in the department. Criteria for this assessment would most likely be the medical school's guidelines for promotion and tenure, as well as departmental goals.

A final point about formative and annual review sessions deserves emphasis. There is no substitute for a thoughtful career plan based on personal and departmental expectations. However, no plan is perfect and each one should be amenable to revision. When expectations or circumstances change, having a compatible relationship with the person you are working for is critical. Here, compatible means a relationship in which faculty and administrators understand each others' pressures, preferred work style, strengths, weaknesses, and ambitions. A relationship of this nature serves both individuals, as well as the total organization.

Establishing a PDC for the Coming Year

After receiving results from the annual evaluation, a new professional development contract can be established for the coming year.

Performance appraisal is a method for planning and managing one's career. Regardless of whether or not one's department uses such a system, the technology of performance appraisal is available, and the concepts can be applied in a variety of settings.

Conclusions

At the outset of this chapter, it was stated that academic medical faculty, particularly new faculty members, should take control of their own careers. Although this advice may seem like common sense, some of the mechanisms for achieving control might appear unconventional. In particular, the emphasis on negotiation in determining the duties and forms of evaluation for one's position may seem too pushy. No doubt there are many department heads for whom such negotiations will be a new experience. However, the complex role of academic physician requires that to be successful, faculty members must design their positions in accord with their own needs and strengths and with a clear view of departmental expectations. To allow others to set the agenda unilaterally is to invite frustration and disappointment. Similarly, although some department heads can be relied on to act as mentors for new faculty members, others may not. Department heads themselves are engaged in a demanding management task. In many cases, the academic medical department head is as unprepared for the complexities of his or her role as a young medical doctor is for the complexities of the academic physician role. Therefore, it is often necessary for the academic physician to "manage the boss" when it comes to setting the agenda for an academic position (14).

The issues at stake in deciding to actively manage an academic career are apparent. The difficulty for most people is arriving at "how" this can best be accomplished. What are the necessary career management skills? In many ways the physician is already equipped with these skills, as evidenced in patient encounters. The doctor-patient relationship is basically a process of negotiating mutual responsibilities. It requires that the doctor possess the ability to diagnose problems, form a treatment plan, determine expectations, and specify outcome criteria.

One's comfort and success in applying patient management skills in a clinical situation is in part influenced by previous experience with the presenting problem. The same is true for managing a career. Most physicians possess the needed management skills in some form. Several concrete suggestions are contained in this chapter for applying the skills. All that is needed is practice. The patient deserves the effort.

References

1. Weisbord MR. Why organizational development hasn't worked (so far) in medical centers. In *Organizational Diagnosis: A Workbook of Theory and*

Practice. MR Weisbord (Ed). Reading, Massachusetts: Addision-Wesley, 1978.

2. Blackburn RT. Academic careers: patterns and possibilities. *Current Issues in Higher Education: No. 2, Faculty Career Development.* Washington, D.C.: American Association for Higher Education, 1979, Pp. 25–27.
3. Samson GE, Graue ME, Winstein T, Walberg HJ. Academic and occupational performance: A quantitative synthesis. *Am Educ Res J 21*:311–321, 1984.
4. Super DE, Hall DT. Career development: exploration and planning. *Ann Rev Psychol 29*:333–372, 1978.
5. Stephens GG. On becoming a teacher of family medicine. *J Fam Prac 4*:325–327, 1977.
6. Geyman JP. Career tracks in academic family medicine: issues and approaches. *J Fam Prac 14*:911–917, 1982.
7. David AK. The selection of an academic position in family medicine. *J Fam Prac 9*:269–272, 1979.
8. Baldwin RG, Blackburn RT. The academic career as a developmental process. *J Higher Educ 52*:598–614, 1981.
9. Sheehy GM. *Passages: Predictable Crises of Adult Life.* New York: E.P. Dutton, 1976.
10. Levinson DJ. *The Seasons of a Man's Life.* New York: Knopf, 1978.
11. Baldwin RG. Adult and career development: what are the implications for faculty? *Current Issues in Higher Education: No. 2, Faculty Career Development.* Washington, D.C.: American Association for Higher Education, 1979, Pp. 13–20.
12. Mathis BC. Academic careers and adult development: a nexus for research. *Current Issues in Higher Education: No. 2, Faculty Career Development.* Washington, D.C.: American Association for Higher Education, 1979, Pp. 21–24.
13. Gjerde CL, Colombo SE. Promotion criteria: perceptions of faculty members and departmental chairmen. *J Med Educ 57*:157–162, 1982.
14. Gabarro JJ, Kotler JP. Managing your boss. *Harvard Bus Rev 58*:92–99, 1980.

4
Managing Academic Committees

DAVID DILL

> . . . when committees gather, each member is necessarily an actor, uncontrollably acting out the part of himself, reading the lines that identify him, asserting his identity. This takes quite a lot of time and energy, and while it is going on, there is little chance of anything else getting done. Many committees have been appointed in one year and gone on working well into the next decade, with nothing much happening beyond those extended, uninterruptable displays by each member of his special behavioral marks.
>
> If it were not for such compulsive behavior by the individuals, committees would be a marvelous invention for getting collective thinking done. But there it is. We are designed, coded it seems, to place the highest priority on being individuals, and we must do this first, at whatever cost, even if it means disability for the group.*
>
> Lewis Thomas

Academic life involves participation in committees: membership on committees, chairing committees, and for administrators, deciding when to use and how to staff committees. Faculty members in major public and private universities report spending on average 17% of their professional time in committees and administrative work (1). The reaction of faculty members to this activity is uniform: they find it boring and burdensome. But the complexity of academic organizations makes committees essential devices for integrating specialized fields and activities.

As Lewis Thomas suggests, committees are collections of individuals *trying* to work together. Often these individual personalities and the parts they play result in dysfunctional behavior. Since committee chairpersons in academic organizations seldom receive training for their role, dysfunctional behavior may persist or worsen. The first issue of concern for an administrator facing a decision is whether a committee can make a useful contribution. Guidelines for this choice are developed in the first section

---

*From *The Medusa and the Snail* by Lewis Thomas. Copyright © 1974, 1975, 1976, 1977, 1978, 1979 by Lewis Thomas. Most of these essays originally appeared in *The New England Journal of Medicine.* Reprinted by permission of Viking Penguin, Inc.

of the chapter. The second section introduces skills of value to committee members who are *not* serving as chairpersons. These individuals also have responsibilities for committee effectiveness and are usually unaware of how to contribute positively. Skills in running and chairing committees are given special emphasis in the third section. Finally, for many department heads, the "politics" of selecting and charging an ad hoc committee has a great deal to do with the committee's functioning and eventual contribution to the department and organization. The fourth section will therefore focus on understanding when to use such committees and how to construct them.

The Nature of Academic Committees

It is important to stress that although committees are critical to academic life, they are often overused. Frequently a committee is formed because an administrator or faculty member does not wish to deal with an uncomfortable problem. Alternatively, a committee may be formed out of reflex action whenever an issue arises. The first rule of committee behavior, therefore, is to determine whether the problem at hand cannot be better dealt with through administrative action or by a single individual consulting others by individual discussion, phone call, or memo. Several minutes spent with four separate individuals is often more productive than an hour-long meeting with the four together.

Extensive research on problem solving suggests that most decision-making situations can be classified into four categories, each of which demands a different mode of decision making (2). As Figure 4.1 suggests, organizational decisions may be conceived as having two dimensions.

The first is a quality dimension that relates to the "correctness" of a decision. The second is called decision acceptance because it depends on subjective aspects of the decision (i.e., people's feelings). Decision *quality* depends on the decision maker(s) having enough reliable information and professional expertise to make the decision. Medical staff opinions about the desirability of alternative solutions are not relevant to decision quality. That type of information is relevant to decision acceptance, the other dimension of the model. Decision *acceptance* is particularly important to those settings where implementation of the proposed decision solution requires the active involvement of medical faculty members or where performance cannot be closely monitored. Both of these conditions are common in academic medical centers.

A Type 1 problem or situation is one where quality is not a key issue (i.e., expertise is not critical) and where wide acceptance is not pivotal to effective implementation. Decoration of the dean's office is an example. Problems in this set can be decided easily by the personal judgment of the leader. A Type II problem has a high quality requirement, but accep-

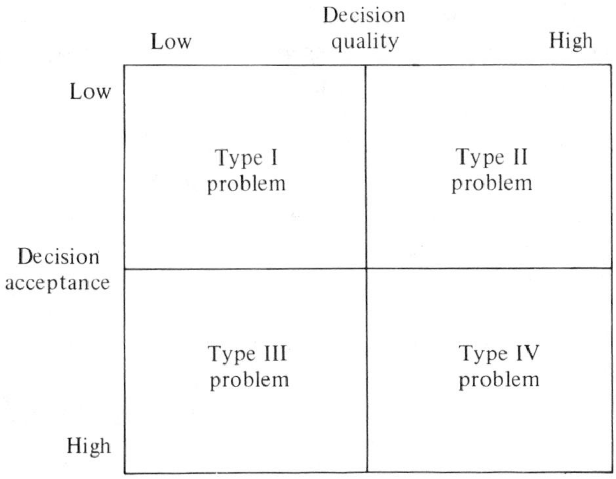

	Decision quality: Low	Decision quality: High
Decision acceptance: Low	Type I problem	Type II problem
Decision acceptance: High	Type III problem	Type IV problem

FIGURE 4.1. Dimensions of organizational decisions

tance is likely to be gained easily (e.g., appointment of the associate dean). In this case, the medical faculty are apt to feel that the leader has the expertise (information) to make the decision, and will comply with the leader's choice. In most such cases the leaders should make the decision themselves. But if the leader feels he or she has insufficient knowledge to make a high-quality decision, he or she can solicit advice and recommendations from other individuals. In contrast, Type III problems do not have a high quality dimension, but acceptance is critical to their success (e.g., distribution of parking places). These decisions (or policies) are best left to group decision-making. A Type IV problem has a high quality dimension (a substantial amount of diverse information is necessary), and high acceptance is necessary for implementation (e.g. a new medical curriculum). These problems are most effectively approached through committee decision-making.

This simple but powerful scheme suggests that many problems currently allocated to academic committees might be solved with equal effectiveness and greater efficiency by an administrator acting alone or in consultation with individual faculty members. In situations where substantial expertise and information are important to the decision and are in the collective minds of the staff, or where broad acceptance is critical to implementation, a committee process would prove most effective.

Given that an administrator has decided to appoint a committee to address a particular problem, what are the special characteristics of these groups that are critical to their successful functioning?

Most notably, committee meetings are, as Thomas suggests, a form of theater. Committee members frequently respond at two levels: to the public agenda, which represents the formal business of the group, and to the hidden agenda, which represents the aspirations, desires, and motives—the emotional life—of the group members. In short, a committee meeting is a "status arena" (3). Because committee meetings are often the only time when individualistic faculty members have an opportunity to discover their relative standing in the medical school, the "arena" function is inevitable. When a committee is newly created, appoints a new chairperson, or is composed of formal authorities such as department heads who are in competition for resources and power and who do not work together outside the committee,"arena behavior" is likely to become even more significant. In contrast, in standing committees that meet regularly it may be hardly noticeable. As a consequence, the successful management of committees requires attending to *two* agendas, the formal agenda and the hidden agenda.

Effective Behavior for Committee Members

It is often observed that what makes effective leadership is good "followership." In the same vein, successful teams and committees are characterized by both effective chairpersons and by participants who are well-prepared and who play constructive roles within the group. Thus the first issue of committee management is defining effective behavior for committee members.

The key to effective committee work is preparation, both in regard to the subject of decision and in regard to skills for effective participation. Committee participants must come as well prepared as the committee chairperson. They should be well versed in the problems and issues to be discussed, should have read the background papers, and should have given thought to how they will contribute. Although preparation for the task may be obvious, it is also necessary to prepare for the hidden agenda of committee meetings, the "arena behavior."

Extensive research on the behavior of small groups suggests that committees are characterized by predictable, destructive or constructive, forms of behavior. For example, the "arena" behavior previously discussed is frequently exhibited in the following individual behaviors (4).

- *Aggression:* Working for status by criticizing or blaming others; showing hostility against the committee or some individual; deflating the ego or status of others
- *Blocking:* Disagreeing and opposing beyond "reason;" resisting stubbornly the committee's wish for personally oriented reasons; using a hidden agenda to thwart the movement of a committee
- *Dominating:* Asserting authority or superiority in order to manipulate the committee or certain of its members; interrupting the contri-

butions of others; controlling by means of flattery or other forms of patronizing behavior

- *Out-of-field behavior:* Making a display in an ostentatious fashion of one's lack of involvement (e.g., reading a newspaper during the meeting); "abandoning" the committee while remaining physically with it; seeking recognition in ways not relevant to the committee task
- *Special-interest solicitation:* Introducing or supporting policies that best meet one's own individual need; trying to induce the committee to be sympathetic to personal misfortune or problems; using the committee time to plead one's own case.

Occasionally one individual may consistently behave in a dominating manner. However, any committee member may exhibit these behaviors depending on the composition of the committee or the issue under discussion. When observing "arena" behavior, it is important to guard against the tendency to blame an individual who resorts to such tactics. It is more useful to see this behavior as a symptom of the ineffectiveness of the committee, particularly the committee's inability to satisfy the individual's need for status through group-centered activity. Indeed, not all participants will perceive the behavior in the same way. For example, the person who appears to be "blocking" may be trying to "test the feasibility" of a proposal ("devil's advocate"). The positive perspective is therefore not to focus on eliminating dysfunctional behavior, but to emphasize the functional behaviors that lead to member cooperation and cohesion.

Research on small groups has shown that effective committees are staffed by people who emphasize task completion and group maintenance (4). Task-related behaviors include initiating activity (e.g., proposing a solution), giving information, clarifying, summarizing, and reality testing (e.g., testing an idea against some data). It is important to stress that task behavior, such as the maintenance behaviors that will be discussed next, may be exhibited by any member of an effective committee. For example, some committee chairpersons may be especially skilled at group maintenance activities, and allow other committee members to assume task completion roles.

Committee building and maintenance behavior provides the "glue" that binds the members of a committee together. Members can contribute to effective group functioning by developing and using several skills when they encounter dysfunctional behavior (4):

- *Harmonizing:* Attempting to reconcile disagreements; reducing tension by using humor or placing a tense situation in a wider context; getting committee members to explore differences
- *Gate keeping:* Facilitating the participation of others, e.g., "We have not heard from Dr. Smith yet;" suggesting procedures that permit sharing

remarks, e.g., limit the talking time of each person so that all have an opportunity to be heard

- *Consensus testing:* Asking for group opinions in order to discover whether the committee is nearing consensus on a decision; sending up a trial balloon to test a possible conclusion
- *Encouraging:* Being friendly, warm, and responsive to others; indicating by facial expressions or remarks the acceptance of other's contributions
- *Compromising:* When your own idea or status is involved in a conflict; offering a compromise that yields status; admitting error; modifying your position in the interest of group cohesion or growth.

We have all encountered and appreciated committee members who exhibit these behaviors. Unfortunately, we have also all experienced committees where these skills were missing. There are two reasons why these behaviors may be absent in a committee. First, the committee chairperson may not effectively fill the needed role of leading the committee in its task functions and overseeing maintenance behavior. Given this abdication of leadership, dysfunctional behaviors may begin to predominate. Even those participants whose natural behavior is that of maintenance may be lured into dysfunctional behavior out of fear that their "ox will be gored" unless they become more aggressive. Even worse, effective maintenance people may drop off the committee. Under these circumstances, the most appropriate action for the committee participant is to meet privately with the chairperson and candidly discuss the problems in the committee's functioning. If the committee chairperson continues to be ineffective, an alternative to dodging meetings or resigning is to confront the administrator or group who appointed the committee and request (with specific alternative suggestions) that the chair be replaced. Again the focus should be on constructive, positive actions that can be taken to improve the committee's work and to prevent meetings from turning into a negative experience.

In summary, effective participation in an academic medical committee involves thorough preparation regarding the material to be discussed. It also involves attending to the status concerns of committee members through appropriate group maintenance behaviors. The following suggestions summarize the points made earlier (3):

- Come with an open mind, but do not hesitate to comment, criticize constructively, or disagree.
- Do not engage in distracting behavior. Prompt attendance, no side conversations, no shuffling of papers, no personal interruptions can all contribute to an orderly and more efficient meeting.
- Do not personalize differences of opinion, but rather try to maintain an objective attitude.
- Speak up when you have something to say, but do not hog the floor. If you restrict your comments to the most meaty item, this should help to

keep the discussion on target and fix your key points in the minds of others.

- Listen actively to others, be tolerant of their points of view, and help them develop their ideas.
- Help the chairperson stick to the agenda, and control wayward members by asking for clarification or making constructive criticism or other appropriate remarks.
- Take notes for reference. Note-taking helps maintain your concentration on the subjects under discussion, helps to fix thoughts in your mind, and can help in the after-stages of a meeting when it is time to implement a decision.

Role and Function of the Committee Chairperson

One of the major sources of frustration in academic life is to be a member of a committee whose head does not perform the necessary functions of the chairperson. Unfortunately, this experience is all too common because: (a) chairpersons are often selected for reasons other than their administrative skills; (b) successful faculty members often lack the interpersonal skills for effectively coordinating the work of others; and (c) skill at committee work is rarely taught or rewarded in academic settings. For these reasons we will outline in some detail the activities important to the role of committee chairperson.

Preparing for the Meeting

The first issue is, how should the committee chair prepare for the actual meeting? One effective means is to sit down with a blank sheet of paper and role-play the meeting: what decisions need to be made, what issues are likely to arise, what needs to be accomplished after the meeting, and what preparation is necessary before the committee meeting can take place. This preparation involves three specific activities:

1. *Plan the agenda.* An agenda is valuable because it directs the chairperson's effort *and* that of the committee. Circulating an agenda prior to a meeting helps individuals prepare for the session and prompts pre-meeting discussions among participants that can promote effective group work. Most important, an agenda helps establish group expectations for orderly procedure and places boundaries between relevant and irrelevant topics. It also provides the chair with a means of coping with wayward discussions. Some critical rules for creating an agenda are (5):

- Sequence items so they build on each other.
- Sequence issues from the easiest to most difficult.

- Limit the topics to a reasonable number.
- Separate topics into "for information," "for discussion," or "for decision."
- If possible, list an ending as well as a starting time.

2. *Do your homework.* Preparing for a meeting not only involves thinking about it, but collecting all relevant information, both written and unwritten. The most important information on controversial topics is apt to be in other peoples' heads. Therefore, contacts with committee members prior to the meeting are an important part of preparation. These contacts can help the chair anticipate potential issues and disagreements as well as the introduction of extraneous material. Pre-meeting contacts are also an opportunity to discover the "hidden agenda" of each participant. Further, the chairperson can use these meetings to prepare the participants. For example, the chairperson can ensure that an important point is raised from the floor rather than from the chair. Most important, these pre-meeting contacts provide an opportunity to encourage participants to do their homework, thereby lowering the odds that committee members will be unprepared to discuss or decide an issue. Finally, preparation involves circulating the relevant background materials and agenda a day or two before a meeting. If circulated too far in advance, the less organized members will forget it or lose it.

3. *Determine the time.* The timing of meetings can be a critical feature of their success. For example, many meetings drag beyond the point of effective discussion. Very few committees achieve anything of value after 2 hours, and 90 minutes is sufficient to cover most issues. The time of day is also important. If a meeting is scheduled first thing in the morning, members may be fresh but anxious to proceed to other engagements. Late afternoon meetings, in contrast, can be more leisurely because committee members are less likely to have competing commitments. Careful timing provides a means of controlling meetings that go on too long. These sessions might be scheduled 1 hour before lunch.

Conducting the Meeting

The greatest single barrier to the success of any meeting is "the chairman's self-indulgence" (3). The clearest danger signal is when the chairperson dominates the discussion. From the standpoint of committee effectiveness the most desirable role for the chair is that of servant rather than master, assisting the committee toward the best deliberations or decisions. If the committee members perceive the chair as motivated by a commitment to their common task, then conducting the meeting becomes a much less onerous job. The major challenge confronting the chairperson during the meeting is to create and reinforce a healthy, problem-solving environment. This can best be accomplished by attending to the following six activities (5).

1. *Begin by previewing the meeting.* The most effective way to establish teamwork is to be task-oriented. Start the meeting with a review of the agenda and session objectives. It is important that this preview permit others to make useful contributions, e.g., to ask questions, or to sharpen the focus of a problem. By indicating that the agenda is open to revision, the chairperson may also discover unsuspected sources of disagreement which can then be placed on the public agenda, rather than being restricted to the hidden agenda.

2. *Encourage problem-solving behavior.* The behavior of the chairperson is a powerful influence on the behavior of other committee members. If the chair focuses on facts, on comprehending points of disagreement, on reducing personality clashes, other members will adopt a problem-solving stance. An important part of a problem-solving orientation is balancing conflicting opinions. If one individual tends to dominate, the chair may ask quieter members for their point of view. Silence often indicates disagreement with the prevalent theme of the discussion.

During this phase, the chair again needs to be conscious of the power of modeling, creating a climate in which *anyone's* assumptions and arguments can be questioned or challenged, even those of the chairperson.

3. *Keep the discussion on track.* The major reason for digression is controversy. Since conflict is uncomfortable, many people prefer to smooth over problems, to avoid confronting the real issue by concentrating on less important or irrelevant topics (6). At these points, the chair must intervene to bring the committee back to the topic. The agenda, previewed at the outset, is a useful point of referral for adjusting behavior, as is attending to the clock. Another useful device is to summarize where the discussion had been and where it appears to be going. Also valuable is the introduction of a short break. Two minutes of stretching and standing can redirect drifting concentration and also serve to interrupt repetitive discussion, creating the opportunity for a fresh contribution.

4. *Control the discussion.* There are many ways to exercise formal control of the discussion. These include requiring speakers to be recognized by the chair, having the chairperson comment on each contribution, using a flip chart or blackboard to summarize ideas, and using parliamentary procedure. The effect of each of these techniques is the same—to reduce the number of direct interpersonal confrontations. These techniques are particularly useful if tension is at a level that precludes rational discussion. On the other hand, many of these techniques are inappropriate at the small committee level or are apt to *produce* tension and earn the chair the enmity of the members. Under these circumstances, informal, personal control is more effective. This includes the use of group discussion techniques often practiced in the classroom. For example, indicating impatience with the garrulous by fixing eyes on the speaker, leaning forward, or nodding to indicate the point is taken. The chair can also suggest "the need to move on" by the tone of a reply. For the partic-

ularly long-winded, it is wise to seize any pause as an opportunity to shift the discussion to another speaker. Similarly, responsible discussion should be reinforced by appropriate intonations and expressions, and by indications that there is ample time for that type of contribution. This is particularly important in order to encourage members to make suggestions, and to prevent the development of a "squashing-reflex" in which new ideas are dampened automatically (3).

5. *Be explicit about decisions and decision procedures.* Committees make decisions by voting or reaching consensus. It is important for each committee member to have a clear understanding and agreement about how they are working, because the members may have different expectations about decision-making. Voting is used when the decision is deemed critical or when the committee is deadlocked. Voting guarantees a decision. It also creates negative reactions because it requires public commitment and creates a win-lose situation. Losers may attempt to balance their position on the next decision, withdraw their commitment to the committee, or entrench their position by insisting on a minority report. In short, voting can destroy the cohesion needed for effective committee work. In contrast, achieving a committee consensus is usually more effective, if more time consuming. Seeking consensus involves listening to all points of view, and this usually results in greater individual commitment to the committee decision. This can be quite important when committee members will be responsible for helping implement the decision. Even if members do not fully support the decision, the fact that their postion had a complete hearing means they are less likely to try to sabotage the decision.

6. *Ending the meeting.* The most critical task at the close is to be explicit about what will happen next. If the committee does not have a regularly appointed meeting time, the time and place of the next meeting should be scheduled before disbanding. It is most critical to ensure that there is agreement on decisions made, on who is responsible for implementing them, and when the work will be completed. It is also useful to close on a note of achievement. Even if the final agenda item is unresolved, an earlier positive decision can be underscored and the members can be thanked for their contribution.

Following the meeting, the minutes should be distributed. They should include (3):

- The time and date of the meeting, where it was held, and who chaired it.
- Names of all present and apologies for absence (listing those who were late as "late comers" will help control the starting time).
- All agenda items (and other items) discussed and all decisions reached. If action was agreed on, record (and underline) the name of the person responsible for the assignment.

- The time at which the meeting ended (important because it may be significant later to know whether the discussion lasted 15 minutes or 6 hours).
- The date, time, and place of the next committee meeting.

These collected suggestions for improving academic committee meetings may seem like a great deal of work. Relative to the time expended in committees and their potential impact on academic effectiveness, the time requirement is slight. The techniques are simple, yet by applying them carefully to the people and problems facing the chair, academic medical committees can be made both more effective and interesting. The key is systematic preparation prior to the meeting, and sensitive observation and intervention while the meeting is in progress.

Management of Ad Hoc Committees

The appointment of ad hoc committees and task forces is a frequent occurrence in academic medical centers. The existence of many separate departments and divisions, the tradition of faculty governance, and the relatively small numbers of administrative staff (compared to nonacademic organizations) create the need for special committees composed of diverse institutional members. Ad hoc committees are often appointed to address unexpected changes in the medical center's external environment (e.g., government requirements for ethical reviews), changes in the internal environment (e.g., reorganization of departments), or anticipated internal/external changes (e.g., a new curriculum). Because these committees are usually established outside the existing committee system to examine issues not easily addressed by traditional decision making processes, ad hoc committees pose special problems of establishment and management.

Ad hoc committees are a powerful management tool, but they have particular costs and benefits (7). An example of benefits is that an administrator can quickly appoint a problem-solving committee with handpicked participants who have the authority to move across organizational lines and regulations. The committee can examine alternatives without requiring the administrator to make a fundamental commitment of the organization. These committees can also be used as a training mechanism to develop and assess potential middle managers as well as to create ties between academic units.

Ad hoc committees also create liabilities. Creating a powerful committee outside the traditional system, even if temporary, can be disruptive. Does the creation of the task force suggest other units are incompetent to deal with the problem? Will the decisions of the ad hoc committee (e.g., a new governance structure) change the distribution of power in the organization and thereby elicit attacks from threatened departments or units?

Given these issues, there are four considerations that should govern the formation of an ad hoc committee or task force. First, what should be the charge to the committee? Second, what criteria should be used for the selection of the committee members? Third, what should be the relationship between the initiator(s) and the committee? Fourth, how should the committee operate and render its report?

A major source of difficulty for a special task force or ad hoc committee is that of determining its task. In a standing committee, or a committee such as a student promotion committee that carries out repetitive activities, the nature of the task is obvious because of the experience of continuing committee members, or because of explicit procedures or regulations. This is less clear in an ad hoc committee. Therefore, the *charge* to the committee by the initiator(s) is significant. Often the charge is vague or inferential. An important first step is therefore a joint clarification of the charge between the initiator(s) and the chairperson. This negotiation often occurs not only at the beginning but at critical points in the committee's work. A wise and experienced faculty member will negotiate the charge, and determine whether there are any covert reasons for forming the ad hoc committee, prior to accepting the role of committee chair. An effective charge addresses both *what* must be done and *how*. For example, what should be the form of the committee's output (e.g., a report, interim presentation, or demonstration)? Does the initiator seek one solution, or multiple solutions? An example of a "how" issue is the committee's jurisdiction. Will it have access to necessary information and the capacity to question others in the organization? An important aspect of negotiating the charge is to provide an adequate balance between support and freedom for the committee to operate. This will be discussed further in a later section.

A second critical issue is the composition of the committee. There are four typical selection methods: a random method, a hierarchical approach, a representative method, and the "diagonal slice" (8). An example of a random method would be to allow whomever wishes to join the committee. The effects of this selection process are, of course, unlikely to be random. Since one of the major values of the ad hoc committee approach is to maximize diverse points of view, a random selection method has little merit. Further, random selection effectively neutralizes the influence of the initiator(s) over the work of the committee. Therefore, although random selection is a possible alternative, there are minimal benefits and substantial costs associated with this approach.

A more typical strategy is what will be termed a hierarchical approach (Fig. 4.2). The hierarchy of the organization—deans, department heads, professors, associate professors, and so on—is used as a basis for committee selection. For example, an ad hoc committee on budgeting and planning procedures for a medical school might include each department

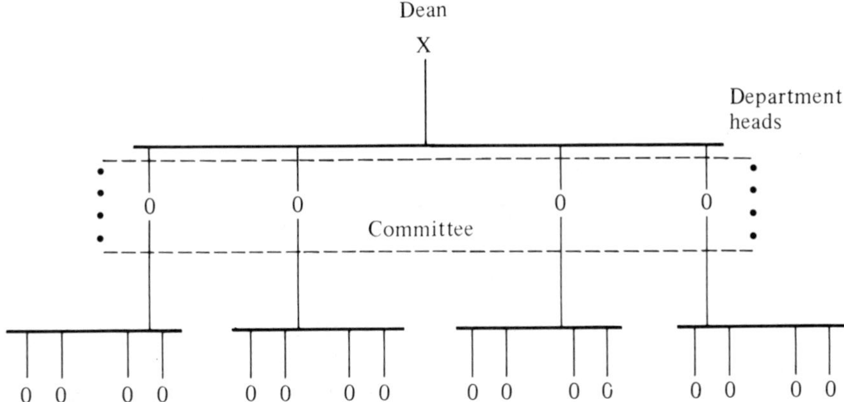

FIGURE 4.2. Hierarchical approach. Shown is an example of an ad hoc committee on budgeting and planning procedures for a medical school.

head in the school. A task force to reorganize a department would include representative full professors, and so on. This strategy for selection is particularly relevant when the problem to be pursued requires the concurrence of formal authority figures, or when negotiation along authority lines is necessary for implementation of the decision. A hierarchical approach can be ineffective for problems that demand high-quality solutions, i.e., those where diversity of information and insight can be important. Also, junior people are apt to become alienated if hierarchically based committees regularly define their conditions of work.

A third, and more prevalent, selection criterion would ensure a committee composed of representative types of people. For example, a committee might be constructed to reflect the demographic composition of the staff: doctors/nurses, tenured/nontenured, older/younger, men/women, whites/blacks. Or individuals may be chosen because they collectively represent the variety of issues relevant to the committee's mission, or because they represent interest groups (e.g., researchers and clinicians). There are obvious problems with a strictly representative approach, most particularly that of ensuring "adequate" representation. That is, once a representative technique of selection is perceived to exist, the potential "representatives" expand geometrically.

The fourth approach is termed the "diagonal slice" (8). This method meets the need for a representative group, but representation is along hierarchical dimensions in one relatively small group of people (Fig. 4.3). In this method, the committee combines both different levels of authority *and* different functional or interest groups. The committee can be carefully selected to represent the senior dimension (i.e., those with tenure, authority, tradition, and age) as well as the junior dimension (i.e., those who are untenured, newer to the organization, less committed to the

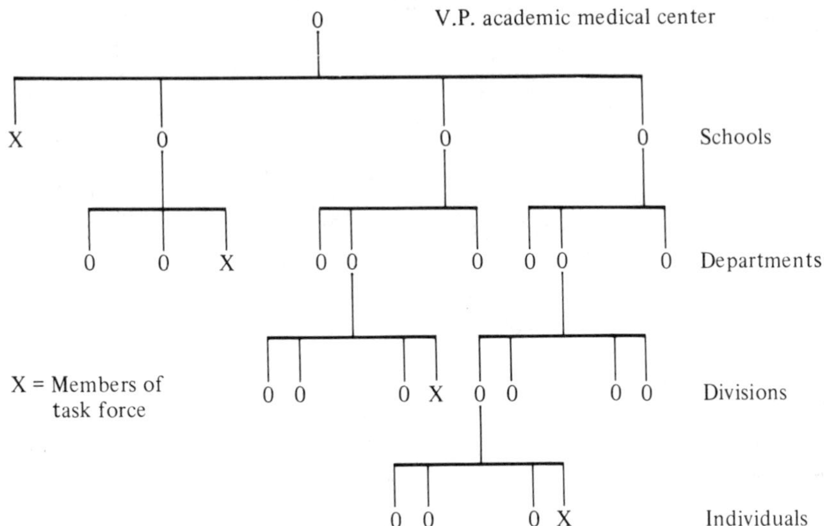

FIGURE 4.3. Diagonal slice approach. Shown is an example of a planning task force for an academic medical complex.

organization's prior traditions and possess less authority). Each individual on such a committee must be artfully chosen. For example, the senior people must be tolerant of junior people, and the junior faculty members must be individuals not easily squelched by participating with those higher in the organization.

A carefully selected diagonal slice committee has several advantages. First, information gathering is likely to be both more adequate and accurate. What, for example, are the points of view of doctors and nurses in satellite clinics or patient care facilities? A carefully selected diagonal slice provides routes of access to information needed for the committee to function, and also provides intracommittee mechanisms for cross-checking the reliability of the information. Second, the diagonal slice provides greater organizational legitimacy than the random or hierarchical approach. If change is to be an outcome of the committee's deliberations, organizational members must have trust and confidence that individual points of view were reflected in the committee discussion. This legitimacy is particularly important to junior academic staff who may justly feel that their views and values are ignored. Finally, the diagonal slice increases the probability of getting organizational commitment. Committee recommendations need to develop a broad base of acceptance. The variety of communication paths available to a diagonal slice committee is also useful in "selling" the committee's recommendations.

A third issue in the management of ad hoc committees is the relationship between the initiator(s) and the committee. If the committee is appointed by a particular person such as the dean, it is best for the dean not to be a committee member, but to be an advocate of the committee and its work. Because the committee is fulfilling a task for the initiator, it is necessary for both the initiator and chairperson to attend to the central issues of support and control (7).

Support relates to whether the initiator will back the committee with personal power and authority. If the work of the committee becomes controversial, will the initiator come to the committee's defense? Will the initiator back the committee's need for information and cooperation from others? Support is also expressed in terms of tangible items: release time from other academic obligations, a committee budget, and secretarial assistance. Not only does this support assist the committee in its operation, but it also demonstrates to the committee members the significance of their work.

Control, in contrast, relates to the restrictions placed on the committee's activities. The specificity of the charge to the committee represents one such restriction. Another is the degree of access to privileged information and key organization members. Control may be ongoing, as when the initiator reserves the right to attend meetings or insists on interim reports. The amount of control is likely to be greater if the task is critical and there is substantial variance in possible outcomes. In contrast, if the task is quite specific and there are few potential outcomes, control is less important.

Similar to negotiating the charge, negotiation of support and control is an ongoing process between the initiator and chairperson. For example, strong public support may be necessary at the establishment of the committee, but greater freedom may later be necessary in order for the committee to attend to its task with high legitimacy.

The final issue is the management of task forces. As should already be obvious from the prior discussion, the skills of the chair in communication and negotiation are much more critical for this type of committee. Because the legitimacy of an ad hoc committee must be developed (unlike, for example, a standing committee of the faculty), the chairperson's academic influence can be critical to establishing the committee's power and reputation. Further, the chair of an ad hoc committee will spend correspondingly more time *outside* the committee seeking information and support, mending fences, and attempting to maintain an independent stance. The chairperson's interpersonal skills in negotiation and communication therefore must be effective. Further, because of the temporary nature of the committee and the inevitably diverse backgrounds of its membership, correspondingly more time must be spent *within* the committee building cohesiveness and commitment (7).

The previous section of this chapter that addressed the meeting management skills necessary to build cohesive committees will be of equal relevance to temporary committees. But the unique characteristics of ad hoc committees lead to some additional suggestions about their management (9).

- Full task force meetings should be held frequently enough to keep all members informed about group progress.
- Dividing into subgroups will be necessary unless the task force is of small size (five to seven members or fewer).
- The chairperson must be cautious not to align himself too closely with a particular position or subgroup too early.
- Interim project deadlines should be established and met.
- Attention should be given to the conflicting loyalties task force members will experience by virtue of being on the task force.
- Communicating information among task force members and between the task force and the outside organization is the chairperson's most critical role.

The operation of an ad hoc committee or task force must be similarly altered, particularly if the group is addressing a critical problem that will require broad support to solve. A major criticism of the work of task forces is their use of time. They may spend up to 90% of their time gathering data, and less than 10% of their time deriving results, formulating alternatives, and promoting recommendations. An alternative strategy is to work "iteratively" (8). That is, produce the complete committee report several times, each time with greater refinement. This would require issuing a first report as soon as possible, approximately one-third of the way toward the committee's deadline. This first report should be a complete report, addressing all issues, if lightly. The task force should then seek criticism from the initiator or decision-making group. Then the committee should write another complete report, addressing objections, adding needed data and analyses, fleshing out the recommendations. The committee would then present the report again and go through another iteration. This process will not only help to build legitimacy and support for the committee's work, but will also help the decision-making groups understand the reasoning that led the committee to its conclusions. This process of developmental discussions and responsive modifications in the committee report is most likely to lead to successful implementation of results.

Conclusions

The organization and collegial traditions of academic medical centers lead to the frequent use of committees in the decision-making process. The large number of specialized schools, departments, and individuals as well as the traditional value of faculty control of academic programs mean that membership on continuing and ad hoc committees will be a necessary part of a faculty member's life. This chapter has reviewed the dynamics of committee functioning, the relevant skills for effective participants, the tools appropriate to chairing academic committees, and finally the issues associated with managing ad hoc task forces or committees in academic medical settings. The skills of effective committee work are simple. They are easily learned through careful observation and immediate experience. Since the role of the academic medical professional is quite complex, however, attention to committee skills is unlikely to be a high priority to the new faculty appointee. By summarizing these points in written form, we hope to shorten the learning curve of the new faculty member and improve both their professional effectiveness and use of time.

References

1. Baldridge JV, Curtis D, Ecker G, Riley GL. *Policy Making and Effective Leadership.* San Francisco: Jossey-Bass, 1978.
2. Field RH. A critique of the Vroom-Yetton contingency model of leadership behavior. *Acad Mgmt Rev 4*: 249–257, 1979.
3. Jay A. How to run a meeting. *Harvard Bus Rev 54*: 43–57, 1976.
4. Bradford LP. *Making Meetings Work.* LaJolla, California: University Associates, 1976.
5. Ware J. A note on how to run a meeting. Cambridge, Massachusetts: *Intercollegiate Case Clear. House* (9–478–003), 1977.
6. Weisbord ML, Lawrence PR, Charns MP. Three dilemmas of academic medical centers. *J Appl Beh Sci 14*: 284–304, 1978.
7. Bradford DL, Bradford LP. Temporary committees as ad hoc groups. In *Groups at Work.* R Payne and C Cooper (Eds). London: John Wiley, 1981, Pp. 121–138.
8. Roberts EB. Strategic Planning—a framework for managerial decision making: organizing for strategic planning. *Management Advancement Program Videotape No. 4.* Washington, D.C.: Association of American Medical Colleges, 1979.
9. Ware J. Managing a task force. Cambridge, Massachusetts: *Intercollegiate Case Clear. House* (1–478–002), 1977.

5
Academic Medical Organizations

JOHN ALUISE

> Both universities and health care delivery systems present complex unsolved organizational problems which limit their effectiveness and their ability to respond to changing circumstances.(1)

Chapter 1 points out that academic medical centers are unique organizations. They are complex and often disorganized systems as a result of the multiple duties of academic physicians, a tradition of individual professional autonomy, and the weak link between academic work and its rewards. This chapter argues that such circumstances highlight the need to manage and integrate better the different features of academic medicine. The intent is to show that improved management and integration will *support,* not undermine, academic productivity and satisfaction. The focus is on medical departments because that is the organizational level at which better management can frequently yield tangible benefits for individual faculty.

Medical faculty are more often engaged in administrative chores than their colleagues in other University departments. The reasons for this range from the complicated organizational patterns just cited, to the need to supervise nonphysician health personnel, and staff turnover. The upshot is that research, teaching, and sometimes even clinical responsibilities are sidetracked to deal with administrative problems or to "put out brushfires." Robert L. Friedlander, Dean of Albany Medical College, New York, argues that such changes in organization and funding have brought with them a "new reality" in which academic medical professionals must place as much weight on their organizational role as they do on their academic role (2). Friedlander gives examples of what this organizational role can involve:

- Medical schools and health centers will have to be more selective in what they do.
- Faculty will have to forsake "academic purism" and get involved in marketing, fund raising, and political activities.
- As institutions become more businesslike, faculty will have to become more accountable for their effort, mission, and income.

- A greater pragmatism about research. . .and its immediate translation to new knowledge for clinical problem-solving will be necessary.
- Medical practices of faculty members will be more efficiently controlled, with new sources acquired for financial support.
- Faculty members will learn more about sharing resources and services.
- Physicians and scientists will need formal preparation in management.
- The use of voluntary faculty may come back into vogue.
- Tenure might be redefined and awarded less frequently.

These points are dramatized and express a bias toward tight management. Some would say they betray an insensitivity to traditional academic values. Nonetheless, the organizational changes to which Friedlander's ideas are addressed are occurring. Academic physicians, especially those who are new in their institutions, should not be indifferent to these issues.

The purpose of this chapter is to introduce academic physicians to organizational concepts that can help them respond to problems that arise in academic medical centers. The following sections will introduce the reader to concepts of organizational health, stages of organizational growth, and organizational problems in academic medical departments. Given this introduction, academic physicians should be better equipped to understand the environment in which they work and better prepared to accept the managerial responsibilities that are an inevitable part of their professional role.

Organizational Health

A business firm defines its effectiveness in terms of its profitability for investors—the "bottom line." Nonprofit organizations such as universities, research laboratories, and symphony orchestras are also concerned with the bottom line but there is no convenient measure such as return on investment to help guide short-run managerial decisions. Consequently, organizational psychologists have adopted the subjective concept of "health" from medical practice as a way to make judgments about organizations and to guide possible interventions.

Academic medical centers, and their departmental units, are different from business and industry because their principal capital is human. Medical centers rely on the skill and knowledge of highly trained professionals to function effectively. Thus, the health of academic medical centers depends on several features that are unique among professional organizations.

Coordination

Academic medicine is not a "single product" enterprise. Instead, departments operate as a collection of separate, sometimes autonomous, divisions that address patient care, education, and research. The coordination of these activities, and the professionals who perform them, is a constant organizational challenge.

Unhealthy symptoms of the coordination problem are observed from strained relationships among the tasks of teaching, research, and patient care. Are these tasks conducted at different sites under the supervision of different people? Is there any form of supervision in operation or are the academics left alone? Are there conflicts regarding the allocation of people and money for education, scholarship, and clinical work? In addition, academic physicians will sometimes admit that their skills are weak in areas such as research despite the expectation that they should obtain grants and publish articles.

Healthy professional organizations, by contrast, have developed mechanisms for dealing with the coordination issue. For example, more frequent communication between individuals and groups is observed in departments that effectively manage faculty behavior. Expected outcomes and accountability for the tasks each professional performs are clearly stated, or at least clearly understood. Healthy departments accept the inevitable conflicts that arise between professional priorities. They create opportunities such as meetings between division chiefs where short-run conflict is addressed before it deteriorates into long-run crisis. Healthy departments also use performance appraisal techniques such as those discussed in Chapter 3 to provide means of integrating faculty objectives with departmental goals.

Integration

Studies of professionals who work in organizations indicate that the more autonomous and highly skilled the individual, the more likely it is that the professional's work values will oppose organizational criteria (3,4). In a frequently cited study of academic medical centers (5), it was reported that physicians do not work well in joint activities, and they resist authority from others over their clinical, teaching, and research responsibilities. Integrating academic physicians into working groups is further complicated because they seldom use effective conflict management skills. Avoiding conflicting issues—"smoothing"—is preferred to confronting problems directly. Another outcome of the academic medical center study was an "anarchy index," a measure of professional resistance to authority and accountability. Except for administrative tasks, few academic professionals acknowledge that they are accountable to a superior for their performance. In short, resistance to integration is a common symptom of unhealthy academic organizations.

The integration of work at some level is clearly a key to organizational health. Healthy organizations allow individuals to weave their personal and professional priorities. They encourage individual initiative and promote nontraditional forms of faculty behavior. These include a wider acceptance of frank communication and even confrontation as appropriate academic conduct. Healthy professional organizations value the independence of professionals by providing them opportunities to exercise judgment and expertise. However, professionals in healthy organizations accept the responsibility to use resources efficiently, promote collegial relationships, and work as a team member.

Leadership

Since professionals are self-directed and prefer to work independently, directive styles of leadership are often inappropriate. Research has revealed three traits of effective leaders in healthy professional organizations (6). The first trait is technical competence. Professionals accept expertise-based authority. The technical competence of leaders provides a basis for recognizing good ideas, helping define significant problems, and providing technical support. The second leadership characteristic is promoting group decision-making. The leader's ability to motivate professionals depends on the involvement of professionals in planning and making decisions. This is especially true for decisions about goals and resources. Professionals are more apt to accept a decision if they understand the logic behind it and have had a chance to evaluate critically different courses of action. Participative decision-making does not imply decisions are totally in the hands of the group. Effective leaders emphasize that goals must be specified and achieved, but allow individuals leeway in how they reach them. The third leadership characteristic is providing work challenge. Academic physicians enjoy hard work. Leaders who are sensitive to work challenge and how to match individuals with tasks are apt to be more effective. In short, leadership symptoms of a potentially unhealthy professional organization are a lack of technical competence, nonparticipative decision-making, and an insensitivity to creating opportunities for challenging work.

Collaboration

Recent thinking about professional organizations offers an amendment to the traditional hierarchical authority structure. A combined horizontal and vertical system of programs, relationships, and accountability is considered more suitable for professionals in organizations (7–9). The organizational description of this structure is called a "matrix" system (10). A matrix system is composed of multiple programs with organizational members working in two or more units. Program leaders plan projects, negotiate activities, and resolve conflicts with professionals in a team manner. A collaborative form of planning and decision-making is critical for this arrangement. Academic organizations with few mechanisms to encourage horizontal collaboration would be diagnosed as less healthy than those where collaboration is routine.

In summary, the characteristics that tend to promote the health of professional organizations such as academic medical centers include a means of coordinating the multiple roles of patient care, education, and research; integrating autonomous professionals into working groups; leadership based on expertise and recognizing the professionals' needs for self-direction; and collaboration that promotes the horizontal coordination of education, research, and clinical work. If these attributes exist in a professional organization, academic physicians will likely find it a "healthy" place in which to work.

Organizational Growth Stages

Studies of organizations suggest they move through cycles, or stages of development, as they mature and expand. Each stage requires a different style of leadership and a different organizational structure.

A model depicting stages of organizational growth (Table 5.1) has been developed to illustrate how organizations mature (11). The four stages of growth have been termed creativity, direction, delegation, and consolidation. Each stage contains a stable growth period that typically ends in a crisis. In some instances, solutions in one phase may later produce a crisis as the organization evolves into its next stage.

TABLE 5.1. Phases of organizational growth.*

	Stage1 Creativity	Stage 2 Direction	Stage 3 Delegation	Stage 4 Consolidation
Growth period	"High commitment by a few"	"Decision-making is centralized"	"Distinct programs or divisions are created"	"Cooperative system of working/cross-sectional teams"
Crises	Leadership "Informal system begins to break down; formal leadership is needed"	Autonomy "Members desire more freedom and input into decisions"	Control "Resources spread too thin; turf issues evolve"	Red tape "Proliferation of planning and reporting to coordinate diverse activities"

*Adapted from Larry Greiner, "Evolution and Revolution as Organizations Grow," *Harvard Business Review 50*(4): July–August 1972, p 41.

Creativity

The initial stage of development is often led by a founder who provides intense energy and commitment. The management is informal, with a small group of dedicated individuals working cohesively to establish a meaningful program. A typical academic department in this stage would be described by the following passage:

A small group of faculty and staff work as a close-knit group. The chairman knows all members personally and has trust in their abilities, and they in the chairman's. There is high *esprit de corps* with a strong sense of mission. There are no layers of management or decision-making. Everyone including the chairman works on all tasks. External agencies may be providing resources and political support. New ideas are welcome and flexibility and change are commonplace. Titles and status are irrelevant.

Stability and organizational health in this stage are difficult to sustain as the number of members increase and personal communication becomes less effective. Newcomers are not necessarily as motivated as the originators. The founder discovers that charisma and dedication alone are insufficient to solve the organization's expanding problems. In time, management responsibilities mount and it is obvious that things are not like they were in the "good old days." A *leadership* crisis often occurs at this point as a consequence of reliance on informal mechanisms of communication and coordination. Leadership is now needed to provide structure and resolve problems. External involvements may demand more of the leader's time, pulling the head away from the department. Several departments have moved through this crisis stage by hiring a business administrator to conduct internal financial and personnel operations, and appointing a vice-chairperson to coordinate internal academic programs. The chairperson is then free to continue work outside the department.

Direction

Following the initial stage of evolution the next phase is characterized by a management focus that emphasizes efficient operation. Policies and procedures are written. Organizational decision-making becomes more centralized. Tasks and projects are assigned to specific individuals or subgroups. Communication becomes more formal and top management provides direction. The following vignette describes a department in the direction stage of development.

The department is led by a chairman and a few senior faculty. These individuals initiate and coordinate all activities. Considerable time is devoted to establishing policies and strategies. There is a need to "get organized." Department members are occupied with program development. There is a clear distinction between decision-makers and functional specialists.

Although the department has evolved and become more efficient, a new crisis can develop as a result of the rigidity of centralized management. The hierarchy becomes impersonal and removed from core problems. Policies once established for a single function become too inflexible to apply across multiple functions. Initiative is stifled when decisions are delayed and new ideas are stymied in bureaucratic channels. The second developmental problem is a crisis of *autonomy.* Faculty members demand greater freedom and participation in decision-making. This crisis is often prolonged because senior leaders are reluctant to relinquish control. If the chairperson and senior faculty are unwilling to share their leadership, the department may be stalled in the direction crisis phase. This crisis is often resolved by a change of leadership.

Delegation

As departmental expansion increases and faculty become more proficient in managing their programs, the organization moves into a decentralized management phase. The delegation stage of an academic department would be described as follows:

> Distinct programs or divisions are created with directors appointed to organize and lead. Faculty and staff are assigned to one or more functional groups. Space, budget, and communications are allocated to each division. Division heads run their units as their "own little department." Grants and resources are obtained for decentralized units. Communication from the top is infrequent, or via division heads.

The problems that evolve in the delegation phase are those of *control.* As the department becomes highly diversified, "turf" issues develop. Communication among divisions and programs breaks down. Resources are spread too thin. Duplication and redundancy may occur. Sectionalism can cause the department to divide into multiple tracks that obscure its primary mission. Parochial attitudes may be prevented by reaffirming the departmental purpose and coordinating divisional activities so that goals are met, and resources are used efficiently.

Consolidation

Organizational maturation should ultimately lead to a final stage called consolidation. A consolidated organization is characterized by a cooperative system of interrelated activities. Members integrate their work by planning and communicating as a team. Collaborative working groups enhance consolidation. A department that reaches the consolidation stage could be described as follows:

> The department has become very global. Extensive planning and coordination are required. Management of information is critical. Divisions operate under dual

accountability—to their unit and the department. Negotiations are necessary to keep individuals' needs aligned with department goals. Interdisciplinary teams are used for planning and implementing departmental programs. Faculty are equally satisfied by individual and group work.

The problem or crisis that occurs in the consolidation stage could be one of "red-tape" with the proliferation of planning and reporting. A consolidated system requires a generalist approach to management, which may create conflicts for professionals who enjoy the uniqueness of specialists. Attrition may also be a natural occurrence at this stage, not from incompetence, but from incompatibility.

A useful means of summarizing these four developmental stages is to relate them to the previously discussed characteristics of a healthy organization. Table 5.2 analyzes the four stages of unit development and suggests the typical mechanisms of coordination, integration, and collaboration at each stage. In the creative stage leadership may be charismatic and entrepreneurial; technical competence may be less critical. The structure may be informal, with frequent communication and easy collaboration being used to solve most problems. Integration is self-evident because the founders believe in what they are doing. As the unit grows in number of faculty and complexity of tasks and programs, predictable strains emerge. In the direction stage these strains are usually dealt with through rule-based, centralized management. At this point "smoothing" becomes a frequent response to conflict. In the delegation stage, complexity is addressed by a decentralized leadership style that attempts to create smaller vertical units as a means of redeveloping collaboration and integration. In the consolidation stage, complexity is recognized, and a matrix structure emerges that combines vertical and horizontal forms of management.

This discussion suggests that organizational health, like physical and psychological health, is a function of both internal growth and environmental change. Progress toward organizational development requires as a first step an appreciation of the dynamic quality of organizational change. Progress can also be enhanced from three types of knowledge: (a) about the stage of development of one's unit; (b) of the leadership and structural mechanisms currently being used; and (c) about the leadership and structural mechanisms potentially available.

Every department and its divisions are at different stages of development. Each stage has an evolutionary and a revolutionary feature. It is unlikely that either feature can be bypassed. Chairmen and faculty need to recognize that leadership requirements change. Organizational crises are inevitable. Conflict should not be avoided, but confronted directly so that appropriate solutions can be reached. Organizational audits and management consultations may enhance awareness of problems and assist faculty members in developing strategies for professional growth.

TABLE 5.2. Organizational health characteristics at different phases of growth.

	Stage 1 Creativity	Stage 2 Direction	Stage 3 Delegation	Stage 4 Consolidation
Leadership:	Leader does everything	Leader as bureaucrat	Leader as mission definer	Leader as supervisor of faculty careers and coordinator of program heads
Structure:				
Coordination	Fewer tasks and activities Every person does everything	Rule-based More specialized	Decentralized vertical structure either by program or subspecialty	Program heads coordinate teams for various programs drawn from unit as a whole
Integration	Shared norms	Contractual	Strong identity with program or specialty unit	Dual identification. Negotiate commitment to unit on specialty knowledge; negotiate commitment to program on tasks
Collaboration	High face-to-face communication More confrontation "Collegial" decision-making	Less frequent face-to-face communication More smoothing	High communication and collaboration *within* decentralized unit	More informal, frequent communication. More confrontational and participatory, particularly in programs

Organizational Problems in Academic Medical Departments

The previous sections introduced some characteristics of healthy organizations and the concept of organizational stages of development. Although these issues are important to understand how academic medical organizations work (or don't work), a fine-grained analysis is needed to help faculty apply the concepts. In this section, ideas about the leadership and structure of academic medical departments will be presented, closely following the organizational health categories introduced earlier. The analysis is based on a study of academic family medicine departments using the organizational health categories (12). Some of the research questions used to gather information are included here along with representative findings. The data should give readers a sense of organizational issues in medicine and a method of inquiry should they wish to conduct a diagnosis of their own department. As described earlier, four general categories are useful for understanding organizational health: coordination, integration, leadership, and collaboration. Each of these four topics is used to discuss the research findings from the study of academic family medicine departments. Observations are also added about two other categories that the academic medical departments reported to be useful in addressing organizational problems. Consequently, a fifth category describes environmental (institutional) constraints that can affect the health of medical units and a sixth category identifies "helpful mechanisms" for the improvement of departmental operations.

Coordination

During interviews with chairmen and faculty the following inquiries were made to analyze departmental goals and objectives. What are your department's goals? What are specific objectives in education, research and patient care? How clear are the goals? What is the level of faculty and staff commitment? How have goals changed?

The departments generally operated under three missions: first, to train physicians to meet shortages in primary care; second, to develop patient care and educational programs to convey appropriate knowledge and to acquire necessary skills; third, to establish family medicine as a credible academic discipline. These goals are widely acknowledged; some have even been legislatively mandated. Educational and patient care objectives were much more clearly stated than research objectives. Faculty members often had different views about the importance and means of achieving goals. Physician faculty were committed to patient care, whereas nonphysician faculty emphasized their work in education and research. The recognition of primary care as an accepted academic field is generally seen as an unmet goal.

As previously discussed, delegation and decision-making in professional organizations are best applied in a decentralized structure. This horizontal structure requires leaders and followers to participate in a collegial or collaborative fashion. Department chairmen tend to function as benevolent autocrats (sometimes not so benevolent), especially with regard to financial and personnel decisions. Delegation to vice chairpersons, administrators, and senior faculty occurred in a few departments, but not in a systematic manner. A decentralized structure was evolving slowly, but faculty found it difficult to negotiate differences and resolve conflicts at a peer level. Assignment of responsibilities occurs in reaction to crisis, not with any prearranged plan or scheme. Some faculty wanted chairpersons to make the difficult decisions whereas others strongly advocated participative management.

Integration

Accountability in academic medical organizations is impeded by a lack of uniform performance standards, especially in patient care, teaching, and research. Faculty struggle with the accountability-reward process. Even though departments have a general view of mission and goals, specific task expectations have not been instituted to the satisfaction of faculty. The academic advancement process has not been developed systematically because of unclear promotion criteria, lack of performance standards, and frequently changing work demands.

The study of academic family medicine departments found only one objective productivity factor for physician faculty—clinical income. Even though this represents only a small portion (10–20%) of faculty performance, patient care activity was shown to be growing increasingly important as subsidized financial resources diminish. Institutional promotion criteria were often criticized by family physician faculty because of their emphasis on research and scholarship. Questions were also raised about faculty evaluation, performance appraisal, equitable compensation and benefits, and institutional acceptance of departmental and faculty performance. Generally, the departmental promotion and reward systems were found to need much attention. Written policies were rarely available.

Leadership

Leadership in professional organizations is problematic because academic professionals pride themselves on independent action and setting their own goals. To assess the nature of the leadership, the following questions were posed: How and when do chairpersons perform multiple roles of educator, clinician, and manager? How do they fulfill their administrative responsibilities? In what way are chairpersons effective or ineffective leaders? What could they do to improve their leadership?

Chairpersons were typically described as innovative, creative, and dedicated. In most departments they were founding fathers. Chairpersons performed multiple duties as clinicians and educators, even though administrative responsibilities were increasing. Only one chairperson of the 10 studied stated that his job was that of an executive. He met regularly with program directors to plan and evaluate department-wide activities and did *not* see patients. Generally the effectiveness of chairpersons was perceived as quite positive in regard to "past" activities, but many concerns were raised about the lack of clear direction for the future and uncertainty regarding departmental priorities. Several leadership needs became evident. They included setting and ranking departmental goals, improving working relationships with other medical school departments, establishing meaningful affiliations with community hospitals, developing efficient financial and administrative systems for departmental management, and negotiation of faculty roles and responsibilities.

Collaboration

Managing conflict is often the difference between high and low organizational performance. The family medicine faculty often reported many conflicting relationships and associations. The most difficult external relationships were with faculty in other departments and administrators within the medical school. Intradepartmental conflicts usually involved chairpersons and faculty, especially when chairpersons maintained tight control. A more subtle conflict surfaced when faculty spoke of difficulties encountered when they ask for peers to assist them with administrative responsibilities. Relationships appeared to be strained because of multiple roles and lack of time-management skills. Even though academic centers are believed to be collegial, teamwork in teaching, research, and management was seldom observed.

Environmental Constraints

Several research questions addressed environmental issues. What institutional policies directly affect the department? What is the department's financial status? What are the major limitations in resources, facilities, and personnel? What external factors or agencies influence achievement of goals and acquisition of resources? The most severe constraint was declining financial resources. Interdepartmental conflicts were also mentioned frequently. Several of the departments studied were also frustrated with rigid accreditation standards because of unique institutional and departmental circumstances. Another constraint was the rapid growth of departmental programs into such areas as geriatrics, nutrition, sports medicine, alcoholism, and other fields of endeavor despite limited numbers of faculty and resources. Pressure also arose from the need to affiliate with community hospital residencies to economize resources. These affiliations had not been a priority for some departments but seem to rank high in future plans.

Helpful Mechanisms

Each organization develops mechanisms that promote attainment of its goals. They can be classified into three categories: (a) policies, procedures, and meetings that help people work together; (b) informal devices and communications that help solve problems; and (c) traditional management functions such as planning, budgeting, and information systems (13). Questions were asked about each category. Very few departments used formal policies. In most cases the family medicine faculty found informality counterproductive—especially regarding promotion, compensation, and personnel supervision. General faculty meetings and one-to-one meetings with the chairperson were the most frequent communications. Even though faculty enjoy freedom to define their professional roles, they often asked for direction and consistency at the departmental level. Management information systems were antiquated, with very little use of computers and monthly or annual reports. Yearly planning and budgeting existed to a slight degree, but in no department did it involve all faculty. Each department could have benefited from a management audit to pinpoint specific administrative problems.

An organizational diagnosis of an academic medical department may detect important questions and concerns that affect professional roles within the department. Major topics and specific diagnostic questions can be raised during interviews that can later become the basis for department policy formulation. It is impossible for any institution or department to avoid organizational problems. However, it is important for a new member of an academic organization to learn about potential difficulties in advance. An organizational diagnosis could at least identify the "blips on the radar screen," so that more detailed investigation can be undertaken.

Conclusions

The American college or university is a prototypic organized anarchy. It does not know what it is doing. Its goals are either vague or in dispute. Its technology is familiar but not understood. Its major participants wander in and out of the organization. These factors do not make a university a bad organization or a disorganized one; but they do make it a problem to describe, understand, and lead. (14)

Academic Health Centers (AHC's) consist of myriad factors which contribute to complexity. AHC's try to accomplish three distinct missions which frequently conflict and lead to confusion, AHC's perform service functions for many kinds of clients without clearly defined performance measures. AHC's are vulnerable to a powerful environment in terms of changing technology, consumer demands, and governmental regulations. AHC's are comprised of highly trained professionals who often experience conflict between professional values and organizational expectations. And finally, influence is issue-specific which limits the control of one person or group. (15)

These two comments provide concluding descriptions of the nature of academic medical organizations. Academic professionals are both the culprits and victims in the organizational problems that they face. Even though academic physicians find many reasons to criticize their institution or department, faculty time and energy should be channeled into resolving problems caused by the complexity of the system. Before organizations (and their leaders) can be changed, they must be understood. The concepts and strategies presented in this chapter can provide a basic understanding of the realities and problems of academic medical organizations. As always, it is the faculty and its leaders that must act to solve them.

References

1. Evans J. Organizational patterns for new responsibilities. *J Med Educ 45*: 988–999, 1970.
2. Friedlander RL. Cited in Korcok, M. Medical News Editorial. *JAMA, 249* (1): 15–16, 1983.
3. Kerr S, Von Glinow MA, Schriesheim J. Issues in the study of "professionals" in organizations: the case of scientists and engineers. *Org Beh Hum Perf 18*: 329–345, 1977.
4. Andrews F, Farris G. Time pressure and performance of scientists and engineers: a five year panel study. *Org Beh Hum Perf 8*: 185–200, 1972.
5. Weisbord MR, Lawrence PR, Charns MP. Three dilemmas of academic medical centers. *J Appl Beh Sci 14*: 284–304, 1978.
6. McCall MW. Leadership and the professional. (Technical Report) Center for Creative Leadership, Greensboro, North Carolina, 1981.
7. Kraus WA. *Collaboration in Organizations.* New York: Human Sciences Press, 1980.
8. Sayles LR. *Leadership.* New York: McGraw-Hill, 1979.
9. Mintzberg H. *The Structuring of Organizations.* Englewood Cliffs, New Jersey: Prentice-Hall, 1979.
10. Davis SM, Lawrence, PR. *Matrix.* Reading, Massachusetts: Addison-Wesley, 1977.
11. Greiner L. Evolution and revolution as organizations grow. *Harvard Bus Rev 50* (4): 37–46, July-Aug. 1972.
12. Aluise J. A Study of the Organizational Structure and Leadership of Academic Family Medicine Departments. Unpublished doctoral dissertation, University of North Carolina at Chapel Hill, 1982.

13. Weisbord MR. *Organizational Diagnosis.* Reading, Massachusetts: Addison-Wesley, 1978.
14. Cohen MD, March JG. *Leadership and Ambiguity: The American College President.* New York: McGraw-Hill, 1974.
15. *The Organization and Governance of Academic Health Centers, Presentation of Findings, Vol. 2.* Washington, D.C.: Organization and Governance Project of the Association of Academic Health Centers, 1980.

Suggested Readings on Professional Development

Bennett JB. *Managing the Academic Department.* New York: The American Council on Education and MacMillan Publishing Co., 1983.

This book of case studies is a rich source of practical ideas and techniques for running today's academic department or division.The first two sections explore the nature of the chairperson's role. Dealing with conflicts is the subject of Section 3. Section 4 covers performance counseling and educational program management. The final sections examine management issues involved in departmental goals, changes, and decision-making and explore such emerging issues as changing attitudes, politics, and personal and professional loyalties. The conclusion outlines an agenda for the future, showing how the chairperson must act as academic entrepreneur, politician, and custodian of academic standards.

Bradford LP. *Making Meetings Work: A Guide for Leaders and Group Members.* LaJolla, California: University Associates, 1976.

Research on small-group behavior has identified a number of mechanisms for making group activities more effective. Unfortunately, this research is rarely connected to the groups that are most frequently used in academic settings: committees and task forces. Bradford, one of the leading researchers and consultants in the field, presents the relevant research in a practical and pertinent handbook that describes group dysfunction, mechanisms for increasing effectiveness, and the role of the group leader. The use of consultants, and the planning of large meetings and work-group conferences are also discussed. The appendices include several helpful questionnaires for evaluating group activities.

Keller G. *Academic Strategy: The Management Revolution in American Higher Education.* Baltimore, Maryland: The Johns Hopkins University Press, 1983.

Academic Strategy is based on the author's visits to campuses across the country and interviews with leading educators and planners. Its description of the special requirements of academic management and its revealing portrait of the latest developments in planning theory make this book a powerful tool for today's academic leaders. The book emphasizes the need for strategic planning in contrast with more mechanical, long-range planning. Strategic planning deals with an array of factors: the changing external environment, competitive conditions, strengths and weaknesses of an educational organization, and opportunities for growth.

Lakein A. *How to Get Control of Your Time and Your Life*. New York: New American Library, 1973.

This is one of the most practical and direct of the many available books on time management. The author addresses such everyday problems as establishing priorities, planning, list-making, organizing paperwork, saying "no," and delegating. Lakein argues that time management problems are frequently symptoms of job ambiguity and overload as well as an individual's search for perfection. Thus, means of dealing with stress, boredom, and fear-of-failure are discussed. The book also contains a chapter on career and life planning.

Levey S and McCarthy T (Eds). *Health Management for Tomorrow*. Philadelphia: Lippincott, 1980.

This book is a collection of invited essays by authorities in the field of health care administration. It addresses the problems and prospects for creating effective health care organizations, including academic medical centers, in an era of shrinking resources and environmental uncertainty. The book examines educational programs for health managers, underscoring the need for closer linkages between the worlds of practice and academia. Community hospital and academic health center governance and internal organizational control also receive attention. The book is an encyclopedia of health care administration with a pronounced future orientation.

Levinson DJ. *The Seasons of a Man's Life*. New York: Ballantine Books, 1978.

Although it is generally accepted that development continues throughout a person's life, in-depth studies of adult development are rare. This book represents a pioneering effort to create a theory of adult development, from the entry into adulthood until the late 40s. A view of the life cycle as a whole is examined by use of the seasonal metaphor. Detailed accounts of the lives of selected individuals provide clear, realistic grounding for the theory.

Starr P. *The Social Transformation of American Medicine*. New York: Basic Books, 1982.

This book maps the historical origins of contemporary medicine and describes the social context of medical care and the profession's role in society. It has received widespread acclaim and has been praised by Dr. Lester King of the Morris Fishbein Center of the University of Chicago as ". . . the best available book on the social problems of medicine." The volume will not make an immediate, practical contribution to the professional development of academic physicians.

However, medical professors will find intellectual enrichment from the scholarly insights it provides about the profession and practice of medicine.

Weisbord MR. *Organizational Diagnosis: A Workbook of Theory and Practice.* Reading, Massachusetts: Addison-Wesley, 1978.

This is a very practical workbook for analyzing one's own unit or organization. The book leads the reader through specific exercises involving organizational diagnosis, assessing purposes, organizational structure, leadership, and power. It also includes a selected set of useful readings. The workbook is particularly useful for academic physicians because its contents reflect major studies of academic medical centers.

Wilson MP and McLaughlin CP. *Leadership and Management in Academic Medicine.* San Francisco: Jossey-Bass, 1984.

The leaders and managers of academic medical centers have a monumental task in dealing with the variety and complexity of their organizations. This book examines the critical role of the leadership cadre in academic medicine. The authors analyze the settings in which leaders work, processes by which decisions are made, and the qualities of executive leadership. The stimulus for this book grew out of the work of the Department of Institutional Development of the Association of American Medical Colleges (AAMC). The book is divided into three major sections: Academic Medical Center and Its Leaders, Critical Managerial Functions, and Leadership for the Eighties and Nineties.

II Medical Education

Education is one of the academic physician's key responsibilities yet rarely is it done with the kind of enthusiasm and attention to detail that doctors give to patient care. Many varieties of learners—medical students, residents, fellows, allied health professionals, colleagues who receive continuing education—are the focus of our educational endeavors. Members of each group require instruction and evaluation that takes account of their present fund of knowledge and experience which is then enriched from teaching and rigorous assessment. Done well, medical education at any level is both deliberate and cyclic. It is deliberate because sound educational activities are thoughtfully planned and managed with great care. It is cyclic because planning always starts by estimating the learner's readiness to receive instruction and proceeds to map a course for learning. Later episodes, in turn, begin where current work ends.

Chapter 6 sets forth some basic principles of educational planning and also offers an opportunity to apply the principles in a simulated curriculum development experience. While the chapter's focus and examples are drawn from undergraduate medical education, they are also directly applicable to postgraduate and continuing medical education.

Chapter 7 addresses the topic of clinical instruction, teaching that occurs in the patient care environment. Seasoned academic physicians know that clinical instruction is very different from instruction in a classroom or lecture hall, yet few have considered the basic elements that contribute to success in clinical teaching. This chapter identifies those elements and points out how clinical instructors can increase their skill and effectiveness.

Chapter 8 is concerned with the evaluation of learners. This is a basic educational responsibility of nearly all academic physicians although it is a duty that few approach with enthusiasm or lasting interest. However, like teaching, the evaluation of medical learners at all levels needs to be thoughtful and systematic to be effective. Learners, teachers, the medical profession, and the public all benefit from evaluative practices that contribute to the accuracy of important decisions that are made about individual physicians; for example, medical school promotion and graduation, licensure, and certification. This chapter describes how mechanisms for learner evaluation can be designed to strengthen decision making and to make the process more acceptable to the persons involved.

The last chapter of this section, Chapter 9, covers material that is likely to be new for most academic physicians. The evaluation of educational programs, in contrast with the evaluation of individual professionals, is almost never discussed in academic medical circles. Instead, program evaluation is usually done implicitly ("our graduates *seem* to be practicing good medicine") rather than according to an explicit plan. Chapter 9 shows how program evaluation can be prepared and implemented and how evaluative information can be used to improve educational programs.

Medical education, including teaching *and* evaluation, requires careful planning and administration to be fully effective. The chapters in this section aim to introduce the academic physician to some basic principles of education and evaluation. Their purpose is to reinforce the idea that educating the next generation of clinicians is one of the academic physician's basic professional responsibilities. As such, education not only deserves time and attention, but should also be viewed as a priority activity within the academic community.

6
An Experience in Curriculum Development

CHARLES P. FRIEDMAN AND RICHARD M. BAKER

The focus of this chapter is a simulation exercise initially designed by the authors and two colleagues in 1980. The exercise has been used routinely as part of a fellowship program for junior clinical faculty members and has been offered on several occasions at national meetings. Following each use, minor modifications in the exercise have been made. Several educators have taken the basic theme of the exercise and adapted it for different audiences.

This chapter has two distinguishable but interrelated purposes, and has been prepared with two potential audiences in mind. First, for the reader wishing to learn about the process of designing the medical curriculum, this chapter provides an immersion into that intricate topic through some background material and an opportunity to work through the simulation. Second, for the reader interested generally in simulations as an instructional strategy, this chapter describes the structure of one such exercise and various ways the exercise may be used. Although it is catering to two potentially distinct audiences, the chapter will not be explicitly subdivided into sections of interest to particular groups.

Readers may generalize from the content of this chapter in two directions. Issues arising in the specific topic we use as an example—predoctoral curriculum development in family medicine—are applicable to the evolution of virtually any curricular area in professional education. Similarly, the format of the simulation exercise is applicable to education around any topic complex enough to require a case study or simulation experience for meaningful learning to occur.

The chapter includes in appendices all materials required for actual use of the simulation exercise. Readers are encouraged to use the exercise, as it appears here or in modified form, in carrying out their own faculty

Acknowledgments: The authors are grateful to Sandra Putnam and Lisa Slatt for their help in developing the simulation exercise, and to Frank T. Stritter who suggested the general perspective for describing curricula employed here.

responsibilities. Because the materials remain under copyright, the authors request acknowledgment when the materials are employed for any formal instructional purpose.

Perspectives on Curriculum

Another World

Although many educators have an intuitive understanding of curriculum, most are comfortable dealing with educational matters on a smaller scale. Most educators focus their attention on day-to-day activities, leaping perhaps from individual meetings of their courses to consideration of courses as a whole. There are relatively few occasions to contemplate the big educational picture which is the world of curriculum. Thinking at the level of the curriculum requires attention to some special issues. Therefore, this section will introduce several new concepts, potentially to make possible a more rigorous approach to curricular thinking and to set the stage for a more productive experience with the simulation exercise.

The word "curriculum" comes to us from Latin, and means "the course of a race." An interesting contrast is found with the Latin "circus," which also can refer to the course of a race but clearly connotes a round course inevitably leading the racers back to the point from which they start. "Curriculum"does not as sharply carry the connotation of circularity; thus one who transits a curriculum can progress to a destination different from where he/she started. It is amusing, therefore, that "curriculum" came into use in education whereas "circus" implies a form of entertainment.

In popular use in education, "curriculum" refers to many different things. Any collection of educational experiences related in some logical way can be viewed as a curriculum. Curriculum carries a connotation of magnitude, so that the term is usually reserved in higher education to refer to a combination of many courses. Educators often use the term "content" to describe the material included within one course. Generally, the term curriculum refers to courses of study taking one or more years to complete, although this is by no means universal. Further fogging the picture, a curriculum can be developed from several component courses that all run simultaneously; so one need not have a temporal sequence of experiences to have a curriculum. Whereas the term curriculum implies an emphasis on the subject matter to be mastered by learners, "content" and "process" in education are very difficult to separate. A statement of curriculum may, explicitly or implicitly, say as much about the methods of education employed as about the knowledge or skills to be mastered.

A Definition

For purposes of this discussion a curriculum can be considered as:

A deliberately designed combination of educational experiences with goals that transcend the goals of the component experiences.

This definition makes three major assertions:

The first assertion embraces the term "deliberately designed." This implies that all curricula derive from some underlying purpose. Even though many curricula have been in existence long enough to blur the original intents, every program of study was developed with a goal in mind. Moreover, in all probability the designers shared a set of beliefs about how education should be structured and what the fundamental purposes of education are. These shared beliefs provide a logic for including or excluding subject matter and a framework for determining its sequence of presentation. Every curriculum articulates its basic beliefs in some way: catalog statements and recruiting literature are two of the most common.

The second assertion relates to the phrase "combination of experiences." Here "experiences" is used generically to refer to the components or building blocks of a curriculum, including but not restricted to: formal courses, clerkships, independent studies, seminars, and preceptorships. A complete specification of a curriculum will include the required, elective, and optional experiences, successful completion of which may be prerequisite to successful completion of the curriculum itself. The complete curriculum specification will also include a statement of how the content of these component experiences is interrelated.

The third key assertion in the definition addresses the goals of the curriculum "that transcend the goals of the component experiences." This is a singularly important statement. A curriculum is, to use a familiar phrase, synergistic: the whole is greater than the sum of its parts. For example, the undergraduate medical curriculum may seek to endow students with the values and ideals of the physician or to inculcate a research-based methodical approach to patient care. However, few medical curricula contain individual courses or other experiences with these as stated goals.

Curriculum Planning

How do curricula come into being? It is useful to present two models: the first is what professional educational planners would lead us to believe is the "rational ideal." The second is the so-called "rumpled reality," which perhaps more clearly resembles the way planning is carried out in our imperfect world. These sharply contrasting conceptions have roots in the formal literature on planning (1).

The rational ideal model is consistent with conceptions of the curriculum as a system (2). As illustrated in Figure 6.1, planning begins with a statement of *curricular* goals and, even more powerfully, a priority ranking of those goals. Assigning priorities to goals helps planners make difficult decisions to include some aspects and not others when resources to implement the curriculum become scarce.

From the goals and their priorities evolves an ideal curriculum design, which specifies the component experiences and their relationships over time. A key aspect of this model is that the component experiences (courses, clerkships, etc.) derive from a clearly defined sense of the curriculum as a whole, expressed in terms of *curriculum* goals and practices.

Following specification of the ideal design, the planners employ surveys, interviews, document analysis and other data collection tools to determine:

- The resources available to implement the curriculum (student, capabilities and expectations, faculty expertise, space, operating funds, support staff, etc.);
- The constraints on curriculum design (maximum time allowable to completion, rules established by the larger university, etc.);
- The organizational climate (what features of programs make them acceptable in the larger context of the university, certifying and accrediting bodies, professional and community organizations).

The results of this three-part assessment process are guidelines to help

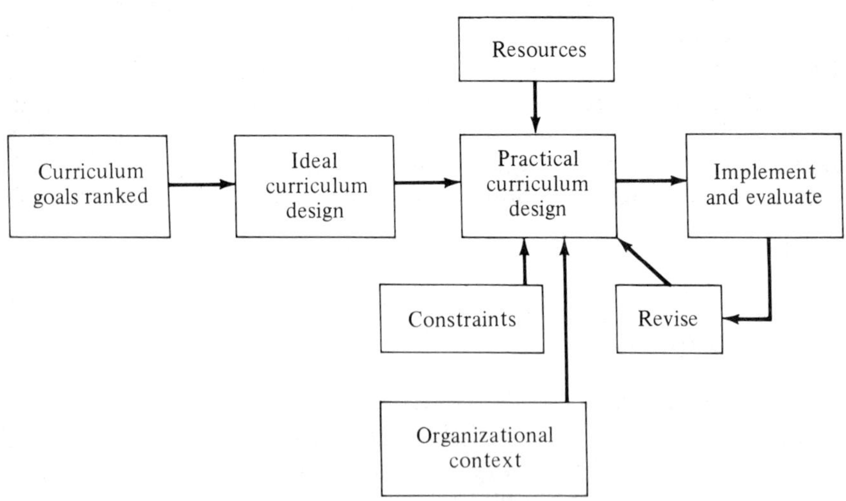

FIGURE 6.1. Systematic planning model (the rational ideal).

curriculum planners shape their ideal design into a practical design, a curriculum that "fits" its context and is realistic in terms of resources and constraints. Following the generation of the practical design, the curriculum can be implemented; students can begin running the metaphoric race. During its use, the curriculum is evaluated. On the basis of evaluation results the curriculum can be modified and improved for the benefit of students who follow.

It is difficult to quarrel with the logic of this rational model. However, one does not require an extraordinary amount of cynicism to recognize the difficulties of actually planning in the real world according to the dictates of such a model. Perhaps it is more realistic to say that new curricula come into being incrementally (3), with planners doing what they can under changing and unpredictable circumstances, as portrayed in Figure 6.2.

According to this scenario, and in direct contrast with the ideal depicted in Figure 6.1, planning does not proceed from a prior conception of the entire curriculum. In this case the curriculum is derived from component experiences. For example, when resources for component experiences 1 and 2 are available, they are put to use designing 1 and 2, which are then implemented and evaluated. Later in this hypothetical example, resources for two more components may become available; but in the meantime the resources for a component developed earlier are lost. When a set of components stabilizes at least long enough to allow the planners some time to think, the curricular goals are inferred from what exists.

The stark contrast between the two models carries some clear implications. The rational ideal assumes that the planners face a predictable future and hold a great deal of control over how resources are allocated. What is out of their direct control can be empirically investigated, with the results of these investigations later reflected in the ultimate curricular design. The "rumpled" model implies instability and relative helpless-

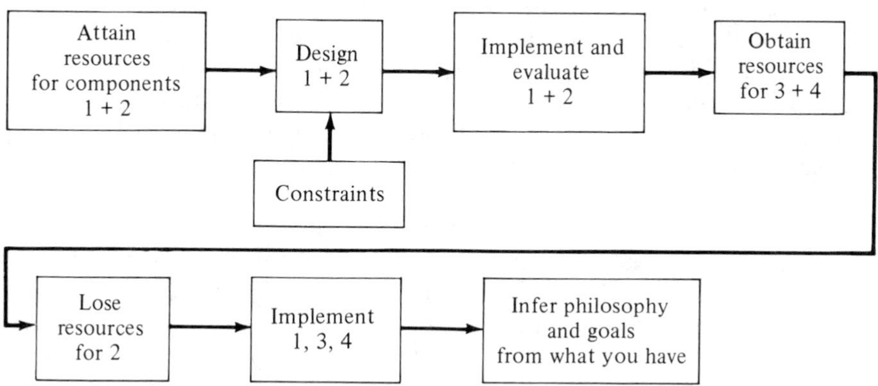

FIGURE 6.2. Pragmatic planning model (rumpled reality).

ness in the face of external influences dictating, among other things, how resources are allocated. It requires planners to react to a rapidly changing set of circumstances.

The reality of curriculum planning usually does not assume either of these extreme postures. Planners probably should aim to follow the rational ideal but at the same time prepare themselves for rumples along the way. The simulation exercise that follows falls somewhere between these extremes. Planners in the simulated situation find themselves able to determine their own goals and priorities for these goals; however, they find their curriculum design to address the goals sharply limited by arbitrary action of an interdisciplinary curriculum committee, of the type known to faculty members of most medical schools.

Describing Curricula

To whatever extent the rumpled reality intrudes itself on the planning process, some ideas are useful to characterize any curriculum. These concepts have meaning only at the curriculum level (they *don't* apply to individual component experiences); they provide a language for curriculum planners to communicate efficiently. As such they shape curriculum design and evaluation activities.

Breadth (4) refers to the number of different subject areas touched upon in a complete course of study. For example, a medical curriculum with six required clerkships has more built-in breadth than an alternative curriculum with five required clerkships.

Depth (4) refers to the amount of learner time and energy devoted to a particular topic area. A curriculum with both a clerkship and subspecialty electives in pediatrics exhibits more depth than an alternative curriculum with only a clerkship.

For a curriculum with fixed length, depth and breadth are complementary issues. One can increase only at the expense of the other. Finding a proper balance of breadth and depth is one of the great challenges of curricular design. Each has an advantage. Often the tension is resolved by externally imposed norms that curriculum planners must accept. For example, in England early specialization is the norm in higher education. Baccalaureate curricula there exhibit much more depth than is customary in the U.S., but much less breadth.

Diversity refers to the mix of types of component experiences built into the curriculum. The basic science aspect of medical education is generally low in diversity, with emphasis on lecture/didactic instruction, with few subject matter options, and with all students moving lock-step through the curriculum. In general, diversity in curriculum is desirable because there is documented variation in learning style, cognitive style, and personality—to complement variations in the background knowledge and personal experiences that students bring to any curriculum. The more diversity, the greater the chance that each student will be intellectually captivated by at least one part of the curriculum.

Vertical integration (5), another desirable aspect, addresses the extent to which component experiences that are sequential in time are logically related. In medical education, vertical integration can be realized if the pharmacology course in the second year explicitly builds on the microbiology presented in the first year—to so great an extent that the connection is clear to the students. Vertical integration is generally easier to realize in a lock-step curriculum. It is difficult to realize a high level of vertical integration in clinical instruction in medical school since students typically take their clerkships in varying orders. For example, planners of a pediatrics clerkship cannot base their clerkship design on assumptions about the particular clerkships the students have completed previously.

Horizontal integration (5,6) is a companion to vertical integration; but it is a concept more often neglected by curriculum planners. In most curricula, students are engaged in studying several different topics simultaneously. Horizontal integration refers to the degree to which these parallel courses are logically interrelated. In medical schools, microbiology and biochemistry are often taught simultaneously in the first curricular year. Horizontal integration could be increased by creating a set of clinical correlation cases common to both courses, which would be discussed from the microbiological viewpoint in one course and from the biochemical viewpoint in the other.

Context integration refers to the degree that the curriculum is consonant with prevailing values and philosophies of the "host" organization and society as a whole (7). It is almost axiomatic that mismatch between a curriculum and its context will generate pressures on the curriculum to adapt. A clear example is the greater prominence of the primary care disciplines in curricula of "community based medical schools," as compared to curricula of older schools with clinical instruction typically focused at tertiary referral centers.

In the simulation exercise that follows it will become clearer how these six descriptive factors can be used as tools for curriculum analysis and planning.

The Simulation

The exercise forming the core of this chapter follows the "research game" format developed by Parker Small (8).

Small developed the "research game" to provide a structured experience for medical students in use of the classic scientific method: formulating hypotheses or abstract models to explain an observed phenomenon, then collecting data that can be interpreted to confirm or refute these hypotheses. Using the rational model, the process of curriculum planning displays several properties analogous to the scientific method (Table 6.1). Just as the scientist must be trained to rigorously collect and then use data to evaluate hypotheses, the "rational" curriculum planner must use data regarding resources and contraints and organizational context to shape an ideal curriculum into one that can function in the real world. The danger in science is that an investigator will go beyond his/her data, or ignore them, especially when findings are counterintuitive or discordant with preconceived beliefs. The curriculum planner is similarly susceptible to ignoring data, being seduced by anecdote, or blindly following intuition. These temptations must be resisted if the value of the rational planning model is to be realized.

The theme of the simulation exercise in this chapter—the process to be modeled and thereby learned—is rational curriculum planning with, wherever possible, a level of rigor and logical consistency approaching that of the scientific method. The exercise is, however, structured so that the rumpled reality will rudely impose itself in the planning process. What follows is a description of the game, which may be addressed by the reader at several levels depending on needs and interests. The next section describes the alternatives and explains how the game works. The actual game materials are in the appendices to this chapter.

Working Through the Simulation

Those playing the game take the role of curriculum planners in a department of family medicine at fictitious but realistic PDQ School of Medicine. (In contemporary medical education, the emergence of family medicine departments raises several key themes in curriculum planning. Although these curricular themes are generic, family medicine is a "natural" setting for a simulation exercise.) All players are first given the case description (Appendix A), which outlines the state of affairs at PDQ regarding family medicine in the predoctoral curriculum and also gives six curricular goals that the department of family medicine has developed. As outlined in detail in the case description, the Curriculum Coordinating Committee (CCC) at PDQ School of Medicine has decided to allow additional curricular time for family medicine in the curriculum.

TABLE 6.1. Analogous features of scientific method and rational curriculum planning.

Scientific method		Curriculum planning
Observed phenomenon	←------→	Goals and priorities
Hypotheses/models	←------→	Ideal curriculum
Empirical data	←------→	Assessment of resources, etc.
Accept/reject hypotheses	←------→	Create practical curriculum

However, the CCC is not going to give the family medicine department unlimited hours and complete flexibility to implement its goals. Instead, the CCC is going to offer the department three alternative and highly specific curricular options. The department must adopt one of these options without modification. Thus the end-point of the game is the selection of one most desirable option.

There are several steps leading up to this end-point. Each may be viewed as an element of the rational curriculum design model.

1. This first step occurs before the options are handed down. While the CCC continues its deliberations, players are asked to take some time to rank-order the six goals. This step is in active anticipation of the fact that the CCC will not give adequate curriculum time to address all six goals, and that as a result some compromises will be necessary. These goals cannot be modified, since they were generated earlier with departmental consensus.
2. After goals are assigned rankings, players are given the approved and binding resolution of the CCC (Appendix B), which contains the three discrete curricular options. Like the departmental goals, these options cannot be modified in any way.
3. The next step gives players the opportunity to collect data about resources, constraints, and context. In terms of the rational curriculum development model, this is the most important step because in real academic life it is the step most often omitted. To get this information in the simulation, players ask "Mother Nature" for it. "Mother Nature" has a treasure chest of data (Appendix C) but *she will respond only to a specific question.* When the game is played in groups, a leader acts as "Mother Nature," circulating about the room responding to each group's question. Individual readers as players can do this by making selective reference to Appendix C, trying to be diligent about not peeking at extra information; that is, information they did not spontaneously decide to seek. Note that "Mother Nature" always gives the same answer to a particular question, so the data bases of different players will differ only according to the questions they happen to ask. In actuality, no players obtain all information available; the average is about 50%.
4. This next step is a formal analysis of each option in terms of the data elicited from "Mother Nature" *and* in terms of the six descriptive curricular variables presented earlier in this chapter. A worksheet is included (Appendix D) to lend structure to this process. Understanding how each option differs with regard to the curricular variables should help players make a final choice. Also on the worksheet is an additional factor, "concordance with curricular goals," which clearly will have impact on the selection.
5. The last step is reaching the decision. When there are enough players to allow several groups to work in parallel, a very interesting discussion can ensue where groups compare their final choices in view of the vary-

ing priorities assigned to goals and the particular data elicited from "Mother Nature." In a completely rational world, groups with the same goal priorities and the same data base would reach the same final decision. When the game is actually played there is, of course, great diversity in priorities and data obtained; however, similar priorities and data bases generally coincide with similar decisions.

As mentioned earlier, there are several ways to use the simulation based on the information provided here. To get the most out of this chapter, it is recommended that you choose some mechanism of involvement with the simulation materials. In order of roughly increasing immersion, the options are:

- *Voyeur*: This entails reading the procedures and the materials.
- *Solitaire*: You can play the game yourself by following the procedures as described in the appendix materials. Playing by yourself will require some perseverance and self-discipline to obtain data from "Mother Nature" strictly according to the rules. Alternatively, you can have a colleague play the role of Mother Nature for you. If you plan to be a leader for a group administration of the game, you may find it useful first to work through the exercise independently.
- *Groups as Players*: As described previously, the exercise is actually designed to be completed by one or several small groups served by one or more "Mothers Nature." Players clearly benefit from having peers with whom to discuss each stage of the exercise. The ideal size for each group is five to six; however, groups as large as eight have been employed with success. One "Mother Nature" can attend to the needs of up to three such groups. (With more groups several "Mothers" will be required.) Having several groups work the exercise simultaneously allows inclusion of a summary discussion comparing and contrasting final decisions and the bases for these decisions. For group play, the "Mother Nature" data should be copied and several copies of each datum stapled onto a manila folder. "Mother" can then rip off the answer to a question from a group. The group thereby retains a paper copy for immediate review, and for reference at later stages of the exercise.

Substantive Curricular Issues

Although the simulation format raises generic curricular issues, and indeed has been adapted to domains other than family medicine, the exercise in its present form raises several considerations related to family medicine particularly. The generic challenge to academic family physicians is to establish the specialty in an appropriate curricular role alongside the other specialties practiced by significant numbers of physicians. The ultimate goals of such efforts should not necessarily be to support the same types of educational programs as other specialties, but rather to design a set of experiences uniquely tailored to the body of knowledge undergirding family practice and the clinical activities family physicians undertake. The motivations behind these goals are manifestly educational. There is a general belief that curricula of most medical schools do not routinely address topics such as disease prevention and continuity of care that are central to family medicine (9), yet at this writing family medicine remains the second to internal medicine in attracting graduating American medical students. Faculty members must impart knowledge, skills, and attitudes which, even if not unique to Family Medicine, have come to be recognized as characteristic of family practitioners.

The history behind this challenge is relatively clear. Family medicine was introduced to the academic medical center primarily at the graduate level. First priority for most departments was to establish residency programs that would train board-eligible physicians. Perhaps because of this initial emphasis, the first decade of family medicine (1969–1979) emphasized course offerings for medical students skewed toward later years of the curriculum, and featuring the clinical preceptorship in a family practitioner's office (10). It was also typical, but by no means universal, for family medicine faculty members to involve themselves in presenting courses on basic clinical skills for medical students in their pre-clinical years. These are clearly worthwhile patterns of involvement but they very likely fall short of presenting principles of family medicine in a systematic way to all medical students. Many medical schools are currently grappling with the issue of just how to do this. The example of PDQ, although fictitious, is nonetheless typical. Options handed down by the curriculum coordinating committee at PDQ in their totality embrace most of the strategies that are being tried nationally. One of the options (A) showcases the required clinical clerkship which some believe is the optimal curricular route for family medicine to take. However, in the simulation, the choice of this option comes at the expense of long-term exposure to family physicians with concomitant potential for vertical integration and coverage of topics in depth. Each option in the simulation has unique strengths and weaknesses. The curricular dilemmas facing academic family physicians are dilemmas with which any academic physician should be able to identify.

Conclusions

Curriculum planning in higher education is a challenging and demanding process for all involved, but especially for the inexperienced. This chapter has sought to cast curriculum planning in manageable terms by presenting a definition, two idealized planning models, a number of concepts useful for analysis of a curriculum, and an exercise that provides occasion to apply these various tools. The hope behind all of this is, of course, to bring an increased level of rationality to the process—reasoned decisions based on data connected to consensual goals and priorities. This is distinguished from the probably exaggerated picture of curricular decision-making as a "smoke-filled room" political process heavily influenced by anecdote and the freely articulated biases of influential faculty members and administrators. Deliberately not treated here are several other skill areas important to curriculum planning. They are addressed in other chapters of this volume. First is the process of collecting valid and reliable data, a process central to rational curricular planning. Several chapters of this book (especially 11–15) are devoted to data collection and analysis. Although not dedicated to curricular concerns, these chapters treat generic issues that bear directly on the processes discussed here. Second are the political dynamics of academic medical centers and their impact on decision-making at all levels, including the curricular. This topic is treated to a significant extent in Chapter 5 and is particularly important because, rational models notwithstanding, curriculum planning *is* political. Third, although revision of a curriculum design based on evaluation is a part of the rational model, this chapter did not explicitly address curriculum evaluation. This is considered in Chapter 9. Finally, for those planning to offer a group administration of the curriculum simulation, skills related to group leadership may be found in Chapter 4.

Ultimately, sound curriculum planning can be boiled down to two principles: (a) plan at the curricular level, recognizing that the curriculum is indeed more than the sum of its parts; and b) plan rationally, based on good data but anticipating the limitations that will be imposed by scarce resources and unpredictable factors. Attention to these principles, in ways outlined in this chapter, seems particularly important to academic physicians in an era of changing knowledge to inform clinical practice, changing needs for numbers and types of physicians, and changing perceptions of the role of the physician in society.

Appendix A: Case Study Description

PDQ School of Medicine*

At PDQ School of Medicine, your group has been appointed to spearhead the development of educational opportunities in family medicine for medical students. The school-wide Curriculum Coordination Committee

---

 PDQ School of Medicine is a fictitious institution. Its characteristics have been tailored to make this case study as challenging as possible; however, PDQ is realistic in many ways. The curriculum that appears on the following page is from a real medical school but is not current.

(CCC) at PDQ has agreed to restructure the entire predoctoral curriculum to make possible the implementation of a limited number of family medicine experiences. This step was taken in philosophic recognition that, as a member of the community of academic departments, family medicine has an unique body of knowledge that medical students should master. PDQ graduates 100 medical students each year; this number is not expected to change in the foreseeable future.

At the present time, there are no *required* curricular experiences in family medicine or courses taught by Family Medicine faculty for medical students. Four years after its creation, the department developed an elective preceptorship for fourth year students, of 6 weeks duration. The preceptorship is currently offered at four different community-based group practices and at the model family practice at PDQ medical center. In the academic year just completed, 21 students took the preceptorship.

The faculty includes 10 full-time physicians, a clinical psychologist, and a social worker. There are 24 residents in the training program at PDQ medical center. Each physician faculty member has a primary area of responsibility, as detailed below:

Number of faculty members	Primary responsibility
1	Chairperson (administration)
2	Medical student teaching
6	Residency teaching
1	Research

However, the chairperson has publicly stated his support for further development of medical student teaching programs and has indicated that all faculty members will be expected to do some teaching in these courses regardless of primary emphasis area.

Overall, the curriculum of PDQ is as follows:

First academic period begins: August; duration: 36 weeks Unscheduled hours per week: 10					
	Hours of				
Required course titles	Lecture	Conference	Lab	Other	Total
Gross Anatomy	65		116	107	288
Introductory Biochemistry	114	15	51		180
Introduction to Clinical Medicine	136			62	198
Introduction to Human Behavior	54				54
Histology	45		57	24	126
Human Physiology	78	16	45	46	185
Neural Sciences	39		40	60	139

Second academic period begins: August; duration: 36 weeks
Unscheduled hours per week: 10

General and Systemic Pathology	134			161	295
Microbiology	31	10	80		121
Dermatology	12		24		36
Introduction to Clinical Medicine	130		86		216
Tropical Medicine	24		36		60
Preventive Medicine	24	48			72
Clinical Pathology	49	24	25	12	110
Pharmacology	80	10	40	10	140
Psychiatry	30	22		20	72

Third academic period begins: August; duration: 36 weeks

Required clerkships	Weeks
Medicine	9
Obstetrics/Gynecology	9
Pediatrics	9
Surgery	9

Fourth academic period begins: variable; duration: 36 weeks

Medicine	6	Each student takes three 6-weeks electives.
Neural Sciences*	6	
Surgical Specialties†	6	

Total weeks of instruction: 144

Reprinted from 1975–76 AAMC Curriculum Directory published by the Association of American Medical Colleges, Washington, D.C.
**Neurology, Neurosurgery, Psychiatry*
†*Ophthalmology, Orthopedics, Otorhinolaryngology, Urology*

The CCC has agreed to modify this structure to allow additional time for Family Medicine. It is currently deliberating the details of a schoolwide curriculum change to accommodate these new courses and will present a detailed proposal soon. For its part, your committee has been working steadily in anticipation of the resolution by the CCC, drafting curricular goals for the predoctoral program. The broad goals that are listed below have been ratified by the department as a whole, having been previously shared with the 10 "community" faculty members who teach in the elective preceptorship. The goals are listed in no particular order.

1. All medical students should acquire skills in *interpersonal communication.* They should be able to use effective interviewing techniques to establish rapport, obtain information, provide information, and counsel.
2. All medical students should gain knowledge and skills relevant to *common problems.* They should be aware of the relative prevalence of specific patient problems in the primary care setting and acquire basic abilities to diagnose and manage common problems.

3. All medical students should learn the principles, skills and attitudes of *preventing health problems.* Emphasis should be on health maintenance in primary care, anticipatory guidance, risk determination, and effective behavior change for health (nutrition, smoking, substance abuse).
4. Medical students interested in family practice should learn principles of *family-oriented care* and of *family dynamics, emphasizing life cycle, stresses, impact of illness, and other family crises.*
5. All medical students should learn to *integrate* physical, psychological and sociological components in caring for their patients. They should understand psychophysiologic mechanisms, illness behavior, stress effects, and sociocultural variation—all as applicable to primary care.
6. All medical students should observe and *participate in the specialty of family practice* to the extent that their training time will permit, in order to understand the family physician's role. This recognizes that almost all physicians will have ongoing contact with family physicians as part of their routine practice. Students tentatively interested in a family practice career should gain adequate exposure to allow an informed specialty choice.

Some Additional Useful Information

1. On the PDQ curriculum chart, the category "other" refers to laboratory demonstrations, small group experiences that are informal, and assigned independent study.
2. PDQ is a state-supported institution, one of two state-supported medical schools in its state.
3. The PDQ department of Family Medicine has two affiliated residencies: one in the city of 250,000 in which PDQ is situated; the other in a smaller city of 80,000 that is 55 miles away from PDQ.

Appendix B: Discrete Curricular Options Handed Down by the CCC

NOTES: These three options cannot be altered in any way by your committee. You must choose one of them. However, you may assume that any aspect of the options not specified by the CCC can be interpreted by your group as it sees fit. For example, the precise content of the clerkship in Plan A is not specified and therefore could be freely determined by your department should you opt for Plan A.

Plan A: Family Medicine can offer a required 4-week clinical clerkship in the third year. This would be a structured clinical rotation analogous to those offered by other departments. Each of the other four clerkships reduces to 8 weeks, to maintain the total length of the third year. This could then be Family Medicine's *only* mandatory curricular input; the elective preceptorship could remain as an option for a limited number of students. Family Medicine could negotiate with Internal Medicine for a role in Introduction to Clinical Medicine in years 1 and 2; but no such role would be mandated under this plan.

Plan B: In the second year, Introduction to Clinical Medicine would be lengthened and then divided into two subcourses: one stressing basic techniques of physical diagnosis (80 contact hours), to be the responsibility of the Department of Internal Medicine; the other stressing basic principles of primary care (80 contact hours), to be Family Medicine's responsibility. In addition, a mandatory 6-week, fourth year clinical preceptorship in Family Medicine would be implemented, to replace some previously existing elective time.

Plan C: Family Medicine would be responsible for a 10-lecture-hour piece of Introduction to Clinical Medicine (ICM) in the first year, the content of which would be negotiated with Internal Medicine, where primary course responsibility would remain. The second year ICM course would conclude with 20 hours of small group seminars. Ten concurrent seminars covering different topics would be offered with 10 students in each, and family medicine faculty would be allowed to teach up to three of the seminars. Students would then have the option of delaying the Ob-Gyn clerkship until the fourth year in order to take a 6-week preceptorship in the third year, or they could take the preceptorship in the fourth year as in Plan B. Either way, the preceptorship would be required of all students.

Please note:

1. You cannot rank the goals again in the context of the plans. Goals as ranked have already been approved by departmental faculty.
2. The "hard facts" that Mother Nature can give you include some medical personnel projections, interest surveys, and statements by school officials for the record. If you ask for some information of this type that simply has not been collected or is no longer available, Mother Nature will so inform you.
3. You may assume that all options are feasible in terms of faculty and finances. This is because the CCC, in generating the options, undertook a detailed feasibility study in conjunction with the department of Fam-

ily Medicine. The medical school is prepared to underwrite all additional instructional costs that accrue from implementation of any of these options.

Appendix C: "Mother Nature" Information

Instructional support facilities in the medical school:	PDQ has a very small Office of Medical Education (2.5 FTE staff) that helps in the design and evaluation of educational programs. There is also an Instructional Media Center that has a small TV studio and, in addition, three portable videotape recording/playback units. Seminar and lecture rooms are ample to accommodate virtually any course design for medical students.
Content of related courses in the PDQ curriculum: (catalog descriptions)	1. ICM, year 1: Organization of the health care system, medical economics, medical ethics, patient interviewing skills, death and dying (organized by Department of Internal Medicine) 2. Introduction to Human Behavior: Basic principles of psychology, especially as they relate to illness; psychosomatics, behavior change strategies, the "sick role," patient compliance and patient education (organized by Department of Psychiatry) 3. ICM, year 2: The screening physical; physical diagnosis, medical problem solving strategies, advanced patient interviewing skills (Department of Internal Medicine) 4. Introduction to Preventive Medicine: Epidemiology and Biostatistics (Department of Community and Preventive Medicine) 5. Psychiatry: Diagnostic categories of psychopathology, methods of assessment, resources for psychiatric care (Department of Psychiatry)

PDQ admission policy:

1. Catalog statement:
   PDQ school of medicine admits, by charter, 80% of its students from the pool of applicants judged to be permanent residents of the state. Students are selected on the basis of potential to succeed in a rigorous course of study, as well as potential to be competent, humanitarian physicians. The applicants' academic record, other achievements, and interview performance are all taken into account when judgments by the admissions committee are made.
2. Statement by the Dean of Admissions: This school has made an effort in recent years to select a class that is diverse in ethnic and demographic background, provided, of course, that we are convinced that every student admitted can perform at an acceptable level of academic achievement. We have great difficulty, however, in predicting the type of career that our students will choose on the basis of information available at the time of admission.

State legislature:

1. Funding History:
   In 1972, the medical school requested funds from the state legislature to create a department of family medicine and a residency at PDQ medical center. These funds were appropriated in 1973 and the department admitted its first group of residents in 1975.
2. Priorities:
   On appropriation of the funds, the legislature stated clearly that this measure was taken to increase the number of physicians who practice in areas of the state, both urban and rural, that have been underserved by the medical community. The

legislature funded the department for a period of 12 years. After that time, it will commission a study of the effects of the undergraduate and graduate program prior to a decision to fund the department permanently.

Medical student attitudes:

1. The Office of Medical Education keeps records of evolving career preferences, as summarized in the following table:

Percentage preferring family medicine as a career choice.

Year of entering class	On entry	At end of 2nd year	On graduation (% entering FP residencies)
1978	19	18	9
1979	22	16	9
1980	25	18	12
1981	37	27	—
1982	38	27	
1983	39	—	
1984	42		

2. Yearly evaluations of the preceptorship suggest that students feel it is giving them high quality exposure to community-based family practice. Students placed at family practice centers consistently give the experience lower ratings.

Present exposure to family medicine:

1. Over the past 4 years, family medicine faculty have been called on to give periodic lectures on selected topics in the Introduction to Clinical Medicine courses, but these have invariably been single lecture, one-shot interventions.
2. There is a Family Practice Club at PDQ that has had stable membership of about 40 over the past 3 years. Average attendance at their bi-monthly meetings is 16.

3. In recent years the elective preceptorship in Family Medicine has attracted the following numbers of students:

Year	Number at community sites*	Number at PDQ Family Practice Center
1980–1981	6	9
1981–1982	8	8
1982–1983	12	6
1983–1984	16	5

*Only six students per year could be accommodated in physician shortage regions under the prevailing arrangements for these years.

Mode of compensation for preceptors:

Under the elective preceptorship in place beginning in 1975, preceptors in community practices receive no direct compensation for their time, aside from a clinical faculty appointment in the department. Money would be available ($100 per student per week) for students placed in the community sites for a mandatory preceptorship, or in the event the clerkship were organized to use some community practices as the bases for clinical experience.

Decision-making in the CCC:

Statement for the record by CCC chairman: All departments were represented at the meeting at which the resolution was passed. Some reservations were expressed by members of major clinical departments, especially with regard to Plan A. However, in light of the fact that family medicine at PDQ is a new department that can be said to be on trial, consensus held that they should be given exposure. Clinical departments having third year clerkships felt, overall, that shortening their clerkship time by 1 week would not be significantly detrimental to the students' interest.

Possible new preceptorship sites:

Under the feasibility study, it was determined that a preceptorship could

	be put in place for 100 students per year. This determination was based on statements of consent to participate obtained from 15 sites, including the affiliated family practice centers and the community practices that participated under the elective program. The feasibility study committee assessed the sites that would be new to the program and determined that they were of equal quality as potential teaching loci to those currently available under the elective program.
Results of feasibility study pertaining to clerkship:	The feasibility study determined that the clerkship could be based at the three affiliated residencies. If structured to comprise one-third inpatient experience, one-third ambulatory care experience in the family practice centers, and one-third community practice experience at teaching practices, then it would be possible to implement a clerkship serving 100 students per year.
Content of existing clerkships:	No primary care except 1 week of ambulatory care. No mandatory community hospital rotations.
Number of residents in affiliated residencies:	Affiliated community hospitals have residency sizes of 12 (total) in the smaller city and 18 in the other program in PDQ city.
Location of preceptorship sites:	Eight are "commuter" sites and seven require student housing.
Undergraduate teaching responsibility of family medicine faculty:	Chairperson dictates that all must devote at least 10% of their time to teaching medical students.
Statement by dean on curriculum policy:	He fully supports the state's mandate to increase medical manpower in

	underserved areas of the state. He is personally committed to doing whatever is necessary to meet this goal.
Relationships with other clinical departments:	Family medicine residents rotate on to all services now. This has been a generally congenial arrangement. The family medicine residents' ratings have been average for all house officers at PDQ hospital.
Family medicine inpatient service at PDQ hospital:	Established in 1978: it has 10 beds.
Instructional formats of preclinical courses:	Lecture method dominates but other formats are used. There is no formal requirement that courses be taught by a large group, lecture format.
National boards at PDQ:	Parts I and II are required for graduation. Average scores have traditionally been in the 450–500 range, only rarely falling outside that range.
Correlation between existence of a clerkship and number of students entering family practice residencies	Black and Stewart found a strong positive correlation between the presence of a required clerkship in family medicine and an increased percentage of students entering family practice.
Other state medical school	It has a new, fast growing Department of Family Medicine with 10 full-time faculty and 36 residents. That department was established 3 years ago. Although there is as yet little contact with medical students, the school is reviewing its curriculum to include family medicine.
Statewide medical manpower needs:	A recent study by a leading consulting firm conducted at legislative request concluded: (1) That the state population will likely not increase dramatically in the 1980s (2–3% increase projected); (2) That the state will have a sufficient absolute number of physicians in the 1980s; (3) That,

unless major incentives for rural and inner city practice are put into place, a severe physician maldistribution problem will persist, wherein 25% of the population will live in regions of significant physician shortage.

Appendix D: Analysis of Curricular Options

Instructions: Rate each plan according to each criterion. Use any rating scheme you (or your group) find convenient.

	Plans:		
	A	B	C
Consistency with goals and their priorities			
Potential for breadth			
Potential for depth			
Potential for vertical integration			
Potential for horizontal integration			
Potential for diversity			
Consistency with organizational context			

References

1. Braybrooke O, Lindblom CE. *A Strategy of Decision.* New York: The Free Press, 1963.
2. Saylor JG, Alexander WM. *Curriculum Planning for Schools.* New York: Holt, Rinehart, and Winston, 1974.
3. Lindquist J. Obstacles to curriculum development. In: *Developing the College Curriculum.* GH Quehl (Ed). Washington DC: The Center for the Advancement of Small Colleges, 1977.
4. Levine A. *Handbook on Undergraduate Curriculum.* San Francisco: Jossey-Bass, 1978.
5. McNeil JD. *Curriculum: A Comprehensive Introduction.* Boston: Little, Brown, 1981.
6. Tyler RW. Specific approaches to curriculum development. In *Strategies for Curriculum Development.* J Schaffarzick, DH Hampson (Eds). Berkeley, California: McCutchan, 1975.
7. Hoyle E. How does the curriculum change? Part 1: A proposal for inquiries. In: *Curriculum and Evaluation.* AA Bellack, HM Kliebard (Eds). Berkeley, California: McCutchan, 1977.
8. Small PA, Jr. Science education: simulation methods for teaching process and content. *Fed Proc 33* :2008–2013, 1974.
9. Shahady EJ. Teaching the principles of family medicine. *NZ Fam Phys 10* :24–26, 1983.
10. Baker RM (Ed). *Predoctoral Education in Family Medicine.* Kansas City, Missouri, 1981: Society of Teachers of Family Medicine.

7
Clinical Instruction

FRANK T. STRITTER, RICHARD M. BAKER,
AND EDWARD J. SHAHADY

A Definition

This chapter suggests a model that can inform the clinical instruction of medical students and house officers. It also provides exercises useful in teaching that model to clinical instructors.

Two definitions are essential. *Clinical instruction* is defined as the interaction between an instructor/practitioner and a learner which normally occurs in the proximity of a patient encounter, focusing either on the patient or a clinical problem associated with the patient (1). The number of learners with whom an instructor deals at any one time is small, generally not exceeding three or four, and the "patient" can be viewed in the context of an individual, a family, or a community. Such instruction has significant consequences because of the responsibilities for health of a population eventually assumed by those being taught. The particular characteristics of ambulatory care education, i.e., precepting, direct supervision and role-modeling in ambulatory clinical settings, support clinical instruction, in contrast to lectures or large group presentations, as the predominant teaching method. *Professional development* is the process of growth in which learners are engaged as they gradually assume the knowledge, skills, and attitudes of physicians. Those are characteristics developed during an educational program under the auspices of a school of medicine, a training program, or a remote training site affiliated with an educational program.

Clinical instruction focuses on learners endeavoring to become the type of professional they are observing or whose role they are practicing. Learners have greater responsibility for management of their educational endeavors than they would in a classroom setting and have specific patient care assignments as the major instructional vehicle. Those assignments are carried out under the supervision of an attending physician who maintains final responsibility for the care that is delivered. They are intended to assist the learner in putting the theory learned in the classroom into practice, in gaining new professional skills, insights, and attitudes, and in being evaluated by the attending physician.

Inpatient or outpatient clinical instruction is often carried out in medical teams with specific patient care responsibilities. Learning in such settings is fundamentally experiential, stimulated to a large extent by a medical problem presented by a particular patient and the subsequent care that is provided. Teaching is responsive to events that are often beyond the instructor's control or influence. Several roles exist on any medical team, but the dynamics underlying the hierarchical structure are often unclear. Attending physicians and preceptors may be viewed as outsiders. Although they typically control how time will be spent on rounds and in precepting encounters, they have little control over the patients whose conditions determine the content of their teaching. In addition, attending physicians cannot control their learners' activities during the majority of the day. The nature of their leadership responsibility often makes them appear intimidating and leads attendings to offer what they believe house officers and students want, emphasizing detailed medical information, rather than discussion of management of complex problems. Clinical teachers are too often narrowly focused, offering more information than a particular problem or group of learners require.

The nature of the learners' roles makes the clinical teaching encounter a particularly difficult one. Senior house officers can serve as barriers between learners assigned to a particular clinic and the attending physician. As team leaders they are expected to facilitate rounds and other instructional discussions about patients, often in a manner not clearly defined. Residents act according to their perceptions of the wishes and capabilities of the attending physician. Both the resident and the attending physican can experience confusion and frustration in such a setting. Beginning house officers are apprentice professionals frequently overwhelmed by the medical responsibilities facing them. Anxious to obtain and demonstrate medical information necessary to deal with their patients' immediate problems, they actively pursue what they feel is organically useful and become frustrated when diverted from it. Concentration on the "right medical answers" and the biomedical aspects of disease often causes young residents to lose sight of important psychosocial patient management issues. Medical students, also members of the team, are enthusiastic about their exposure to real clinical experiences, but tire of performing routine functions. They want more meaningful responsibilities, but can feel overlooked by house officers and intimidated by attending physicians. They feel they cannot take the initiative and often become passive observers.

The content of clinical instruction in the ambulatory care setting represents a unique blend of biomedical and psychosocial knowledge and skills that varies according to the particular primary care specialty. In addition, more work is usually required in preventing disease, promoting health, counseling, coordinating team work, consulting with others, working with community resources, and managing a practice than in a typical inpatient setting.

Observations of both clinical instructors and learners indicate that the blend of competencies required for ambulatory care requires clinical instruction different from other medical specialties. For example, the third-year medical student in a tertiary care setting is often taught the diagnostic process through delineation of a comprehensive differential diagnosis by consulting subspecialists. This same diagnostic process applied in the primary care setting has the potential to be inappropriately expensive, time consuming, and possibly even dangerous for the patient. In ambulatory care, patient problems must be diagnosed using personal knowledge of the patient's history and condition together with probabilistic thinking based on epidemiologic principles. Therapeutic trials and monitoring progress over time assist the diagnostic and therapeutic process. "The Diagnosis," itself, can often be of limited importance in many primary care situations and is often characterized by uncertainty. Reassurance and attention to the psychosocial aspects of the patient's problem can be sufficient and often do not depend on developing a definite diagnosis.

Learning in ambulatory care settings should expand the learner's fund of knowledge just as will involvement in patient care on a clinical rotation with appropriate supervision and learning resources. The particular concerns of ambulatory care, however, demand that learners experience an involvement with and participation in the function of the health care team, office management and relationships with community resources. Therefore, ambulatory care education at any level requires experiences with appropriate clinical instructors and patients. An endocrinologist in a university hospital's diabetic ward will emphasize approaches to managing hospitalized diabetic patients, but often does not adequately emphasize or even understand the primary care component stressed above. In contrast, the primary care physician precepting in an office or ambulatory care unit can complement the specialist's information by explaining the complexities of diabetic patients within the context of the family and the community.

When one considers these issues, the difficulty of clinical instruction and, in particular, ambulatory care clinical instruction, becomes apparent. It has been suggested that the classroom teacher has more difficult decisions to make about his learners in a shorter period of time than a physician caring for a patient. When instruction and patient care are combined, as they are in the physician/clinical instructor's responsibilities, that difficulty is compounded. Clinical instruction must, therefore, take place in a difficult context at best. It does, however, have a significant impact on learners. To accomplish this well it must be integrated into an overall clinical curriculum in which the major stimulus for learning comes from individual patients seen by learners.

Clinical instruction should accomplish much for the learner. It should:

1. . . . articulate a rationale, set expectations, and provide an orientation for care of the patient.

2. . . . guarantee that what is presented, studied, and mastered is related to medical practice and is based on the health care needs of a population.
3. . . . relate concepts taught to an overall structure as well as indicate any relationship between concepts.
4. . . . develop concepts and responsibilities from simple to complex.
5. . . . stimulate the learner to be increasingly more responsible for his/her own learning and to be accountable for it.
6. . . . emphasize flexibility so that instructors can serve either as providers of information, facilitators, or as consultants depending on the particular needs of the learner.
7. . . . include opportunities for systematic and periodic evaluation by which the learner uses self, peers, patients, fellow health care team members and instructors to develop an appraisal of his/her own comprehension and performance.
8. . . . specify an evaluation plan in which the learner can contribute to continued analysis and improvement of instruction.

A Developmental Rationale for Clinical Learning

The model suggested in the following paragraphs is based on the assumption that clinical instruction can influence a medical student's or house officer's development as a clinician in a stepwise fashion.

The basis for the model comes from developmental psychology. A developmental model describes an individual's actions as being governed by an internal mediating process that varies according to that individual's experience and stage of development. According to Sprinthall and Thies-Sprinthall (2), the assumptions of a developmental model are: (a) humans process experience through cognitive structures called stages; (b) such stages are organized hierarchically from the less complex to the more complex; (c) growth occurs first within a particular stage and then only to the next stage in the sequence; (d) growth is not automatic but occurs only with appropriate interaction between the learner and his environment; and (e) a learner's behavior can be determined by awareness of his particular stage of development.

Early credit for thinking about learners from a cognitive developmental perspective can be attributed to John Dewey (3).

> The aim of education is growth or development . . . Ethical and psychological principles can aid . . . in the greatest of all constructions—the building of a free and powerful character. Only knowledge of the order and connection of the stages in psychological development can insure this. Education is the work of supplying the conditions which will enable the psychological functions to mature in the freest and fullest manner.

Later credit can be attributed to Piaget (4), Bruner (5), and Case (6). However, the major portion of their writing was in reference to children.

Kolhberg (7) has addressed development from a moral perspective whereas Loevinger (8) defines it as ego. Knowles (9) and Chickering (10) have studied adults.

A model for educational and professional development recognizes that physicians continue to learn throughout their productive years, that learning does not cease on completion of a basic educational program. Further it must recognize that as learners mature into their professional roles they: (a) evolve a learning focus from abstract concepts emphasized in a classroom to the concrete problems faced in the clinical setting; (b) begin to apply learning immediately as opposed to some future time; (c) increasingly use personal patient care experiences as a basis for learning; and (d) increasingly orient learning to what they perceive to be their future professional roles. This is a personal orientation marked by no one best technique of clinical instruction for all learners at all times.

The Learning Vector

The model suggests that the most effective physicians are those who function at higher developmental stages. It posits an approach by which the clinical instructor modifies instruction based on the stage that the learner has achieved. The instruction is, therefore, designed to promote professional development by assisting the learner to reach the highest possible stage. It has been named "The Learning Vector" and is depicted in Figure 7.1.

This model can help in determining conditions or situations in which specific approaches to instruction are optimal for facilitating professional growth. The instructor progressively modifies instruction from one of

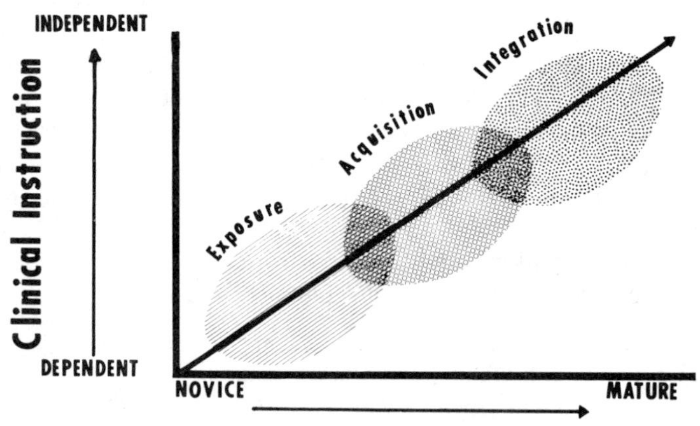

FIGURE 7.1. The learning vector.

keeping the learner dependent on him/her personally to one of assisting the learner to become independent in pursuing his/her instructional needs. This developmental model is somewhat different than strictly humanistic (11) or behavioral (12) educational orientations. The humanist would place responsibility for what is learned heavily on the shoulders of the learner whereas the behaviorist would place it almost completely on the instructor. The developmentalist gradually shifts the responsibility from the instructor to the learner. The developmentalist promotes a gradual professional maturation, collaborating with the learner on approaches to or characteristics of future instruction. Figure 7.1 depicts in two dimensions how this might be accomplished.

Professional Development*

In Figure 7.1, the horizontal axis of the Learning Vector refers to the professional development of the individual learner. Initially, the learner can be considered a novice in the clinical setting. The learner matures as contact with and exposure to the clinical setting increases. That development can be classified in three stages.

Exposure: in which the learner is merely exposed to basic facts, information, concepts, and skills from a variety of sources, e.g., an instructor, a patient, a host of other health professionals and even the environment.

Acquisition: in which the learner tests the information, concepts and skills by applying them to clinical responsibilities. In this stage, the learner begins to focus on decision making and reasoning.

Integration: in which the learner weaves those characteristics into a professional identity that he/she perceives can embark him/her on a professional career.

Six characteristics comprise development or maturation at each of the three stages. First is the *cognitive*—the information or knowledge base coupled with both intuitive and analytical reasoning that allows the physician to use that base (4). Second is the *technical*—the physical or procedural skills that allow the physician to ascertain information about his patients and/or to conduct procedures of various types (13). Third is the *attitudinal*—the interests, values, and ethics that guide resolution of moral problems, action, choice, argument, and rationalization (14). Fourth is *psychosocial*—those characteristics of interpersonal and human interaction that underlie sensitivity to and communication with patients and fellow health care team members (15). Fifth is *socialization*—the

---

*References cited are generic and not necessarily directly applicable to physician development.

internalization of professional values, the commitment to the professional role, and acquisition of norms typical of the fully qualified professional (16). Sixth is *independent learning* which enables the physician to keep abreast of the changing science of his field (9). During *Exposure* and the early stages of *Acquisition*, progress along each of the six dimensions occurs relatively independently of the others. This progress is often jagged and uneven. Toward the end of *Acquisition* and beginning of *Integration,* progress on the dimensions begins to merge and finally becomes inseparable. Development and/or modification of any single dimension influences development and/or modification of all the others. They have become interdependent.

It is possible that the development of these characteristics over time may not be as linear as suggested by the figure. Perhaps they cannot be integrated into a single dimension, but each must be considered independently. Perhaps the characteristics of the institution (e.g., school, hospital, or clinic) in which the instruction and development occur constitute a third dimension of the figure. The model would then appear as a series of models or even a multidimensional model. It seems likely, however, that even if it is more irregular than Figure 7.1 represents, it does display basically stepwise properties.

Clinical Instruction

The vertical axis of the Learning Vector figure refers to the instructional approach adopted by the clinical instructor. Instructional psychologists (17–19) write that certain components of instruction should characterize every teaching interaction. If these components are not always explicit, they should at least be implicit.

Those events are: (a) *orientation,* in which the learner decides what and how much to learn, (b) *learner practice or responsibility,* in which the learner masters knowledge or undertakes a task related to previous expectations, (c) *evaluation,* in which a judgment is made about the quality or worth of the completed task, (d) *feedback,* in which the learner decides how successful he/she was in completing the task, and (e) *closure,* in which the instructor concludes the interaction with the learner and plans new learning.

The problem for educators is essentially one of assessing which stage the learner has achieved and then providing instruction that will stimulate growth to the next highest stage. The Learning Vector model suggests a continuum of occasions at which such an assessment might be carried out. Three of those occasions are depicted in Figure 7.1.

Exposure occurs when the learner is a novice in his development and calls for the clinical instructor to act authoritatively. The learner functions in a highly structured environment, depending on the instructor for setting objectives and expectations. The activities that the teacher provides are also structured, that is, the instructor demonstrates procedures to the learner, encourages comprehension by asking Socratic questions, and asks the learner for questions or comments about the demonstration. The power of the role model is considerable. The instructor provides short-term evaluative feedback to the learner by setting criteria, reporting strengths and weaknesses to the learner, suggesting areas for improvement, and indicating preparation necessary for the next interaction. An example in primary care training might be the third-year clerkship in the teaching clinic. In this setting learners observe primary care practice by faculty and senior residents. Patient care provided is followed by a discussion of the data obtained and the plan of patient management. Objectives and expectations should progress from the data collection techniques of interviewing, physical diagnosis, and laboratory tests to problem-solving, therapeutic skills, and follow-up. Clinical teachers act as role models and provide information to augment learners' basic knowledge of primary care.

Acquisition calls for the instructor to begin functioning more collaboratively with the learner. The instructor adopts an inductive approach and suggests that learners accept more responsibility for their own learning. Roles and the objectives of instruction are negotiated. The learner begins to choose from among options presented to him in addition to addressing objectives that the instructor has set. The learner performs procedures while being observed by the instructor, using mutually derived criteria and standards. Similarly, assessment and feedback become a mutual process involving both learner and teacher. The instructor asks the student to evaluate and comment upon his/her own success in meeting the expectations. Thus the learner develops skill in analyzing his own performance. Examples would be the fourth-year student preceptee or more experienced first-year resident in the clinic. Such a learner would independently gather data about the patient's problem, then present the management plan to the instructor for validation or modification. He or she would carry out clinical procedures, such as sewing lacerations, while being observed by the instructor before the two discuss the procedure.

Integration suggests that the instructor delegate even more responsibility to the learner for the instruction. The instructor now becomes a consultant to the learner. As part of what is taught, the learner might be asked to develop a learning contract. The learner begins to decide what to learn and agrees to be accountable for what is undertaken. The learner now begins to perform the less complex procedures independently, reviewing a self-analysis with the instructor on completion of the task. In the evaluation, the learner is now responsible for compiling a systematic analysis based on evaluation information derived from self-analyses, instructors, peers, colleagues and patients. The instructor then discusses the integration of the evaluative information with the learner. Second- and third-year residents are examples of appropriate learners in this phase. They can function relatively independently in clinical care settings, with a clinical instructor available as a consultant. Patient management and clinical procedures for which the resident has been appropriately trained can be carried out through chart monitoring or reviews at the end of the clinic session. More complex skills, such as counseling or management of a health team should be demonstrated to experienced clinicians or consultants. In novel or complex situations, however, the resident must be able to recognize the need for further learning and establish goals for acquiring more knowledge and skills.

The underlying theme of this model is that the learners are progressively more responsible for the conduct of their own learning as the instructor determines they are able to accept it. Instructors must gradually modify their role from provider of information to facilitator to consultant. If this process is begun with learners early in their clinical training, then it is likely that their professional development will be more effective and efficient. Learners will likely be able to more effectively use available resources in becoming life-long learners.

Preparation for a career as a physician imposes requirements. The profession itself—as represented by licensing boards, certification and hospital privileges—specifies that certain knowledge, skills, and attitudes must have been acquired as a condition for admission to the profession. Educational institutions specify minimum standards of achievement as conditions for awarding degrees or certificates of completion. By using the Learning Vector model as a guide, it is possible for clinical instructors to assist learners in reconciling these requirements imposed from the profession with the developing professional's need to become increasingly self-directed. Instructors can assist in matching these requirements with learners' personal goals, choosing ways of achieving those goals, and determining when and how those goals have been satisfied. The quality of clinical instruction becomes a function of the instructor's ability to teach in response to the individual learner's stage of development. Clinical instruction should not be static or unchanging, but dynamic and situational.

Organization of Clinical Instruction

Clinical instruction is difficult but learning can be accomplished systematically in a clinical setting. The experience of clinical teachers and the conclusions of educational research can be forged into a set of guidelines for clinical instruction. The guidelines have, therefore, been developed from that experience, review of the literature, on clinical teaching (20–23), and many discussions with both clinical teachers and clinical learners. Reflection on one's teaching in light of these guidelines can help instructors better realize the learning potential that exists within any interaction with a learner. It can help the instructors achieve greater self-awareness and increased congruence between intentions and actions.

Diagnose the Learner

The diagnosis of the learning problem requires recognition of the stage of the learner's professional development as well as the reason for the consultation. The stage of development will usually, although not necessarily, be recognized by the learner's level of training. The necessary instruction will differ according to that level of maturity—a first-year resident is less developed than a third-year resident.

Recognizing the reason for the specific consultation is somewhat more difficult. Medical students and residents are generally taught to present material in a traditional manner, reflecting their impressions or diagnosis of a patient, not the questions they have about that patient. Students and residents are also aware that they are constantly being evaluated and they want to make favorable impressions on their teachers. As a result, determining the learning problem may be difficult. The following example illustrates this issue.

While you are precepting, a third-year resident asks your advice about a choice of antihypertensive medications. After a brief traditional presentation of the case, you are tempted to suggest a treatment plan, but instead you ask selected questions that reveal that the real problem concerning the resident is patient compliance. Too often, instructors accept the reason for the encounter on face value. Further questions of this specific resident revealed adequate understanding of antihypertensive medications but poor knowledge, experience, and literature base about compliance. You can now reinforce the learner regarding the knowledge of antihypertensive medication and provide information about compliance.

Set Expectations for Learning

Successful learning and teaching take place when expectations are mutually agreed on and understood by both learner and teacher. Conversely, most ineffective and uncomfortable teaching encounters do not have a clear learning contract or expectations for the encounter.

The following example helps to describe how to set expectations:

A first-year resident asks you to see a patient with her who has knee pain. She believes the patient has a torn medial meniscus and should be referred for orthopedic evaluation. Her expectation is that you confirm the diagnosis and management decision.

After listening to the presentation you are not certain of the diagnosis. To set the tone of the encounter, you suggest that perhaps you could go through a knee exam with her and review the anatomy of her knee rather than that of the patient's. You also ask how much she wants you to discuss in the presence of the patient. Agreeing to these activities and procedures before entering the patient's room is vital for an effective encounter. In this particular case, the resident had already informed the patient of her impression and suggested management. The learner agreed to the procedures outlined but asked that she be the one to tell the patient if you did not agree with the diagnosis and/or management. Your exam revealed the presence of a patellar-femoral syndrome, which you then carefully demonstrated while avoiding embarrassment for the learner.

The discussion about treatment occurred outside the exam room, thus allowing the learner to reenter the exam room and inform the patient of the change in diagnosis. Expectations were set mutually by the learner and instructor.

Match Expectations with Instruction

This straightforward guideline reminds us to evaluate the success of the learning contract. Ask the learner and yourself if the expectations that were set were met. It is also important to evaluate the appropriateness of the learning contract and whether the needs of the situation and the learner are met.

In the previous example, the instruction seemed to match the expectations that were set. However, an outside observer could evaluate the effectiveness and appropriateness of the instruction and whether it met the learner's needs.

Question the Learner in a Problem-Solving Sequence

Not all teaching leads to effective learning. Optional learning occurs through self-discovery, a process facilitated by instructors who value it and are skilled at ensuring that it happens. Clinical teaching is an ideal opportunity for self-discovery. Instructors can lead learners through a problem by posing questions that allow learners to realize what they already know. This process should be structured and organized in a manner that emulates the clinical decision making process. Open and frank discussion with learners of *why* we do what we do is just as important as *what* we do.

If instructors can determine the process they use when making medical decisions and implementing solutions, they can develop appropriate guidelines for teaching. As instructors identify each step in the process of making decisions, they can assist learners to develop their own approaches to the process. Learners can observe as instructors think aloud in applying their particular approach to any given problem. Questions that instructors can use as a basis for discussion when managing clinical problems are:

1. What is the patient's problem? (Is there really a problem—if so, what is it?)
2. What are the likely causes (etiological diagnoses) of the problem?
3. What facts in the case (e.g., history, physical findings) support or mitigate against each possibility?
4. Are there factors or characteristics of the patient's environment that I should consider in eliminating unlikely possibilities and deciding the cause(s)?
5. What additional information do I now need in order to eliminate other alternatives and how will I obtain that information?
6. Now what is my best estimate of the cause(s)?
7. What information do I now need to consider in choosing a treatment?
8. What are the treatment options? Which is the best and why? (e.g., What drugs are contraindicated?)
9. What are the options for preventing this type of problem from recurring?
10. How will I monitor the appropriateness of the treatment and/or prevention plan?
11. What specific qualitative and/or quantitative results are needed to confirm that the treatment and/or preventive plan(s) has been appropriate?

Just as important are the attitudes, values, and ethics that influence both the decisions one makes about patients and one's relationship with those patients. For instructors to teach about these influences, they must be aware of their own attitudes, values, ethics, and the influence these have on their decisions. Just as they do when demonstrating a decision-making process, they can ask themselves questions while managing each clinical problem. Such questions might be:

1. What values and ethical priorities do I have as a physician?
2. What values do I have that might be appropriate to consider in this particular case?
3. What role might my values play in the way I approach this problem?
4. How can I continue to grow and learn about myself and the impact my values have on the care I deliver from this encounter?

If instructors ask such questions and openly discuss the answers with learners, they can legitimately expect their learners to do similarly. This can be a strong contributor to developing self-awareness among clinical learners.

Evaluate Your Learners and Provide Constructive Feedback

Experience indicates that instructors tend not to assess performance or provide feedback in a constructive manner. Part of the difficulty is that it is not pleasant to tell someone they have not done well. We often feel that we should not cast stones unless we ourselves are free of fault. This feeling has to be balanced with the realization that instructors are not discharging their responsibility to patients and learners if they don't evaluate and help their learners understand their strengths and weaknesses.

A few suggestions that might help:

1. Negative comments are usually heard if they are accompanied by positive ones.
2. The instructor's nonverbal behavior and tone of the setting are as important as the words used by the instructor.
3. Make assessment part of an initial learning contract. Learners should be informed about what will be assessed before the encounter begins.
4. Ask the learners to first assess their own performance. Frequently they will be more critical of themselves than you are.
5. Couch statements in a positive, specific manner without value judgments intimating a good or a bad job. Statements such as "Another way you could have asked that question is __________," using questions such as "Are there side effects of this drug that make it hazardous to utilize in the elderly?" are often helpful.

Evaluate the Teaching Encounter

Agree with learners as part of the learning contract that feedback regarding effectiveness of the teaching encounter will be requested. Learners should be asked to give the feedback in specific terms. Statements such as, "You could have asked different questions" are more helpful than "This was a good or a bad encounter." This also helps learners see that professionals must be critical of their own performance as well as that of others.

Closure

There should be an identifiable conclusion so that both parties understand what was accomplished and that the task is completed. A brief summary is a helpful way of coming to closure. Summation should include not only highlights but emphasis on what was learned, how it was learned, and what remains to be learned. Also helpful is agreement on when and how any follow-up will occur.

Although it will not always be possible to adhere to all guidelines in every teaching encounter, clinical instructors should try to address as many as possible. This will ensure that little is left to chance and that a maximum amount of teaching is accomplished. It is then likely that a maximum amount of learning will also occur.

Exercises for Clinical Instructors*

This section offers a series of exercises and instruments that can be used individually or in a faculty development activity to address the topic of clinical instruction. They can be used to help clinical instructors examine their own tacit assumptions about clinical teaching and learning. They can also be used to practice new approaches to clinical teaching and to reinforce older appropriate ones.

Four separate exercises are presented. Each incorporates selected issues and guidelines concerning clinical instruction that can be valuable in conducting varied teaching encounters such as brief hallway consultations with medical students and house officers or multiweek rotations.

Specific instructions are included for the use of the exercises. They were originally developed for and used in a 1-day workshop on clinical teach- (24). They have been modified based on new concepts and information derived from various evaluations conducted over the years. A brief introduction to each of the exercises is as follows:

1. Attitudes Toward Learner Responsibility Inventory (ATLRI)

The ATLRI is a 20-item questionnaire that takes approximately 10 minutes to complete individually. In a workshop setting it can be productively followed by a 50-minute group discussion. Questions addressed in such a discussion could focus on the advisability and feasibility of providing learners progressive responsibility for their own clinical learning and how to accomplish it.

2. The Clinical Learner: A Consensus Exercise

This consensus exercise is an excellent stimulus for discussion. It asks participants to respond and then participate from what they perceive to be a learner's perspective. Workshop participants are often clinical instructors who have generally not viewed teaching from that perspective since they were learners.There are no "right" responses to the exercise, but varied responses and heated discussion often result. The entire activity can be completed in 1 hour.

3. Guidelines for Clinical Teaching

The Guidelines are not an exercise but a set of questions about a clinical teaching encounter with which that encounter can be analyzed. They can be considered criteria for good teaching. Every teaching encounter cannot be characterized by all of the guidelines but can incorporate many

---

*The exercises and instruments can be adapted for use when the original source is cited.

of them. The guidelines can either be used as a basis for undertaking the following Diagnostic Exercise or be used to analyze an actual teaching encounter.

4. Analysis of Case Studies: A Diagnostic Exercise

This exercise is composed of three case studies containing a patient, a learner, and a clinical instructor. A different teaching problem forms the basis for each of the three cases. Each can be completed individually or in a group and the response compared to the appended response of the "experts." The case studies can also be used as the basis for a practice exercise.

Attitudes Toward Learner Responsibility Inventory (ATLRI)

Motivation is a significant determinant of learning. One way to increase the learner's motivation is to increase "ownership of the situation" by giving the learner more control over what is learned and how it is learned. Learners can take responsibility by participating both in setting expectations and in developing the criteria that assess that learning. They can conduct certain learning activities while unsupervised. They can evaluate their own learning and that of their peers as well. An institutional or programmatic attempt to improve the motivation of learners should have as its goal increased individual responsibility for learning. This *Attitudes Toward Learner Responsibility Inventory* (ATLRI) was developed to assess clinical instructors' receptiveness toward increasing learners' control over their own learning. It consists of 20 Lickert scale-type statements, 10 worded positively and 10 worded negatively. Attitudes are sought to each statement with the possible responses being Strongly Agree (SA), Agree (A), Undecided (U), Disagree (D), and Strongly Disagree (SD). Letter responses are translated to numerical scores by using Part II, the scoring portion of the ATLRI. Responses are scored in reverse numerical order for positive and negative statements, with possible total scores ranging from 20 to 100. The higher the score the more likely the instructor is to give learners responsibility for their own learning. Scores below 40 indicate that the instructor is not favorably inclined toward doing so, scores of 40–60 indicate that the instructor is undecided or only somewhat disposed, scores of 60–80 indicate that the instructor is moderately interested and scores over 80 indicate commitment to giving learners responsibility. The average score is 70 (Table 7.1).

The Inventory can be used in two ways. First, it can be administered individually such as prior to and at the conclusion of a faculty development activity for clinical instructors. This will help to assess both the need for and the subsequent impact of the activity. It can also be used by clinical instructors who want to begin rethinking their own approach to teaching. Second, it can be used to stimulate group discussion on the question of the advisability and/or feasibility of providing learners in a clinical setting responsibility for their own learning. Such a discussion can be an integral part of a faculty development process.

The Clinical Learner: A Consensus Exercise

This exercise is generally used to stimulate discussion about clinical teaching from a learner's perspective. Groups of eight require approximately 30 minutes to complete the exercise. First, participants are asked to rank individually all 12 teacher behaviors on Part I from most to least helpful in facilitating clinical learning. They are asked to complete the ranking as they feel residents would rank the behaviors. Next, participants are asked to develop a group consensus ranking of the behaviors by discussing and negotiating. These rankings are contained in Part II and the participants are then asked to compute a difference score between themselves individually, the group score, and the residents' actual rankings (22). A difference score is the sum of the differences between the residents' actual rank and the rank developed consensually in the exercise. Finally, the facilitator processes the exercise by leading a discussion about the differences in ranking and their interpretation. After sharing views from the unique perspective of clinical learners, participants can now undertake a more substantial examination of the issues from the perspective of clinical instructors (Table 7.2).

Guidelines for Clinical Teaching

Observing someone else's and/or analyzing one's own clinical teaching is an important step toward improvement. The guidelines presented here are characteristics of clinical teaching that make clinical teaching effective and beneficial to the learner. They are based on the discussion included in the previous major section of this chapter.

These guidelines can be used periodically by the individual instructor in self-evaluation after an interaction with a learner. They can be used by a peer who has been asked to review a colleague's instruction prior to a discussion between the two individuals. They can also be used as a basis for developing an evaluation instrument to obtain feedback from learners.

*1. Was a *diagnosis* of the learning problem made? How was it made? Which stage of development has the learner achieved in relation to the clinical problem in question?

*2. How were *expectations* developed, communicated, and/or agreed to by the learner? Were they set in a mutually agreeable way?

---

*Primary and should be present in every teaching encounter; others may not be present in every encounter.

TABLE 7.1. Attitudes toward learner responsibility inventory: Part I.

Indicate your feelings about each of the following statements by circling the appropriate response.

Strongly Agree (SA) Disagree (D)
Agree (A) Strongly Disagree (SD)
Undecided (U)

1. When opportunities exist for learners to work and/or study on their own, they will learn more efficiently.	SA	A	U	D	SD
2. Even though attempts are made to give learners significant responsibility for their own learning, most will seldom take advantage of those opportunities.	SA	A	U	D	SD
3. On their own, learners will generally accomplish any goals that they participate in setting.	SA	A	U	D	SD
4. Learners will gain from opportunities to participate in determining criteria for accomplishing goals that have been set.	SA	A	U	D	SD
5. Clinical instructors should closely supervise their students to facilitate efficient learning.	SA	A	U	D	SD
6. When learners participate in setting their own expectations, there is a positive interaction between instructor and learner in the clinical setting.	SA	A	U	D	SD
7. Learners have the highest regard for a clinical instructor who directs them systematically step by step through the learning process.	SA	A	U	D	SD
8. Allowing learners to control their own time will usually result in appropriate use of that time.	SA	A	U	D	SD
9. Allowing learners to assume control of their own learning causes confusion in the clinical setting.	SA	A	U	D	SD
10. Most learners would not understand the rationale for participating in setting their own learning expectations.	SA	A	U	D	SD
11. When learners are given more responsibility for their own learning, they will be more likely to accomplish that learning.	SA	A	U	D	SD
12. Learners can profitably evaluate the learning and/or performance of their peers.	SA	A	U	D	SD
13. Learners usually cannot provide constructive feedback to their peers after evaluating their learning.	SA	A	U	D	SD
14. Learners who are personally committed to goals and expectations set by the instructor and/or the program will require little supervision.	SA	A	U	D	SD
15. Most learners will make decisions about their own learning and their own conduct in a manner that benefits the clinical unit in which they are working, their peers, and their patients.	SA	A	U	D	SD

16. Clinical instructors should generally observe their learners as they fulfill their responsibilities.	SA	A	U	D	SD
17. Learners should be evaluated daily by the instructor to determine whether they are performing adequately.	SA	A	U	D	SD
18. Clinical instructors should plan their learners' schedules and assign responsibilities.	SA	A	U	D	SD
19. Clinical instructors should be the principal source of evaluation for any learner.	SA	A	U	D	SD
20. Learners can constructively evaluate their own learning and/or performance.	SA	A	U	D	SD

Attitudes toward learner responsibility inventory: Part II. Scoring and Interpretation

Circle the response you gave to each item; sum all circled numbers under each column and then sum across all columns for an overall total score.

Item #	SA	A	U	D	SD
1	5	4	3	2	1
2	1	2	3	4	5
3	5	4	3	2	1
4	5	4	3	2	1
5	1	2	3	4	5
6	5	4	3	2	1
7	1	2	3	4	5
8	5	4	3	2	1
9	1	2	3	4	5
10	1	2	3	4	5
11	5	4	3	2	1
12	5	4	3	2	1
13	1	2	3	4	5
14	5	4	3	2	1
15	5	4	3	2	1
16	1	2	3	4	5
17	1	2	3	4	5
18	1	2	3	4	5
19	1	2	3	4	5
20	5	4	3	2	1

Subtotals __ + __ + __ + __ + __ = ____

Plot your total score on the scale below by placing an X at the location of your score.

0–20	20–40	40–60	60–80	80–100
Little or none	Slight	Moderate or undecided	Marked	Extensive

TABLE 7.2. What are the most helpful teaching behaviors? (Part I)

In 1979, literature on teaching behavior research and practice was reviewed with the goal of identifying helpful teaching behaviors of clinical teachers in individual or small group settings. A list of 44 possible items resulted. In addition, a research project resulted in residents' identification of behaviors most helpful in facilitating their learning.

A representative 12 of the items are listed below in random order. Rank order items 1 through 12 as you think the residents ranked them. It is important that you rank them *as you think they are important to residents,* not as you would rank them yourself.

	Scoring			
	Individual rank	Difference	Group rank	Difference
A. Encourages residents to raise questions and ask for help.	______	______	______	______
B. Asks residents about the effectiveness of his/her teaching.	______	______	______	______
C. Treats residents as colleagues.	______	______	______	______
D. Discusses current research and research findings.	______	______	______	______
E. Answer questions raised by residents clearly and precisely.	______	______	______	______
F. Encourages residents to express feelings/values and opinions in relation to particular patients or problems.	______	______	______	______
G. Contrasts implications of various diagnoses or therapies.	______	______	______	______
H. Cites or discusses his/her own participation in research.	______	______	______	______
I. Advises residents of their progress regularly and systematically.	______	______	______	______
J. Identifies what he/she considers important in clinical conditions or problems.	______	______	______	______
K. Has good relationships with patients and their families.	______	______	______	______
L. Asks residents to evaluate quality of their own performance (i.e., residents).	______	______	______	______
		______ Total		______ Total

Table 7.2 What are the most helpful teaching behaviors?* (Part II)

Rank	
5	A. Encourages residents to raise questions and ask for help.
11	B. Asks residents about the effectiveness of his/her teaching.
1	C. Treats residents as colleagues.
8	D. Discusses current research and research findings.
3	E. Answer questions raised by residents clearly and precisely.
6	F. Encourages residents to express feelings/values and opinions in relation to particular patients or problems.
7	G. Contrasts implications of various diagnoses or therapies.
12	H. Cites or discusses his/her own participation in research.
9	I. Advises residents of their progress regularly and systematically.
4	J. Identifies what he/she considers important in clinical conditions or problems.
2	K. Has good relationships with patients and their families.
10	L. Asks residents to evaluate quality of their own performance (i.e., residents).

*Ranks derived from 72 primary care residents (22).

*3. Did the instruction *match* the expectations that were set? Was it appropriate for the needs of the learner and the patient?
4. Did the instructor's questions follow a logical *problem-solving sequence* allowing the learner to "discover" ? Were they organized?
5. Did the instructor *assess* the learner's comprehension and/or performance observed in the encounter?
6. Did the instructor provide or otherwise facilitate constructive *feedback* to the learner?
7. Did the instructor attempt to *evaluate* the effectiveness of his own teaching with the learner?
*8. Did the instructor and the learner reach a mutually satisfactory *closure* ?

Analysis of Case Studies: A Diagnostic Exercise

The three case studies presented in this section can be analyzed by a single individual or by a group of clinical instructors working together. The focus of each case is a resident at a different stage of training with a different instructional need. Each case is meant to be brief and provocative and requires that at least one of the principles of clinical teaching be addressed.

When used individually, a clinical instructor should read each case, decide how to respond, and then answer one or more of the eight questions presented in the Guidelines for Clinical Teaching above. The instructor should respond in writing to the questions that he or she deems most applicable to each case. The responses can then be compared to those prepared by the two expert clinical teachers that are presented at the conclusion of each case. The responses presented are not meant to be construed as correct, as correct answers may not exist. They are, however, intended to reflect thoughtful appropriate approaches to the teaching problems by two experienced clinical instructors.

In a group exercise, groups of three normally have the best discussions. In a 20-minute session one individual plays the role of the clinical instructor, one the learner, and the other an observer, with the roles then rotated for each of three successive sessions. The "instructor" in each group is advised to pick a case that seems valid for his type of teaching, to select two or three of the Guidelines to emphasize as teaching goals during the session, to take a few minutes to plan the conduct of the interaction, and to try out a technique or approach not tried before. The "observer's" role is to evaluate the interaction using the Guidelines and then to provide feedback to the "instructor." To be maximally beneficial the feedback should be behaviorally descriptive, nonjudgmental, and noninterpretive. In other words, it should be specific. The "learner's" role is to act naturally as if he were a learner in that situation. The "learner" should not act as a particularly difficult learner. After approximately 10 minutes of playing the various roles, the "observer" should conduct an evaluation session asking both the "learner" and the "instructor" to point out strengths, weaknesses, and aspects of the interaction that might be improved. The "observer" can conclude by adding observations from notes written during the role play. An advantage of this exercise is the opportunity to hear how various teaching behaviors affected the "learner" and how helpful each might have been in facilitating his learning.

After the case studies have been conducted, a short reflection on the total experience led by a facilitator may be helpful. Questions that might be addressed include the following: 1) What were the aspects of each session that were particularly interesting or helpful? 2) What was learned that may be helpful in future teaching responsibilities? 3) What specific teaching guidelines will be emphasized by each participant as he or she strives to continue improving?

Case Study I: Johnnie Jones

Patient: Johnnie Jones is a 3-year-old boy who is brought into the emergency room by his hysterical mother who suspects that the child may have roundworm infection. She has been told that a classmate has "worms" and she recalls vividly a childhood friend who had "worms" that had to be treated with medicine that "turned her yellow."

Learner: The learner is a second-year resident in primary care/pediatrics who has difficulty in coping with the concern of the mother in this instance. He focuses on the treatment of the infection in the child and neglects the hysteria evidenced in his mother.

Teacher: The attending physician observes the resident in the emergency room and notes the disregard for the mother's feelings. He wants principally to determine if the resident has assessed the treatment he has provided and then to provide feedback that would be helpful to the resident.

Response of Teacher A

The resident must be stimulated with questions that will lead him to realize that he has not addressed the mother's anxiety. Then he must be confronted about his insensitivity to her emotions and his failure to deal with them. Constructive criticism is important, such as: "I felt that the biggest need in the situation was reassurance for the mother. You can provide that, as you did, by directly investigating the child's problem. A more direct approach that includes the mother is also helpful." Techniques for combating the mother's misinformation, fantasies, and fears could be discussed together with appropriate reassurance approaches for parents.

Response of Teacher B

The primary concern for the teacher is one of teaching vs. supervision of patient care. The attending physician may disapprove of the learner's behavior he has observed, become frustrated, and may wish to confront. Confrontation, however, should probably not be the first choice. Rather, the resident should be asked if he needed assistance and how he felt about the encounter. The resident might be willing to share his frustrations about the hysterical mother. The resident most likely feels inept about dealing with the mother and has, therefore, focused on the child's problem. The preceptor should be supportive, empathetic, and helpful in making suggestions. It is possible, however, that the resident will not seek assistance for the right problem. A more direct approach may, therefore, be necessary in the interest of patient care. The attending physician must determine the necessity of intervening to deal with the mother. This could result in modeling of appropriate care for the resident.

CASE STUDY II: MRS. RODMAN

Patient: Mrs. Rodman is a 23-year-old white married woman who has swollen feet, ankles, and hands; morning nausea; and has missed one menstrual period. She is a diabetic who has been married for 8 months and has used a diaphragm for birth control. She is fearful of being pregnant as she and her husband have not discussed having children, and she is aware of the problems of diabetes and pregnancy. The physical examination confirms pregnancy, establishes a blood pressure of 140/95, and edema of 4+ in the lower extremeties and 2+ in the hands.

Learner: The learner is a third-year resident who has been following Mrs. Rodman for the past year and has developed a fairly good relationship with her. The resident would like to recommend abortion, but is extremely concerned about the feelings of the patient and has just participated in a seminar on ethics emphasizing patient rights. The resident has just seen Mrs. Rodman in clinic and wants to seek the advice of the attending physician before giving a final recommendation to the patient.

Teacher: The attending physician in the ambulatory care clinic is concerned that the patient receive the best possible advice, but is equally concerned with helping the resident process the choices he faces in preparing that advice. The attending physician also wants to help the resident integrate recent discussions about the larger philosophical issues and a final goal is, therefore, to help the resident cope with his own feelings. What questions might the resident be asked to emphasize a sound approach to clinical reasoning?

Response of Teacher A

Goals for this resident would be first to plan the final few minutes of this visit and then to plan the follow-up. Asking questions such as "What do you think you can accomplish today with this patient?" will stimulate the resident to set priorities, perhaps informing the patient of her pregnancy and eliciting her feeling about it. The resident should also be questioned "When will you schedule the next visit? What will you accomplish in it?" The likely conclusion is that the patient and her husband need detailed information about the medical situation in the very near future. Finally, the resident should be asked about a decision to recommend for or against abortion, and encouraged to review the case with the attending physician and perhaps with a gynecological consultant. The resident should also be encouraged to take the opportunity to consider a personal philosophy about ethical issues in this case.

Response of Teacher B

The resident would like to recommend an abortion, but is concerned about the feelings of the patient and the ethics of the encounter. These three points represent areas of confusion for the resident. The resident might first be asked to clarify why an abortion is medically necessary. Although a medical necessity probably exists, it must first be established clearly in the resident's mind. The next questions would then be: "Does your medical opinion interfere with the patient's rights?" Many times the physician confuses medical appropriateness with an appropriate ethical decision or lets personal feelings interfere with reasoning. The rights of the patient are clear. It is certainly her baby and her husband's, and the ultimate decision is theirs. The physician's responsibility becomes clear when the physician understands their responsibility and their rights. The instructor can assist the resident by identifying his own feelings and asking the resident to identify his.

CASE STUDY III: MRS. SMITH

Patient: Mrs. Smith is a 33-year-old married woman who has pain in her epigastrium before meals and during the night. She has had no previous health problems of significance but her father has recurrent ulcers. She and her husband have recently moved to the area, and her husband is unemployed. Because of financial pressures, she just dropped out of her graduate studies to take a job. Her physical examination is normal except for mild tenderness in the epigastrium on deep palpation. The stool is negative for blood.

Learner: The learner is a first-year medicine resident anxious to make a definite diagnosis. The resident wants to hospitalize the patient for observation, x-rays, films, and endoscopy. As a medical student the resident learned that the differential diagnosis must be complete and treatment depends upon knowing the cause of the symptoms. The resident is most concerned about the uncertainty in delivering primary care and doing "anything short of rigorous medicine."

Teacher: The attending physician in the Medicine Continuity Clinic wants to help this resident learn an important principle of primary care—that a therapeutic trial and close follow-up can constitute cost-effective management in this setting. He wants to empathize with the resident's concerns about dealing with "uncertainty" but also wants to avoid over-investigation of Mrs. Smith. In his teaching, the attending physician wants to determine the resident's learning need from this particular encounter and jointly set expectations for the encounter.

Response of Teacher A

Unless there was good rapport between the first-year resident and the attending physician, it would not be wise to confront the resident's insecurities about become a practicing physician in the "real world." Instead, the resident should be asked to develop alternative diagnosis and management plans that recognize the patient's ambulatory status. The resident should be asked to propose a therapeutic regimen and list any potential complications that could cause the patient to seek medical attention during the week. Finally, the instructor should ask that the resident establish criteria by which a presumptive diagnosis would be confirmed or other diagnoses could be pursued. Most importantly, the resident should be asked to point out to the patient the costs and risks of hospitalization and further diagnostic testing. A Socratic method should be used to help the resident develop expectations for interacting with the patient.

Response of Teacher B

The attending physician should first ascertain the resident's actual reason for initiating the interaction. What is it that the resident hopes to accomplish? The resident may have had prior experience with a similar patient who required hospitalization. Rather than responding specifically to the resident's desire to hospitalize the patient, he might be asked to establish the differential diagnosis and indicate support for it. It is hoped that the resident would answer his own question, i.e., he might conclude that there seems to be no life threatening indication at this time and that hospitalization would not be necessary. The resident's uncertainty in this case would likely be related to his stage of development. The attending physician should recognize a knowledge utilization problem and help the resident set an expectation for using his current level of knowledge.

Implications

These are guidelines for learning the complex role of a clinical instructor. In conclusion, the authors suggest three they consider paramount: recognizing and using role modeling; understanding the influence of evaluation on learners; and requiring learners to accept both increasing responsibility and accountability for their own clinical learning.

Role modeling is, on the surface, a passive process that consists of an instructor/practitioner demonstrating patient care in front of learners. It is not quite as easy as that, however. According to Jason (25), a role modeled by an instructor is witnessed by learners, whether it is intended or not. It can, therefore, influence learners positively or negatively. Role modeling can help learners develop a sense of direction as they observe a clinician who has already achieved the goal of becoming a professional. Role modeling can also serve as a standard against which learners can compare their comprehension of a concept or performance of a task. Finally, consistent attitudes on the instructor's part can serve to shape a learner's attitude toward learning, toward the practice of medicine, and toward patients. Our learners do, as Jason (25) indicates, "tend to become as we do, not as we say."

From this discussion, however, one might conclude that all instructors must be flawless representations of professional behavior. Stritter, (26) has suggested, however, that few teachers can be the "ideal" professionals they may wish to be. How then can role modeling be most effective in clinical instruction? First, learners could be assigned only to those instructor/practitioners who possess the desired ideal personal characteristics. Second, an attempt could be made to change the characteristics of those instructors who don't fit the ideal model. Finally, steps could be taken to help ensure that teachers are not required to represent all aspects of the ideal practitioner. All three are feasible, but the third alternative is, perhaps, the most realistic, given the complexities of clinical teaching environments and the difficulty of effecting major personal changes in people.

A second guideline for the clinical instructor is to understand the influence of evaluation on the learner. Evaluation involves a judgment about a learner and subsequent communication of some type of constructive feedback that assists that learner in improving comprehension, performance, and/or demeanor. Evaluation can be derived from the clinical instructor, the learner himself, and other individuals in the learner's environment, such as other health professionals, other learners, and patients.

Evaluation is often described as the "hidden curriculum" because of the way the learner responds to it in his learning, rather than a formal description of a curriculum. Jason (27) has listed the characteristics of effective evaluation as: (a) the learner understands the reasons for the evaluation; (b) the learner is receptive to the evaluation; (c) the learner is challenged to do a self-evaluation; (d) the evaluation is in language the learner can understand; (e) the evaluation deals with issues the learner has agreed are important; (f) the evaluation emphasizes one major issue at a time; (g) differences in views between teacher and learner are discussed and resolved; (h) the evaluation is based on actual observation; (i) learners are provided with specific and objective evidence; and (j) evaluation leaves the learner with a desire to continue his learning.

A final guideline is adherence to the principle that the learning occurs in an environment that supports the progressive maturation of the student/ physician. As suggested by the Learning Vector model, this means enabling learners to become less instructor dependent and more independent in their clinical learning. Learners should be given increased responsibility for deciding what they will learn, how they will learn it, and when they have learned it. Even though some learners may not willingly accept increased responsibility, instructors, by asking for it and expecting it, will influence clinical learners to become independently functioning professionals.

References

1. Stritter FT, Flair MD. *Effective Clinical Teaching.* Bethesda: National Library of Medicine, 1980.
2. Sprinthall NA, Thies-Sprinthall L. The teacher as an adult learner: a cognitive-developmental view. In *Staff Development.* GA Griffin (Ed) Chicago: National Society for the Study of Education, 1983.
3. Dewey J. What psychology can do for the teacher. In *John Dewey on Education: Selected Writings.* R Archambault (Ed) New York: Associated Press, 1970.
4. Piaget J. Intellectual development from adolescence to adulthood. *Hum Dev 15*:1–12, 1972.
5. Bruner JS. *The Process of Education.* Cambridge: Harvard University Press, 1960.
6. Case R. The underlying mechanisms of intellectual development. In *Cognition, Development, and Instruction.* J Kirby, and J Biggs (Ed) New York: Academic Press, 1980.
7. Kohlberg L. Moral stages and moralization: the cognitive developmental approach. In *Moral Development and Behavior.* T Likona (Ed) New York: Holt, Rinehart and Winston, 1976.
8. Loevinger J. *Ego Development.* San Francisco: Jossey-Bass, 1976.
9. Knowles MS. *The Modern Practice of Adult Education.* New York: Associated Press, 1976.

10. Chickering AW. Developmental change as a major outcome. In *Experiential Learning.* T. Keeton (Ed) San Francisco: Jossey Bass, 1976.
11. Rogers CR. *Freedom to Learn.* Columbus, Ohio: Charles E. Merrill, 1969.
12. Skinner BF. *The Technology of Teaching.* New York: Appleton-Century Crofts, 1968.
13. Posner MI, Keele SW. Skill learning. In *Second Handbook of Research on Teaching.* RMV Travers (Ed) Chicago: Rand McNally, 1973.
14. Rokeach M. *Understanding Human Values.* New York: The Free Press, 1979.
15. Havinghurst RJ. *Developmental Tasks and Education.* New York: McKay, 1972.
16. Bucher R, Stelling JG. *Becoming Professional.* Beverly Hills: Sage Publications, 1977.
17. Gagne RM. *Essentials of Learning for Instruction.* New York: Holt, Rinehart and Winston, 1974.
18. Glaser R. Instructional psychology: past, present and future. *Am Psychol 37*:292–305, 1982.
19. McKeachie WF. Psychology in America's bicentenniel year. *Am Psychol 31*:819–833, 1976.
20. Stritter FT, Hain JD, Grimes D. Clinical teaching reexamined. *J Med Educ 50*:876–882, 1975.
21. Irby DM. Clinical teacher effectiveness in medicine. *J Med Educ 53*:808–815, 1978.
22. Stritter FT, Baker RM. Resident preferences for the clinical teaching of ambulatory care. *J Med Educ 57*:33–41, 1982.
23. Mattern W, Weinholtz D, Friedman C. The attending physician as teacher: *N Engl J Med 12*:1129–1132, 1983.
24. Stritter FT, Hain JD. A workshop in clinical teaching, *J Med Educ 52*:155–157, 1977.
25. Jason H. Shaping attitudes thru role modeling. *Fac-Sheet 2*(3):3, 1983.
26. Stritter FT. The "ideal" role model. *Fac-Sheet 2*(6):3, 1984.
27. Jason H. Shaping attitudes thru critiquing. *Fac-Sheet 2*(4):3, 1983.

8
Evaluation of Learners

WILLIAM C. MCGAGHIE

The term evaluation frequently has a negative connotation, especially among learners who are enrolled in a program of study. When physicians-in-training are evaluated it often means they are in the spotlight, under close observation, for a long time. They are being judged. Evaluation takes on even greater significance when the judgment to be made is for high stakes, such as whether or not to grant a license to practice. High stakes evaluative decisions require sophisticated data collection and interpretation methods to ensure the decisions are accurate. That is why the medical specialty boards have elaborate certification testing programs. Specialty certification is an evaluative situation where the negative connotation probably develops from a fear-of-failure, a legitimate worry for physicians who have spent years preparing for careers that can't get underway until they pass a test. Evaluation done for such other purposes as documenting a clinical clerk's ability to do a physical examination or assessing character traits among prospective residents is often viewed negatively by learners because it is seen as "a waste of time" or "meaningless." In short, evaluation is a process to which most learners grudgingly submit. It is rarely a process they seek or enjoy.

This prevailing view of evaluation is unfortunate because formally or informally, evaluation is a daily fact of life in medical circles. Done well, it can make a large and useful contribution to the learner's educational experience. A good evaluation plan for a residency program would, for example, include a mechanism for documenting each resident's practice experience, as Thomas McGlynn and his colleagues have done (l). The findings from such an evaluation allow for balanced assignment of patients to residents, assuring each physician-in-training a broad-based clinical education. In addition, documentation of residents' experience is increasingly being requested by hospital privilege committees before they will allow young doctors to practice independently (2).

Learner evaluation also serves the interests of those responsible for administration of educational programs. For example, standardized achievement test scores from medical school classes may suggest to a

Dean that an undergraduate curriculum is strong, or that it has weaknesses that need attention. The same would be true of a residency program director who uses in-training exam results diagnostically, as an aid to show where the program needs improvement. Properly conducted, learner evaluation can provide data that can be used as an educational tool, rather than as a weapon. This chapter aims to describe factors that should be taken into account in designing and operating an effective system for learner evaluation.

The chapter begins with a discussion of four reasons why the evaluation of medical learners should be taken seriously at all educational levels. This concerns the purposes of learner evaluation. Next, a description of the context of learner evaluation is given to provide a framework for viewing individual evaluative events as parts of a larger educational program. The third section presents concrete issues that evaluators must recognize when designing a learner evaluation plan. The technology of learner evaluation then receives attention, including a review of basic terms and a description of various educational measurement tools. The chapter concludes with a short section on evaluation and values, topics that are always linked, yet rarely addressed simultaneously.

Purposes of Learner Evaluation

There are four main reasons why learner evaluation deserves close attention from academic physicians. A sound learner evaluation plan is needed to (1) document experience, (2) provide feedback about educational progress, (3) reach decisions about competence, and (4) judge the effectiveness of educational programs.

Document Learner Experience

Most clinical teachers acknowledge that they exercise very little control over the type of cases seen by medical students or residents in the clinic or hospital. Patients arrive for clinic visits or are admitted to an inpatient service as a result of concerns about their health, not because the patients wish to advance medical education. Individual cases, and the health problems they represent, often present on an uncontrolled or "random" basis. Unless patients having different problems are selectively distributed among clinical learners in a controlled way, it can be argued that clinical medical education is also a random process (3).

We all know better, of course, yet few clerkship or residency program directors have evidence that the educational experiences (cases seen) of their learners provides a representative sample of cases in their specialty. Inspection of patient encounter data can lead to startling insights. In one primary care internal medicine residency program, for example, the clinical experience of 10 residents was recorded for a 1-year period. The investigators kept careful records of the patients seen by the residents and the residents' clinical decisions. The results showed that the practice gave the 10 residents ample opportunities to delivery ambulatory care. However, the data also revealed that, "Individual training experiences varied widely and individual deficiencies are readily apparent" (1). One of the residents saw a total of 36 patients over 127 clinic visits during the year. Of 12 common primary care problems identified by the investigators, this resident was judged proficient in treating only two: hypertension and diabetes mellitus. At the other end of the spectrum, the most active of the 10 residents saw 105 patients over 273 visits; proficiency was established for all 12 patient problems including anxiety–depression, congestive heart failure, arthritis, and headache.

Another program that documents residents' patient management experience concerns resident effectiveness "after hours" on the telephone. This program has been described by Curtis and his colleagues (4).

Documentation of learner experience is an important part of learner evaluation, even though few programs currently keep such records. There is a growing trend, however, that is likely to prompt clerkship and residency program directors to keep patient encounter data about their learners. The trend originates from the specialty boards, such as the American Board of Internal Medicine, that increasingly recognize how board exam scores need to be amplified by practice data to ensure the clinical competence of candidates for certification (5). The widespread introduction of microcomputers in medical education settings (discussed in detail in Chapter 15) will greatly reduce the difficulty of collecting and storing patient encounter data.

Feedback About Progress

A common complaint among medical students and residents is that they rarely receive concrete information about "how they are doing" clinically or educationally. Medical learners are usually eager to discuss their experiences and are anxious to discover ways in which they can increase their fund of knowledge or improve their clinical skill. Performance "feedback" is a term that is widely used to describe information that gives learners knowledge of the results of their study and clinical work. Given *specific* feedback about their progress or deficits, clerks, residents, fellows, and even colleagues can either move to new areas of clinical practice or take steps to improve marginal performance.

An educational program needs to have three basic features before useful feedback can be given to learners. First, the program needs to have a clear set of goals that represent a graduated series of milestones for medical students or residents. It makes sense, for example, to expect an intern to demonstrate mastery of basic cardiopulmonary resuscitation (CPR) before beginning work on advanced cardiac life support skills (ACLS). Without such goals, clinical learning is aimless; there are no benchmarks against which to gauge trainee progress. Second, the program needs to have a means to collect, store, and retrieve routinely data that learners and their teachers can use for educational feedback. Documentation of patient encounters is one useful type of educational data. Other types of data include in-training examination scores, anecdotal comments from attendings, and peer evaluations. Third, the program needs faculty who are willing to take time to discuss candidly the evaluative data with students or residents. Effective feedback about educational progress cannot occur unless a plan is in place that identifies goals to be accomplished, routinely collects data about learner progress, and provides frequent opportunities for trainees and faculty to discuss clinical learning.

Data-based feedback about their readiness for clinical work can be helpful to medical learners from the very beginning of their involvement in a clerkship or residency. Establishing a knowledge and skill baseline for each trainee (pre-training assessment) would clearly identify the difference between their clinical proficiency as novices versus their expected proficiency as reflected in learning goals at various stages in a program. Periodic staged evaluation of trainee skill and knowledge to provide performance feedback is termed formative assessment. Cognitive tests and appraisals of procedural and interpersonal skills are all appropriate for pre-training assessment and formative assessment depending, of course, on the learning goals trainees are expected to achieve.

An innovative approach to providing progress feedback to medical students is seen in the work of Paula Stillman and her colleagues (6). They have trained nonphysician patient instructors to evaluate the students' interviewing skills. Not only do the patient instructors submit to interviews by each student, but they also give the students feedback about the quality of the inteview at a level of detail far more precise than feedback ordinarily received from attendings.

Providing feedback to students and residents has many parallels with the performance appraisal techniques suggested for clinical faculty in Chapter 3. In both situations, individuals set goals whose attainment is assessed at regular follow-up intervals. An individual's educational or professional development can be managed using these methods, and steps can be taken (before it's too late) to improve performance if shortfall is detected.

Decisions About Learner Competence

The most visible form of learner evaluation in medical circles occurs when individuals submit to evaluations designed to measure their professional competence or fitness to practice. These are the high-stakes evaluations mentioned earlier. Results from various tests of learner competence are used to reach a number of key decisions; for example, the readiness of foreign medical graduates to undertake advanced study and clinical work in U.S. hospitals, whether an individual can receive a license to practice, specialty and subspecialty certification, and recertification. Competence tests are typically composed of multiple-choice questions, and are usually long, demanding, and expensive. Careers are shaped by competence test results; medical learners take them very seriously.

Measures of medical competence are created by committees of physicians—usually academic physicians—under the auspices of a certifying board or licensing agency. The committees frequently receive guidance in test planning and question writing from specialists in educational evaluation. The National Board of Medical Examiners and the larger specialty boards use elaborate test development procedures and tightly enforce test administration, scoring, and security to ensure that their evaluations are reliable and valid (7).

Accurate decision-making about the professional competence of physicians is linked directly to the quality of the data that are used for competence evaluation. Sophisticated test development and security procedures, and continuing research on improved ways to measure medical competence, are all used to advance the profession and to protect the public. However, much additional work is needed to develop a "test" or other evaluations to certify clinical competence in the real world of medical practice.

Judge Program Effectiveness

One index of the effectiveness of an educational program is the performance of its current and former trainees on various measures of professional achievement. For example, residency program directors frequently use board exam results of recent graduates to gauge the quality of postgraduate medical education. Of course, such judgments of program effectiveness need to take the backgrounds of the residency graduates into account. Are the residency graduates successful on the examination as a result of good clinical teaching or because they *entered* the program with an extensive fund of knowledge and well-developed clinical judgment?

Learner evaluation data—of which test scores are only one type—can be used in a number of other ways to determine the effectiveness of an educational program. To illustrate, family medicine residents rotate through a number of specialty services during their training. At the end of each rotation it is customary for the residents to be rated by their specialty preceptors in terms of technical skill, clinical judgment, professional working relationships, and other indicators of performance. Consistently low resident ratings from services such as obstetrics, orthopedics, or pediatrics may indicate that the residents need to be better prepared before they undertake future rotations. A conference series or a set of required readings might be needed to ensure adequate preparation. Alternatively, consistently low ratings may reflect a bias against generalists among physicians on a specialty service. Here, learner evaluation data can be used to strengthen the famiy medicine residency program either by correcting the source of professional bias or by scheduling the specialty rotation in a different clinical setting.

There are many variations on the four main reasons why schemes for the evaluation of medical learners need to be carefully designed and managed. The key point is that a thoughtfully designed system of learner evaluation can fulfill several purposes for several different audiences. Medical learners at various levels, their teachers, program administrators, and the public all stand to benefit from thorough and rigorous learner evaluation.

Context of Learner Evaluation

Just as it is important to have a clear sense about one's purpose for evaluating medical learners, it is equally important to understand that individual evaluative events do not occur in isolation. Instead, learners experience different types of evaluation on a continuous basis throughout their medical careers. The context of learner evaluation is also shaped by changes or differences in expert thinking about required skills and abilities. Such changes arise from research data, technological advancements, cultural variations, and the evolution of clinical opinion. Faculty who are responsible for evaluating medical learners are also part of the context. Finally, the views of learners about the need for and benefits from learning assessment affect the context of evaluation.

Continuum of Medical Evaluation

That physicians are evaluated throughout their careers is patently obvious yet rarely considered by most clinicians. The evaluation process begins well before one's admission to medical school, concludes at retirement, and involves a host of major and minor career decisions. Figure 8.1 shows the medical career continuum and identifies key points at which evaluation data are used to reach career decisions. All of the decisions shown are important, although for different reasons. For example, the decision to admit a student to medical school is significant because at least since the 1970s, the medical school attrition rate has been very low—about 2% (8). Consequently, for nearly all students the evaluative decision to admit them to medical school is tantamount to a decision to grant the M.D. degree. The decision to grant a license to practice is also significant because it provides the physician statutory authority to practice medicine without supervision. Confidence in the accuracy of these

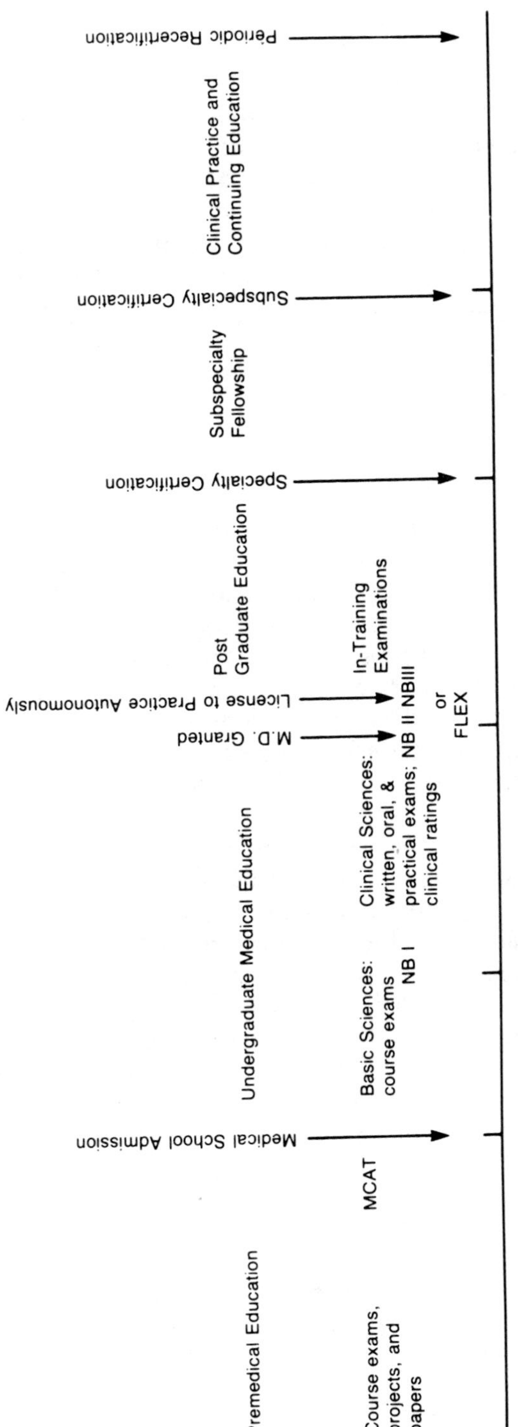

FIGURE 8.1. Contiuum of medical learner evaluation.

and other evaluative decisions stems from the quality of the data they are based on.

Few academic physicians are directly involved in the creation, use, or interpretation of high-stakes evaluations such as admission tests and board certification exams. Instead, their work on learner evaluation is usually confined to the undergraduate and postgraduate years of medical education. At these points on the continuum academic physicians frequently participate in such evaluative activities as test development for basic science courses, giving oral exams and clinical ratings to students in basic clerkships, and supervising clinical care by residents. Despite their apparent isolation, all of these events are links in an evaluation chain that stretches for decades. And each link, each evaluative event, is important at the time it is undertaken.

Settings

In addition to pointing out that evaluation occurs throughout the clinician's career, Fig. 8.1 suggests that physicians are judged in different settings for different reasons. The form and function of the multiple-choice tests that are used for student evaluation in basic science courses are vastly different from the residency program director's subjective impressions which shape letters of recommendation for subspecialty fellowships. Yet the tests and the impressions are both measures; the former quantitative, the latter qualitative. Each has direct consequences for those being assessed. Each is appropriate for use for a focused purpose in a specific setting.

Not only do evaluative settings reveal the purpose of an assessment, they also limit the type of evaluation that can be done. For example, evaluation of residents' procedural skills can, with a few exceptions such as CPR, be done accurately only at the bedside. However, the clinical setting is judged less appropriate than a classroom for evaluating learners' fund of knowledge because preceptors must usually rely on inefficient and potentially unreliable oral reports. Learner evaluations done in different physical settings will serve different purposes and with varying degrees of precision. Thus it is important to have a clear purpose in mind (answer the question: Why bother?) before planning a learner evaluation and selecting a site.

Zeitgeist

The definition of medical competence is neither static in time nor the same for all nations and cultures. The definition evolves with the introduction of new technology, with changes in expert opinion about the skill and knowledge that physicians need to acquire and use, and as public opinion shifts about the efficacy of medical interventions for various problems. The physician of the 1980s is far more resourceful than the 1920s physician and, we suspect, will be much less sophisticated than doctors in the 21st century. Even today medical competence in some cultures means the physician incorporates nonscientific folk remedies and beliefs into practice. In many developing countries the competent doctor may never even see patients, spending time instead supervising a district health team that would include nurses, medical aides, midwives, and sanitary personnel who are responsible for giving care.

Thus technological and professional change and a respect for national and cultural variation are important factors to consider when planning evaluations of medical learners.

Evaluators

Faculty who plan and implement learner evaluations are a key fixture of the educational context. Worthwhile evaluations take hard work and some knowledge about available methods of assessment and technical terms such as reliability and validity (see Table 8.1). Worthwhile evaluations also require faculty commitment, a readiness to devote time and energy to educational activities. Research shows that a strong commitment to learner evaluation is rare among clinical faculty, especially those at academic medical centers (9). Although it is reasonable for faculty who are responsible for clinical evaluations of students and residents to expect colleagues to contribute to the process, it is naive to assume that most academic physicians see evaluation as a top professional priority. Consequently, departmental evaluators are more likely to get the help they need by being well-organized, efficient, and realistic about the time demands they place on colleagues.

Learners

All medical learners are a part of the educational context. Their acceptance or rejection of evaluation as a worthwhile educational activity reflects the climate of student–faculty relationships. Sometimes the climate is stormy, adverserial. Elsewhere it may be open and honest, with faculty making their educational expectations and standards known to learners and giving learners feedback about their educational progress. The medical education literature suggests that the customary behavior of medical learners falls between the two extremes. It is well-established, for example, that clerks and residents quickly learn the importance of managing attendings' *impressions* of their clinical skill (10). Wearing a "cloak of competence" and learning to "cover" oneself by hiding areas of ignorance or uncertainty are common among medical learners in clinical settings. The costs of this behavior are obvious. Learner weaknesses go uncorrected, faculty members are misled into thinking all's well, and patient care may suffer, just to maintain an amiable clinical atmosphere.

The clerks and residents, of course, are only responding to unwritten criteria for what it takes to move on. Thus although the learners are an important feature of the clinical evaluation context, they also influence the context by managing faculty impressions. Faculty who take time to probe the skills and knowledge of learners at a level deeper than clinical impressions are often surprised, as McGlynn et al. have shown (1). Given good evaluative data, faculty can work to ensure, not guess, that students and residents are not just "safe," but also clinically fit.

Issues Evaluators Must Recognize

Given a clear sense of the purpose for and the context of learner assessment, a number of concrete issues should be recognized in designing a plan for learner evalution. Seven issues are especially prominent.

1. The *evaluation system* that operates for a course, clerkship, residency, or any other educational experience *will likely have a stronger influence on learner behavior than any other educational variable* (11). Whether an evaluation plan is structured and formal or loose and easy, medical learners will act to satisfy faculty expectations (often, their *impressions* of faculty expectations) as seen in the way learners are judged. In short, evaluations such as clinical ratings and tests motivate medical student behavior, sometimes in ways that, if left alone, will conflict with faculty intentions (12).

2. The *criteria* for an evaluation are embodied in the procedural and interpersonal skills, knowledge, clinical judgment, and other qualities the faculty expects learners to acquire during a period of study and work. Criteria are outcomes. Some of the outcomes are intentional. These are expressed as learning objectives whose accomplishment is assessed at fixed points in training, like ACLS certification during internship. Other outcomes are incidental, unplanned, and may be highly valuable, such as not "coming apart" under stress, or objectionable, such as labelling patients "crocks" or "gomers." Proposed criteria need to be stated before a coherent evaluation can be undertaken.

3. *Standards* are different from criteria. Standards refer to how well or to what degree educational criteria must be accomplished. Common expressions of educational and professional standards include passing scores on examinations; the minimum number of, say, diabetes mellitus patients a second-year resident is expected to see; or an editorial board's consensus about the acceptability of a subspecialty fellow's manuscript for journal publication. Most faculty find standard setting very difficult

because it is judgmental, and often appears arbitrary. Evaluators can't avoid the standards issue. Instead, they should accept its necessarily qualitative foundation and rely on reason and experience to help them decide the level of achievement expected from students, clerks, residents, or fellows.

4. Evaluators need to worry about the *quality of the data* they use to reach decisions, just as physicians attend to the quality of the data used in making clinical decisions. Analogous to the sensitivity and specificity of laboratory data, physicians responsible for learner evaluation should concern themselves with the reliability and validity of data from educational evaluations. Failure to do so leaves academic physicians in what Richard Friedman has called a "fantasy land" with particular reference to interpreting grade reports and letters of recommendation for housestaff slots (13). Data of poor quality contribute to inaccurate educational (and sometimes, career) decisions just as poor clinical data may be not only useless, but harmful.

5. The skill of evaluating learners is rarely found in the academic physician's repertoire. Consequently, almost without exception, faculty evaluators need to be *trained* to ensure that they collect and interpret data accurately. Such training can take many forms: workshops on test construction and question writing, videotape analyses of student or resident case presentations, mock chart reviews, and reading about evaluation in the medical education literature. To the degree that learner evaluation is held to be an important activity, the development of faculty skill in evaluation will not be left to chance. Instead, it will be practiced.

6. Medical educators at all levels need to recognize the *consequences of evaluation* for learners and for training programs. The consequences for learners are clear. Student admission, promotion, and graduation from medical school; "matching" with a desired residency; licensure and certification; and other hurdles on the medical education continuum all depend on evaluative data. The consequences for patients are equally obvious. They include assurance that doctors are clinically capable, cost conscious, and attentive to patients' personal and family lives. The consequences of learner evaluation for programs are also significant yet rarely considered by most academic physicians. These can include documentation of high-quality teaching, continued accreditation, status in the professional community, and many others. Of course, evaluation poorly done can yield consequences opposite the positive ones listed and also lower student, patient, and faculty morale.

7. A key issue has been implied, yet not clearly stated at several points in the discussion. It is that *many persons and organizations beyond medical learners and faculty have a stake in the evaluation system.* The Bakke case of the late 1970s brought the medical school admission process into public view. For some citizens, the case led to the conclusion that the presumably unbiased academic admission system contains political ele-

ments. A similar case for public interest in medical evaluation can be made about licensure, which is a civil, not a professional, decision. Public laws govern procedures for granting and rescinding a physician's license to practice, despite the fact that measures developed by the medical community (FLEX and National Board examinations) are frequently accepted as a proxy for licensure. The point is, of course, that the evaluation of medical personnel throughout their careers is of interest to physicians, medical education institutions, medical boards and agencies, and the public. Medical learner evaluation is not strictly a guarded professional matter.

Technology of Learner Evaluation

The practice of learner evaluation requires a set of measurement tools, a technology, that can be used with ease and confidence. It also requires familiarity with a few basic terms. Like all academic specialties, educational measurement and evaluation has a technical language that gives practitioners a set of common meanings and allows them to communicate with precision and efficiency. A short list of 20 basic terms in educational measurement and evaluation appears in Table 8.1. Academic physicians with responsibility for learner evaluation will frequently encounter these and other new terms. By understanding the language of educational measurement, or referring to Table 8.1 or other sources (e.g., textbooks) for clarity, academic physicians can enrich their technical knowledge.

Many different tools are available for the evaluation of medical learners, ranging from long aptitude tests such as the MCAT to short bedside encounters. Some evaluative tools such as subspecialty certification examinations are highly quantitative and objective whereas others such as letters of recommendation are qualitative, subjective. Each type of measure has a place in medical learner evaluation. However, the decision to use one of the tools should be based on a clear understanding of one's evaluative purpose and context.

Table 8.2 describes common evaluation methods in medical education. The table also contains a short comment about the advantages of each method and a statement about potential problems associated with using each procedure. A citation is given for each method to encourage further reading by those who seek more detailed information.

No single evaluation method is valid for all purposes. Academic physicians need to think hard about their reason for wanting to assess a student's or resident's knowledge, procedural skill, self-confidence, dependability, honesty, or any other clinically relevant characteristic. Only after identifying the purpose of evaluation (e.g., educational diagnosis, annual promotion, performance on a particular rotation) should the academic

TABLE 8.1. Twenty basic terms in educational measurement and evaluation.

Affective domain. The area of learning and performance related to beliefs, emotions, sentiments, professional values, etc.

Cognitive domain. The area of learning and performance related to knowledge and comprehension.

Composite score. A score that is a combination of several subtest scores, usually derived by addition. Different weights can be applied to the subtest scores to increase or decrease their relative importance in the composite.

Criterion. A set of tasks, body of information, or professional attributes—the content—that test scores are said to represent.

Criterion (content)-referenced test. A test specifically designed to provide information about examinee knowledge or skill in regard to a clearly defined body of content.

Difficulty index. For a single test question, the percentage of individuals in some specified group (e.g., second-year medical students) who answer the question correctly.

Discrimination index. For a single test question, an index that shows the power of the question to separate examinees who score high on the total test (usually the top 27%) from examinees who score low on the total test (usually the bottom 27%).

Evaluation, formative. Periodic assessment of an individual's progress through a program or course of study.

Evaluation, summative. A final evaluation event that is usually done at the conclusion of a long period of education and experience (e.g., specialty board certification).

Norms. Statistics that describe the average or typical test performance of a particular group of examinees.

Norm-referenced test. A test whose results are interpreted by comparing scores obtained by individuals with a group average or norm.

Psychomotor domain. The area of learning and performance related to skilled tasks and motor behavior.

Reliability, interrater. The degree to which the scores or ratings given to the same person or object by two or more independent raters are in agreement.

Reliability, test (single administration). The degree to which scores on a test "spread out" a group of examinees, highlighting their individual differences, and yielding a normal distribution. This type of reliability is a valued property of norm-referenced test scores.

Score, raw. The first quantitative result obtained from scoring a test. Raw scores are usually expressed as the number of correct answers on a test.

Score, standardized. Test scores that result from mathematically adjusting a set of raw scores and recasting them on a scale having standard properties. National Board scores, for example, are former raw scores having a standardized mean = 500 and standard deviation = 100.

Standard. A quantitative or qualitative value, usually expressed as the minimum passing score on a test, that is used to interpret achievement.

Standard error of measurement. A statistic giving an estimate of the amount of error in an individual's test score. Use of the SEM encourages test score interpretation within confidence bands rather than as fixed points on a scale.

Test, standardized. A test that is carefully constructed, administered under controlled, uniform conditions, and whose results are interpreted in a uniform way for an examinee group. Most standardized tests are norm-referenced.

Validity. The degree to which the results from an educational measurement provide useful information for a specific purpose, e.g., diagnosis, prediction, placement, promotion, certification. Validity has different meanings for different types of tests and various forms of validity evidence are needed, depending on one's purposes.

TABLE 8.2. Evaluation methods commonly used in medical education.

Method	Description	Advantages	Problems
1. Records of clinical encounters (1,2,4)	Case-by-case documentation of: 1. clinical problems seen; and 2. decisions made about each problem	Long-run formulation of learner practice profile. Helps identify clinical problems where more experience is needed. Useful for gaining hospital privileges after training	Requires high degree of learner compliance. Cumbersome without computerized data management system
2. Formal (external) examinations (7)	Long, standardized, norm-referenced examinations covering large bodies of medical content; often composed of separate disciplinary subtests	Usually high-quality exams that give a general portrait of an examinee's fund of knowledge	Test content may not match local educational objectives. Not useful to pinpoint specific learning deficits. High monetary cost
3. Local (internal) examinations (14)	Examinations written by local faculty typically for use in courses or clerkships; can be criterion-referenced	Can be keyed to match closely local teaching emphases; exams unite instruction and evaluation.	Quality can suffer if faculty are disinterested or unschooled in test development. Major cost is faculty time.
4. Simulations (15)	Written and computer-based approximations of clinical encounters with patients. Often termed patient-management problems (PMPs)	Apparently realistic approach to evaluating clinical reasoning and problem-solving. Enjoyed by clinicians; excellent for instruction	Technical problems in scoring and failure of performance to generalize across cases makes them questionable measures for evaluative purposes.
5. Objective structured clinical examination (OSCE) (16)	Examinees rotate through a series of stations where, in about 5 minutes each, they are questioned, asked to interpret clinical data, perform a procedure, or otherwise show proficiency with clinical materials.	Concrete, realistic approach to evaluating learners' clinical skills. Requires prompt responses to real clinical material. Bluffing is unlikely.	Faculty cooperation is vital; can create administrative problems unless tightly managed.
6. Checklists (17)	Step-by-step "yes-no" or "right-wrong" protocols used to assess either skill at a clinical procedure (e.g., CPR) or at preparing a clinical product (e.g., a sterile tray)	Useful to evaluate *specific* procedures and products. Little guesswork once checklist items and their order are agreed on	Can appear simplistic unless procedures or products are critical. Use may require much faculty time. Rater training is essential.

7. Rating scales (17)	General assessments, often of learner character or noncognitive professional qualities, based on the rater's memory rather than direct observation of specific events	Allows evaluators to quantify important qualitative factors that underlie good clinical care.	Frequent "halo" effect, meaning low ratings are rare
8. Oral examinations (18)	Face-to-face learner–evaluator encounters where learners are questioned about clinical subjects; sometimes used to see if learners can withstand stress	Historically grounded; have been used for learner evaluation for over 3000 years. Encourage student–faculty interaction.	Notoriously unreliable approach to learner evaluation. Unstandardized; subject to capricious evaluator behavior
9. Anecdotal records (13,19)	Dean's letters, faculty letters of recommendation	Highly personalized approach to description of learner achievement and frequently, learner readiness to pursue more advanced training	"Halo" effect is common. Frequently difficult to interpret as recipients try to "read between the lines."
10. Chart reviews (20,21)	Faculty–learner case discussions based on data contained in patient charts and recent progress notes	High relevance due to grounding in real clinical work. No or low cost, straightforward, immediate feedback about patient management	Cases selected should be representative of learner's practice not chosen because, they are unusual.
11. Patient instructors (6)	Laypersons are taught to function as patient, teacher, and evaluator for students	Very high realism; patient instructors provide students excellent feedback. Especially useful to evaluate skills in physical diagnosis and interviewing	Training of patient instructors takes some time, careful management of the evaluation plan is needed.
12. A-V reviews (22,23)	Learner–faculty review and critique of taped encounters involving the learner and patients	Very high realism; allows mutual assessment of patient management and learner's interpersonal skill and professional qualities	Can be "highly charged." Some learners need time to "desensitize" from seeing or hearing themselves on tape.
13. Educational prescription contracts (24)	Written agreement between learner and evaluator about learner's educational goals for a specified period of time	Clear specification of learner's educational intentions and what support faculty will provide. States educational criteria *and* standards	Some learners and faculty are reluctant to express expectations for one another.

physician select a measurement tool that will produce meaningful data to inform the needed decision.

Congruence

Academic physicians often assume that the measures they use to assess learners closely conform with the goals of clinical education. In a word, there is a belief that goals and tools are congruent. However, evaluator Jon Wergin argues that the congruence assumption may not be valid because clinical settings are often uncontrolled, learners are frequently unaware of what is expected of them, and faculty standards differ (25). Wergin observes that two widespread problems are often responsible for the lack of congruence between educational goals and measurement tools: (a) "Fallacy of False Quantification" (FFQ), and (b) "Law of the Instrument Fallacy" (LIF).

Of the two problems, the FFQ is the most frequently encountered. "Simply stated, the FFQ is the tendency to focus on those skills or objectives that are most easily measured" (25). For example, medical student knowledge is regularly evaluated using multiple-choice questions while student ability to apply knowledge in novel situations is ignored. Once again, focusing on the purpose of evaluation will help avoid the FFQ trap.

The LIF is based on a famous statement made by philosopher of science Abraham Kaplan: "Give a small boy a hammer, and he will find that everything he encounters needs pounding" (26). Expressions of the LIF involve routine and unquestioned use of a measurement procedure without giving much thought to its value in a particular context. Thus today we see patient management problems (PMPs) widely used in medical circles to measure learner skill at "problem-solving," "clinical judgment," and "decision-making," without a clear indication of what those terms mean. The upshot is that clinically relevant skills such as problem-solving are defined by the method of measurement, an example of the tail (PMPs) wagging the dog (evaluation of problem-solving).

When evaluating medical learners, congruence is more likely when academic physicians look beyond skills and abilities that are easily measured and adopt a healthy skepticism about the value of routinely used measurement instruments.

Interpretation of Evaluative Data

Valid interpretation of evaluative data is the endpoint of learner assessment. This is the case for both formative assessment and summative assessment. It is important to underscore the point that *validity is not an intrinsic property of data.* Instead, valid interpretation of data is a consequence of the way an evaluator gives them meaning and uses data for a practical purpose, such as promotion of students from the third to the fourth year of medical school. Valid interpretation comes from human comprehension and judgment about the numbers. It takes skill and care. It depends on linking three elements of learner evaluation mentioned earlier.

1. *Consensual evaluative criteria* (e.g., skills, knowledge, experience, sentiments) need to be identified and used as the foundation for the construction or adoption of measurement tools.
2. Assurance is needed that the resulting tools are *accurate measures* (i.e., reliable) of the criteria of interest. Put another way, the quality of evaluative data should be confirmed, not assumed.
3. Evaluative *standards* (i.e., minimum acceptable performance) need to be established and preferably shared with faculty and learners. This reduces the likelihood of "gamesmanship" by clarifying expectations and making data interpretation a straightforward activity for learners and faculty.

Analogous to the interpretation of data derived from scientific experiments, the interpretation of data arising from learner evaluation always contains uncertainty. In statistical terms, evaluators, like clinical scientists, need to be sensitive to the probability of making false-positive and false-negative decision errors about the progress or competence of learners. However, the probability of reaching accurate decisions is increased as evaluative criteria, measures, and standards are carefully formulated and tightly linked.

Managing Evaluation

The practical business of managing learner evaluation is a direct extension of the conceptual and technical matters that have been discussed. Academic physicians who are directors of basic science courses, clerkships, residency programs, and continuing education programs can benefit from suggestions about how to set up and run their evaluations. Here are 13 practical tips:

1. Obtain a consultation from a specialist in educational evaluation or from an experienced clinical colleague, preferably both. Even a short consult with one who has evaluation experience can help keep program planning and execution on the right track.

2. Become informed about educational evaluation by reading the literature. Articles by academic physicians frequently appear in such journals as *Evaluation and the Health Professions* and the *Journal of Medical Education.* Occasional articles about evaluation also appear in specialized clinical journals. Rely on the published work about evaluation to keep you from "reinventing the wheel."

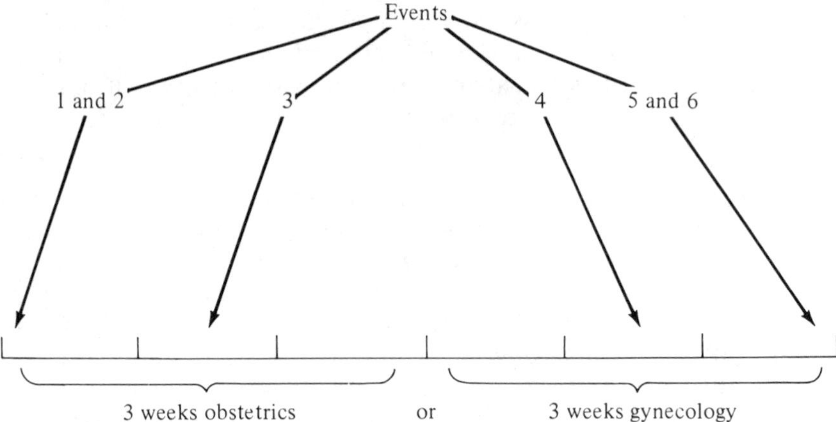

FIGURE 8.2. Timetable for an obstetrics and gynecology clerkship.
(1) Distribution of APGO Instructional Objectives (27). Students are told they are required to take a written examination keyed to the objectives on the last day of the clerkship.
(2) Distribution of Clinical Activities forms. Students are expected to maintain records on patients they work-up and follow.
(3 and 4) Supervised clinical experience. One-half of each student group starts on Obstetrics, the remainder starts on Gynecology. Rotation occurs after 3 weeks.
(5) Multiple-choice examination in National Board format, oral examination.
(6) Grading: clinical performance ratings by attendings and housestaff (50%), written examination results (25%), results of an oral examination on student's patients and a specific, preassigned topic (25%).
Information provided by William D. Droegemueller, M.D., Professor and Chairman and by William N.P. Herbert, M.D., Associate Professor and Clerkship Director, Department of Obstetrics and Gynecology, University of North Carolina, School of Medicine.

3. Start evaluation planning at the earliest possible date, especially if other faculty members will be involved as, for example, test question writers. Busy clinicians tend to put off evaluation until the last minute, often at the expense of learners. Avoid this trap by getting started well in advance of the deadline.

4. Locate your educational offering on the continuum given in Fig. 8.1 (p.131). Next, develop a scaled-down version of the continuum in the form of a timetable that covers the period of the course, clerkship, or other educational offering. An illustration is given in Fig. 8.2 for the 6-week Obstetrics and Gynecology clerkship at the University of North Carolina School of Medicine. Now ask: "What do I know or what can I safely assume about the prior preparation of learners who will enroll in my program?

5. Clarify the purpose for which evaluative data are needed. Illustrations include (a) grade assignments to satisfy institutional requirements,

(b) documenting residents' patient care experiences so that hospital privileges can be obtained after training, (c) reporting to the AMA to obtain continuing education credits (attendance figures), and (d) reporting to the Dean who may subsequently use the information when preparing letters of recommendation for residency positions.

6. No evaluation can attend to or measure all of the possible outcomes from a course of study or period of clinical experience. Physician evaluators need to focus their efforts by making conscious choices about the criteria that will be assessed. Ask: "What can I reasonably expect learners to accomplish given constraints of time and energy?"

7. Carefully match educational measures with educational criteria. Many excellent measures including examinations, rating scales, chart reviews, and educational prescription contracts can be created and used in the local setting to match local learning goals (Table 8.2). "Off the shelf" measures such as National Board examinations or rating instruments developed at other hospitals or clinics should be used with much caution.

8. Using the timetable derived in step 3, lay out an optimal data collection plan in terms of frequency and depth of learner evaluation. Check to see if the institution governs the amount of time allowed for evaluation by a formula that takes account of the number of instructional hours devoted to courses and clerkships.

9. Identify local resources for collecting and analyzing data. For example, is there secretarial, word processing, or computer support for the evaluation? Are statisticians available for consultation about data analysis and interpretation? Be certain that you clearly understand who is responsible for establishing and maintaining learner files (computer or manual), checking the accuracy of evaluative data, and badgering faculty who fail to submit reports on time.

10. Achievement standards should be spelled out *to the degree of precision permitted by the evaluation criteria.* Some standards are quantitative, others are qualitative. It is one thing to assert that all students are expected to achieve a score of at least 70% on a biochemistry test, quite another to assert that "residents should be attentive to the psychosocial needs of terminally ill patients." Physicians responsible for learner evaluation should also identify the consequences of learner failure to meet current standards. Does failure mean more focused reading; additional, selected clinical experience; repeating an entire clerkship or rotation; or dismissal from medical school or the residency?

11. The degree of involvement by other faculty should be addressed, especially if these colleagues are from departments different from one's own. Memos seeking or confirming faculty participation need to be exchanged, most likely at the level of department chairpersons. In addition, faculty training often needs to occur well in advance of planning an evaluation design. This will help colleagues understand the logic and process of the evaluation.

12. Decide in advance exactly who will receive evaluation reports. Learners and the School Registrar are legally entitled to the material. But to what extent, if at all, will faculty members in your or other departments, other schools, local or state medical organizations, or even the public have access to data? Evaluation data can be sensitive and it is wise to establish a dissemination policy before an assessment is undertaken.

13. Conduct a *postmortem* after an educational experience and an evaluation of its learners. Ask: "How can the evaluation be improved first, to serve the interests of learners and second, to serve the interests of faculty, and administration? All of us learn from experience, usually in a haphazard fashion. However, a systematic review of the experience should be a learning process if the academic physician engages in thoughtful advance planning.

Evaluation and Values

The central message of this chapter has been that the evaluation of medical learners is an important professional responsibility, an activity that warrants the academic physician's thought, time, and energy. In academic settings, the faculty's approach to evaluation is the most visible and concrete indication of expectations for the next generation of clinicians. In short, an evaluation plan is a direct expression of faculty values. It makes explicit the ordinarily tacit features of clinical medicine that are observed from faculty role-modeling and personal demeanor (see Chapter 21).

A sound evaluation plan minimizes the probability that chance will shape and maintain the skill, knowledge, and habits of future physicians. A well-designed evaluation for a program or course of study indicates faculty members acknowledge their ethical responsibility to supervise physicians-in-training (9), and to certify learner competence. Evaluation, even for relatively short portions of the continuum shown in Fig. 8.1, is a sign that academic physicians control the clinical and educational environment. Education and patient care thus become deliberate, orderly processes where expectations are clear and standards are known.

Medical students, residents, and more advanced learners should, ideally, understand how evaluation can contribute to their acquisition of skill and knowledge. This means the learners' best interests are served by evaluation and feedback; that similar to faculty, learners should grow to appreciate evaluation and its role in career development. Evaluation should also be seen as an omen. Determination of clinical competence is increasingly becoming a fact of professional life, particularly in the form of specialty recertification. Early recognition by learners that evaluative events can contribute to the maintainence and growth of competence throughout a professional career is itself an expression of values that few academic physicians would dispute.

Conclusion

The evaluation of medical learners is a complex responsibility for academic physicians, a task that occurs throughout a medical career. Done well, such evaluation takes account of the purposes for data collection and analysis; is sensitive to the context of evaluative activities; recognizes persistent issues (e.g., criteria and standards) involved in learner evaluation; and takes advantage of the current technology of educational measurement. Academic physicians should also appreciate that all evaluation plans are grounded in a system of values that tacitly reflect professional consensus about the proper conduct of clinical medicine.

References

1. McGlynn TJ Jr, Munzenrider RF, Zizzo J. A resident's internal medicine practice. *Eval Health Prof 2*:463–476, 1979.
2. Curtis P, Resnick J, Warburton SW. *Resident Inpatient Documentation for Family Practice Training Programs: A National Perspective.* Chapel Hill, North Carolina: Department of Family Medicine, University of North Carolina School of Medicine, 1982.
3. Hainer BL, Curry HB. Selective patient enrollment: a tool for improved residency training. *J Med Educ 57*:835–840, 1982.
4. Curtis P, Talbot A, Liebeseller V, Phillips K. After hours calls: an American study. *Can Fam Phys 25*:284–292, 1979.
5. American Board of Internal Medicine. *Evaluation of Clinical Competence.* Philadelphia, 1983.
6. Stillman PL, Burpeau-DiGregorio MY, Nicholson GI, Sabers DI, Stillman AE. Six years of experience using patient instructors to teach interviewing skills. *J Med Educ 58*:941–946, 1983.
7. Hubbard JP. *Measuring Medical Education: The Tests and the Experience of the National Board of Medical Examiners,* (2nd Ed) Philadelphia: Lea & Febiger, 1978.
8. Johnson DG. *Physicians in the Making.* San Francisco: Jossey-Bass, 1983.
9. Smith AC, McGaghie WC. Student evaluation in clinical education: a field study at one medical school. *Ann Conf Res Med Educ 23*:217–222, 1984.
10. Hass J, Shaffir W. Ritual evaluation of competence. *Work and Occup 9*:131–154, 1982.
11. Frederiksen N. The real test bias: influences of testing on teaching and learning. *Am Psychol 39*:193–202, 1984.
12. Newble DI, Jaeger K. The effect of assessments and examinations on the learning of medical students. *Med Educ 17*:165–171, 1983.
13. Friedman RB. Fantasy land (sounding board editorial). *N Engl J Med 308*:651–653, 1983.
14. Newble DI, Elmslie RG, Baxter A. A problem-based criterion-referenced examination of clinical competence. *J Med Educ 53*:720–726, 1978.
15. McGuire CH, Solomon LM, Bashook PG. *Construction and Use of Written Simulations.* New York: The Psychological Corporation, 1976.
16. Harden RM, Stevenson M, Downie WW, Wilson GM. Assessment of clinical

competence using objective structured examination. *Br Med J 1*:447–451, 1975.

17. Risley B. Principles for developing instruments to assess students' clinical skills. *Respir Care 23*:158–166, 1978.
18. Meskauskas JA, Norcini JJ. Standard-setting in written and interactive (oral) specialty certification examinations. *Eval Health Prof 3*:321–360, 1980.
19. Leichner P, Eusebio-Torres E, Harper D. The validity of reference letters in predicting resident performance. *J Med Educ 56*:1019–1021, 1981.
20. Payne BC. The medical record as a basis for assessing physician competence. *Ann Intern Med 91*:623–629, 1979.
21. Ramsdell JW, Berry CC. Evaluation of general and traditional internal medicine residencies utilizing a medical records audit based on educational objectives. *Med Care 21*:1144–1153, 1983.
22. Lin P, Miller E, Herr G, Hardy C, Sivarajan M, Willenkin R. Videotape reliability: a method of evaluation of a clinical performance examination. *J Med Educ 55*:713–715, 1980.
23. Shepherd D, Hammond P. Self-assessment of specific interpersonal skills of medical undergraduates using immediate feedback through closed-circuit television. *Med Educ 18*:80–84, 1984.
24. Pratt D, Magill MK. Educational contracts: a basis for effective clinical teaching. *J Med Educ 58*:462–467, 1983.
25. Wergin JF. Congruence evaluation. In *Evaluating Clinical Competence in the Health Professions.* MK Morgan and DM Irby, (Eds) St. Louis: C.V. Mosby, 1978, pp. 52–58.
26. Kaplan A. *The Conduct of Inquiry.* New York: Harper & Row, 1964.
27. Association of Professors of Gynecology and Obstetrics (APGO). *Instructional Objectives for a Clinical Curriculum in Obstetrics and Gynecology.* Chapel Hill, North Carolina: Health Sciences Consortium, 1979.

9
Evaluating Educational Programs

GEORGE B. FORSYTHE, JAMES C. SADLER,
AND RUTH DE BLIEK

Having snatched a few moments out of a busy schedule to go through paperwork which had piled up on the desk, Dr. Peffer noticed the report of last week's department education committee meeting. The minutes reminded him of the lively discussion that had taken place during the meeting about the educational goals of the department and whether or not they were being adequately addressed by the residency program. A survey of recent graduates suggested that some of the goals may not be emphasized as much as they should be. As director of the residency program, Dr. Peffer was concerned about his inability to offer more than just his opinion about the program's effectiveness in attaining the goals. At the department chairperson's request, he agreed to undertake a systematic evaluation of the residency program. But how?

Dr. Peffer felt overwhelmed by the complexity of the assignment, especially given all his other responsibilities. How could he develop a rather casual request made at a departmental meeting into an assignment for which he would receive adequate support and recognition? What aspects of the program can and should be looked at? Who would help him conduct this evaluation? How would he gather information and how would he report his findings?

This chapter is written for the academic physician who is faced with questions similar to those posed by Dr. Peffer. Nine considerations are presented that will help make the planning and implementation of a program evaluation manageable and informative. The major theme of this chapter is that program evaluation can and should be a purposeful and systematic process that serves the decision-making needs of the program personnel.

The specialized scholarly literature on program evaluation derives from such disciplines as psychology, economics, and research design. This literature provides a basis for enumerating a set of principles that can help the academic physician plan and carry out an evaluation of an ongoing educational program such as a clerkship, a residency program, or continuing medical education.

Each clinical education program exists within its own context and has its own set of resources and constraints. Because programs differ, any evaluation should take into account the goals, the environment, and the processes unique to a program in order to help decision-makers make adjustments and judgments.

The nine program evaluation considerations presented in this chapter are derived from the relevant literature. Topics have been selected that are most concerned with the essential points of program evaluation. Each consideration should be guided by the character and needs of the specific program. They are:

1. What purposes will the evaluation serve?
2. What is the nature of the assignment?
3. What resources and constraints influence the evaluation?
4. Who should do the evaluation?
5. What are the dimensions of the program?
6. What criteria and standards should be applied in the evaluation?
7. What components of the program should be evaluated and what sources of information are available?
8. What methods of investigation should be used?
9. How should the findings be presented?

What Purpose Will the Evaluation Serve?

The evaluator should specify the purposes of the evaluation in the early planning stages because this will influence other decisions made in the evaluation process. This process is similar to the selection and definition of a suitable research question prior to beginning a research project. Four major purposes of evaluation are presented in this section: informed decision-making, measurement of outcomes, assessment of program effects, and improving organizational processes.

Informed Decision-Making

Evaluation should provide information for decisions regarding educational programs (1–3). Stufflebeam (3) defines evaluation as "the process of delineating, obtaining, and providing useful information for judging decision alternatives." The evaluator must identify alternatives that are useful and realistic for a particular program. Many portions of a residency curriculum are dictated by specialty boards or accrediting bodies, and hence are beyond the control of the program staff. Delineating the "discretionary" portion of the curriculum will highlight areas in which information should be collected to improve the process of making decisions.

Scriven distinguishes two roles of program evaluation: formative and summative (4). This distinction is based on the types of educational decisions the program director will make and the nature of the information required to make them. Formative and summative evaluations are most easily distinguished when a program has a discrete beginning and end, for example, a fellowship program funded by an external agency for a 3-year period. Formative evaluation works to improve or re-"form" an educational program. It explores questions such as: "Is the rotation in the intensive care unit the appropriate teaching environment for our junior residents to learn fluid and electrolyte balance?" "Two interns did not do well on the in-training exam in nephrology. Is this the result of what occurred on the nephrology rotation or is it a learner problem?" "How can we improve the scheduling of our residents?" "How can we improve our residency selection process?" Formative evaluation is developmental. It is undertaken when it is possible to take action to modify or improve the program.

Summative evaluation focuses on accountability. It asks questions such as: Does this fellowship program fulfill the professional development needs of the specialist so well that it should be continued beyond its initial funding period?" "Should we continue to teach this course?" Summative evaluation usually occurs *after* an educational program is completed.

Measuring Specified Objectives (Outcomes)

One of the earliest formal approaches to evaluation emphasized the measurement of program outcomes or the "process of determining to what extent the educational objectives are actually being realized"(5). Here the prespecified outcomes (the objectives) are measured against the actual outcomes to see if the objectives were actually met. For example, an objective for an advanced rotation in neurology or radiology might be to perform a myelogram successfully. A broader objective for a residency program may be to have all residents pass the specialty board exam on their first attempt.

When outcome measurement is the dominant evaluation focus, a program must have clear and measurable objectives. Since such objectives are not always available, the evaluator's first task may be to assist in their development. Although definition and measurement of outcomes are essential in any educational program, the overall usefulness of evaluation is highest when outcome information is combined with other evaluative information. Restricting an evaluation to only the assessment of specified program outcomes may distort what the program is actually accomplishing and how it is being accomplished. For example, a residency in family medicine may have a specific educational goal that residents will develop facility in basic interviewing skills and the recording of affective data. An evaluation effort based solely on the objectives would assess whether residents had mastered these skills, but would likely not deal with the process of how the skills were learned or whether the residency program had a positive or a negative effect on the process. It may be that the majority of residents do master interviewing skills, but do so as a result of peer modelling or instruction by the medical staff. To a residency director, the knowledge of what is fostering or precluding the development of desired competencies may be as important as knowledge about the outcome.

Assessing All Program Effects

In contrast to an objectives or outcome-based approach to evaluation, Scriven argues that evaluation is the process of identifying accurately all the effects of an educational program (6). He points out that the effects may or may not match the intended outcomes stated at the beginning of the program. Any educational program will likely have both intended and unintended effects.

For example, in a residency program many types of effects—cognitive, psychological, and social—may occur that are unintentional. Some unintended results may be beneficial, for example, incidental learning in a content area such as statistics may occur because of influence from a particular set of fellows who are immersed in research issues in their own academic programs. Other unintended outcomes may be counterproductive, such as stress-related illness, marital difficulties, and substance abuse.

Improving Organizational Processes

Evaluation may also be desirable for its psychological effect (7) on program participants. First, it may motivate desired behaviors in those being evaluated. Indeed, it may be argued that the evaluation system in an educational program drives the behavior of participants. For example, medical students tend to study the material that they expect will be on the upcoming exam. Likewise, a formal ongoing evaluation of clinical teaching may direct attendings to examine more consciously the process by which they are facilitating the developing competence of residents.

Evaluation may also serve a facilitative function (8–13) in which the information collected serves to identify and create responses to various needs. This contribution of evaluation emphasizes multiple perspectives of the program, differing values of participants, and qualitative as well as quantitative data. The evaluator plays a more active role in analyzing the dynamics of the program in order to make inferences and propose alternatives for action. Depending on the specific evaluation strategy used, the evaluator may be in a position to ensure that the concerns of all participants in the program are represented in the decision-making process. As information is gathered about a program, communication among the participants and the faculty may be improved, expectations of the varied constituencies may be clarified, and teamwork may be enhanced.

What Is the Nature of the Assignment?

The prudent evaluator can avoid misdirected effort and misunderstandings if early attention is given to organizational needs and dynamics. Three general questions should be addressed: a) What are the circumstances prompting the evaluation? b) Are there underlying issues or decisions that need to be addressed? c) Is a specific product or result expected from the evaluation effort? In answering these questions, the evaluator should communicate extensively with the various members of the audience for whom the evaluation is intended. Differences in opinion between, say, department chairperson and a faculty committee should be resolved before the evaluation is planned.

What Are the Circumstances Prompting the Request?

This question suggests a conscious consideration of the impetus for the evaluation effort and may provide clues as to how to proceed. Is it a short-term ("one shot") or a long term, continuing effort? What forces are directing the evaluation—a specific problem or issue, a forthcoming external review, internal interest in how the program is doing, anticipation of an upcoming change in faculty, changes in organizational demands? These questions focus on the context of the evaluation effort; they constitute the beginning of the process of describing the purpose, scope, and duration of the inquiry.

Are There Underlying Issues or Decisions that Must Be Addressed?

This question refines the previous one to define the needs and expectations of the decision maker and the audiences the evaluation will serve. It relates directly to the purposes of evaluation mentioned in the previous consideration. Is there concern with assessing program outcomes, improving the program, addressing a specific problem, or documenting the program activities for accountability? If the evaluation is prompted by specific departmental issues, the evaluator must clarify them. For example, if certain goals are not being realized by a program, is the faculty more interested in modifying program activities or in redefining program goals?

Is There a Specific Product or Result Expected from the Evaluation?

This question addresses how the evaluation results will be used. Is the evaluator expected to provide specific recommendations or simply detailed descriptive information? Are there existing explicit criteria and standards that must be used in judging the program or is the evaluator expected to develop them? Has the decision maker already made the decision and merely wants information to corroborate that decision? What type of information is expected—quantitative, qualitative, or both? What type of evaluation report is expected, how will it be presented, and to whom will it be addressed? An understanding of the desired product or result helps delineate the subsequent evaluation effort.

What Resources and Constraints Influence the Evaluation?

What resources are available for the evaluation? Consideration should be given to such resources as: the evaluator's time and expertise; the availability and accessibility of support personnel in the department or school (e.g., educational R & D personnel, behavioral science staff); and finances for data collection, data analysis, or consultation. Perhaps the most important resource an evaluator can have is administrative and organizational commitment to the evaluation effort.

In addition, it is helpful to identify the constraints that may limit what the evaluation can accomplish. Constraints include such concerns as the timeframe within which the evaluation must be completed; the amount of time program participants have for information-gathering efforts; the expertise available; and the finances to support additional resources.

After clarifying the assignment and identifying resources and constraints, the evaluator and the department chairperson should formalize the evaluation agreement in writing. This document should include a description of the assignment, time and resources available for the task, and the anticipated product of the evaluation. By circulating a summary of this agreement to program faculty, the department chairperson can facilitate staff awareness and cooperation.

Who Should Do the Evaluation?

Although this chapter assumes that an academic physician will be the primary evaluator, it is important for conceptual reasons to explore the various types of evaluator roles. Two distinctions can be made (7). The first is the distinction between an internal and an external evaluator. The internal evaluator usually works directly for the program administration. In a clinical department, this may be a member of the faculty, the director of residency training, or a professional educator who works in the department. The external evaluator comes from outside the program. An accreditation review committee is a good example of an external evaluation group. Another external evaluator might be a residency director from another institution observing the operation as a consultant in order to provide a fresh perspective. The obvious advantage of using an internal evaluator is that he or she will be intimately familiar with the program and highly responsive to the internal decision makers. The disadvantage, however, is a potential loss of objectivity and autonomy. Conversely, the external evaluator may bring a more objective and independent view to the gathering and interpreting of information on a program but may have greater difficulty in developing an in-depth portrait of an unfamiliar program.

The second distinction is between an evaluator with technical expertise and one with content expertise (4). A technical evaluator is one who has extensive training and expertise in evaluation methodology and whose major responsibility is conducting evaluations. This type of evaluator may be expected to have training and experience in various methodological areas such as evaluation design, statistics, observation, interviewing, and questionnaire design. The technical evaluator should also have expertise in educational methodology and organizational behavior in order to help the program director formulate the precise questions that will serve the decision requirements.

A content-expert evaluator is a person, such as an academic physician, whose professional training has not been in program evaluation and whose job responsibilities are not solely devoted to conducting evaluations. This individual has expertise in the subject matter of the program. For example, this physician should be able to observe an event such as the interaction between a chief resident and a set of medical students and readily assess the appropriateness of the medical content and clinical reasoning involved in a discussion of a particular patient.

These two distinctions may be combined in a typology to distinguish between four types of evaluators: internal-technical, internal-content, external-technical, and external-content (7). Table 9.1 portrays these four types with an example of each. A combination of evaluation expertise, particularly in the form of an evaluation team, is desirable for complex evaluations.

TABLE 9.1. Types of evaluators.

	Technical expertise	Content expertise
Internal	In-house education or evaluation professional	Physician faculty member
External	Outside evaluation consultant	Outside peer reviewer

Technical expertise involves skill in evaluation design, quantitative or qualitative evaluation methodology, data collection, designing instrumentation, and data analysis. Content expertise involves skill in the substantive content of an educational program (e.g., internal medicine, psychiatry, surgery, obstetrics and gynecology, pediatrics, or family medicine).

What Are the Dimensions of the Program?

A thorough description of the program is important for three reasons. First, it helps to establish limits on the scope of the evaluation. Most training programs are part of larger educational endeavors—the residency program is part of a larger system of medical education within the department and the medical school. Is the evaluation effort just to deal with the residency program or is it to cover all department training programs? Second, identifying the program's specific components clarifies the objects of evaluation and the sources of information. For example, if the selection process is included as one of the key components of the residency program, then the evaluator will want to seek information on the criteria used to select candidates. He or she may also wish to relate specific entering characteristics of residents to the attainment of departmental goals as they proceed through the program. Third, a thorough program description may resolve a number of issues simply by clarifying the roles and responsibilities of the program participants.

Program Components

Developing an initial program description is not a perfunctory task, as any given program involves a set of experiences, depending on the perspective of the particular observer. Identification of the purposes and processes of a program, for instance, may vary greatly depending on whether one talks with the participants, the program staff, or the program's funding or administrative agency. The attempt to enumerate the many elements of a program may result in an extensive list of attributes. In no particular order these may include: a program's objectives, philosophy, trainees, clients, faculty, materials, time frame, activities, finances, setting, outcomes, organization, and relationship to other programs.

Table 9.2 lists some of the components that may be commonly found in any residency program. Program components such as these will require some initial description or definition and may be the objects of evaluation.

Program as a System

Program description is made easier when the program is viewed as a system that has clearly defined elements and relationships. In *The Medusa and the Snail* Lewis Thomas defines a system as "a structure of interacting, intercommunicating components that, as a group, act or operate individually and jointly to achieve a common goal through the concerted activity of the individual parts" (14). Medical care, as Engel has noted (15), involves a physician's awareness of a continuum of interacting systems ranging from the molecule, through organ systems and various lev-

TABLE 9.2. Examples of components of an educational program.

Program component	Examples
Learners:	Residents, clerks, fellows
Teachers:	Full-time faculty as attendings and preceptors
	Part-time clinical faculty
	Behavioral science professional
	Pharmacologist
Resources:	Patients
	Other health professionals, (nurses, health administrators, technicians, community workers, social workers, dietitians)
	Funding
	Equipment
	Library
	Texts
	Journals and other references
	Audiovisual material
	On-line bibliographic search capability
	Expert consultative systems
	Personal computer file
Activities:	Attending rounds
	Lectures and conferences
	Preceptorships
	Grand rounds
	Chart audits
	Video-taped encounters
	Simulated patients
	Rotation through other services

els of interacting social systems, to the biosphere. Educational programs may also be thought of as systems since they are comprised of interrelated subsystems and function within the context of a larger sytem. They both act and react within their environment.

From this systems perspective it should be clear that the components listed in Table 9.2 are not static but interactive. For example, the quality of the faculty, the variety of patients encountered, and the entering characteristics of the residents interact to shape the nature of the training program. These program components are also dynamic, evolving over time as residents mature, department heads retire, and budgets expand or decline. Thus, when describing a program, consideration should be given to the relationships among the attributes of a program and their change over time. Figure 9.1 depicts several categories of components that an initial description of a program should consider. Each arrow in the diagram represents an interactive relationship among the major components of the program. Both the components and the interrelationships should receive attention in the description in order to capture the dynamic nature of the program.

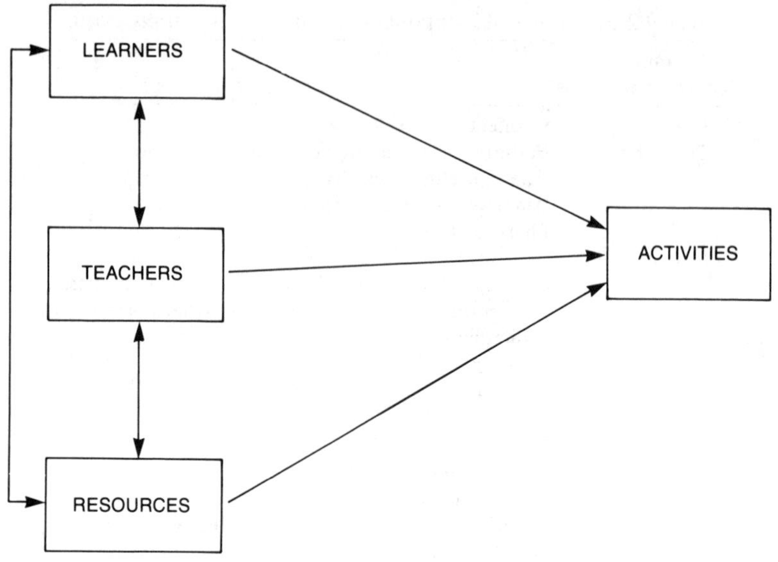

FIGURE 9.1 The program as a system.

The Program and Its Environment

Consideration must also be given to the program's environment, which places demands on and provides resources to the program. The program also makes demands on its environment and produces products that in turn become part of the environment. A residency program is a subsystem within an environment composed of three interacting systems: (a) organizational and community, (b) professional, and (c) societal.

Organizational and Community System

The immediate environment of a medical training program is comprised of the hospital and the community that it serves. The type, size, age, location, and administration of the hospital may affect how a residency program operates. Similarly, the demographic characteristics of the population that the hospital serves determine the types of patient encounters available to contribute to the training of residents. Features of the community such as location, size, and amenities may affect the potential for attracting desirable candidates and the personal satisfaction of the participants. Many of these factors are obviously beyond the control of a program director. An important outcome of the entire descriptive process is to identify those program components that can be changed and those that must be taken as givens.

Professional System

Residency programs exist within the context of a larger professional system. The profession dictates accreditation requirements, which influence the content of a program; it provides resources in terms of trained physicians who serve as faculty members; and it receives the products of the program (new specialists). Interactions with the professional system will vary by program and specialty. Medical school curricula and board requirements of various specialties differ widely on such issues as the amount and type of educational experience required. Directors of some programs have considerable flexibility in designing clinical education experiences whereas others simply schedule rotations based on externally imposed requirements. Similarly, the professional system often dictates how participants in a program are to be evaluated; some residency programs have required in-training exams whereas others view these as optional.

Societal System

The larger societal system also affects the educational program and how it functions. For example, the resources and constraints with which an educational program must work may be affected by such factors as government regulations regarding health insurance; the physician pool available to serve the health needs of a particular region; the expanding biomedical technology supporting medical practice in a particular specialty; and the evolving legal precedents concerning various social issues (e.g., physician accountability, abortion, and informed consent).

Figure 9.2 portrays the educational program set within the environmental systems. The relationship between the program and its environment is dynamic and interactive. Organizational maps and flow charts of program activities may be helpful in isolating the various points where larger systems intersect with the program. As mentioned above, some environmental influences on a program are not susceptible to change by the program staff or director and must be recognized as such in any plan for educational improvement.

In summary, initial program description helps refine the scope of the evaluation. A summary of the program description should be included in the final evaluation report because it helps set the stage for further evaluation efforts. Five elements of the program as a system should be considered in the description.

1. The *boundaries* that delineate the program from its environment (e.g., a specialty residency program to be considered separate from a larger general residency program in which it is placed);
2. The organizational and community, professional, and societal *systems that make up the program's environment* and specifically influence the program (e.g., the hospital; the community; the accreditation board; and financial, technological, and legal contexts in which the program operates);
3. The *inputs* to the program such as the requirements (demands) and resources arising from each facet of the environment identified in (2);

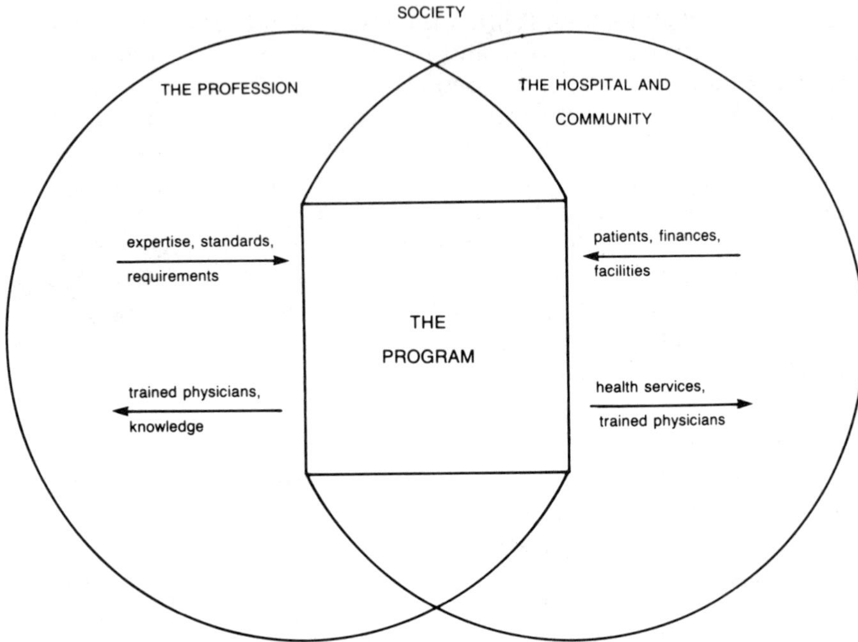

FIGURE 9.2 The program within a social environment.

4. The *components of the program* and how they are interrelated (e.g., the trainees, the faculty/trainee ratio, the patient population, the structured teaching activities, the support staff, the information management resources);
5. The *outputs* or products (e.g., the quality of graduates, the kind of research, the quality of patient care) that the program returns to its environment.

What Criteria and Standards Should Be Applied in the Evaluation?

Evaluation, whatever purpose it serves, involves the process of judgment (16). The process of judging involves the comparison of an object to a criterion (or set of criteria) and assessing the degree of congruence between the two based on a standard (or standards). An important facet of judgment in evaluation is the setting of criteria and standards. In essence, judgment involves: specifying criteria for comparison, adopting and defining standards, and deducing the degree to which the object of evaluation meets the standards for each criterion. Criteria are what is judged; standards are used to determine how much is sufficient. One criterion for a family medicine residency program might be continuity of care. A standard used to assess this criterion may be that 70% of all patient care is handled by the primary physician.

The selection of criteria and standards for judgment depends upon the values underlying the approach to evaluation. For example, an objectives-based approach to evaluation will use goal achievement as the criterion for judging merit and worth (5). This traditional use of goal achievement provides explicit criteria and standards for judgment; however, it fails to consider the possibility that the goal may not be appropriate. A third-year clerkship may have as an explicit objective that students master patient management plans for certain syndromes. Given the novice state of knowledge that a third-year medical student has, it may not be appropriate to expect development of such skill until the fourth year and beyond.

Other approaches advocate implicit criteria and standards including the opinion of those familiar with the content of a program (17) and the opinion of those with explicit technical expertise in evaluation (4). In many cases, decision makers establish the broad evaluation criteria and standards and it is the role of the evaluator to clarify them and collect information accordingly.

Part of the evaluator's task, together with that of the staff and participants, is to establish the bases for judgment that fit the program's circumstances. Often criteria and standards are established at a single point in time and later changes in the program and its environment may make them inappropriate. Or, in another type of mismatch, they are set only after the program has been operating for some time. The latter situation is like starting a game without first learning how to keep score. Thus, the search for relevant standards and criteria should be an ongoing concern in the evaluative process.

What Components of the Program Should Be Evaluated and What Information Is Available?

Components to Be Evaluated

Two themes concerning the objects of evaluation are mentioned in the literature: (1) almost any component of a program can be considered as an object of evaluation; and (2) identifying these objects is a critical precursor to designing a program evaluation (7). Traditionally, learners and teachers have been the focus of program evaluation, as can be seen in the prominent use of examinations, grades, questionnaires, and rating scales. However, restricting the evaluation focus to these two groups (both as objects and sources of information) fails to consider the other components of the educational system that influence learners and teachers and that may be amenable to change. An evaluator should regard participants and teachers as parts of the larger educational system and should consciously plan to collect information on the other components of the program as well. For example, the director of a residency program may want to collect information about the effectiveness of various clinical teaching strategies. Initial component description in the systems perspective will help identify the elements to be evaluated early in evaluation planning.

Sources and Types of Information

A theme of this chapter is that program evaluation should be a systematic, continuous, and purposeful information-gathering effort that serves program decision makers. Evaluation theorists have proposed a variety of conceptual frameworks for organizing types and sources of information (8,9). Earlier in the chapter an educational program was conceptualized as an open system with interacting components. As a system, a program exists within a larger context, accepts inputs from its environment, engages in a variety of processes, and produces products—both intended and unintended. Several evaluation theorists have used this systems model to identify the types and sources of information used in evaluation.

One widely known framework is Stufflebeam's CIPP ("sip") Model (3), an acronym that represents four types of evaluative information: context, input, process, and product. This model illustrates a systems approach to evaluation and is useful in organizing information gathering for decision-making. Stufflebeam points out that this systematic approach to evaluation can be used for the improvement of ongoing programs and the planning of future programs (3).

Context information examines the relationship of the program to its environment and might include learner needs, specialty requirements, available resources, and community concerns. The collection of context information is based on a needs assessment in which general goals and specific objectives are developed for identified educational problems and opportunities. Evidence is gathered to examine the educational setting and to determine a set of objectives that guide program design. For an ongoing program, such information is useful to reaffirm or change the program's goals in light of current demands and resources.

When planning a new program, *input information* is used to match the organization's current and envisioned resources with the previously identified program goals. For programs already in operation, input information assists the decision maker in aligning program goals with currently available resources. As an illustration, a program may have a commitment to intensive monitoring of residents by preceptors. However, the patient load and the limited number of attendings available to facilitate this educational process may preclude satisfactory use of the approach.

Once the program is in operation, *process information* is needed to ensure proper implementation of the educational program and to document and overcome any procedural problems that develop. This is largely a descriptive effort in which activities and events are monitored with quantitative or qualitative measures so that strengths and weaknesses in the program design can be identified. For example, this might involve documenting the attending/resident ratio; the types of patients seen at different times and at different training sites; or the length of time patients spend waiting because of educational requirements.

Product information comes from results generated by the program at all stages of operation. Comparisons are made between documented outcomes and the original expectations or objectives for the program to determine the extent of congruence. For example, departmental expectations or past performance of residents could be compared with current clinical performance and scores on in-training and board exams.

Again, the interrelationships between these various categories of information are important. What are the links among the entry skills, the departmental expectations, the practice and educational environment of the service, and the competencies residents develop? Table 9.3 provides examples of types and sources of information pertinent to a formative evaluation of a residency program. The types of information are conceptualized according to the CIPP model for illustrative purposes.

TABLE 9.3. Sources and types of information in a residency program.

Information	Program components	Sources
Context	1. Needs of residents 2. Professional association requirements 3. Expectations of department and hospital 4. Community needs	1. Candidates, residents 2. Professional associations 3. Faculty, hospital administration 4. Community leaders
Input	1. Resident/attending ratio 2. Nature of patient care demands 3. Financial resources	1. GME standards 2. Hospital records 3. Government, private foundations
Process	1. Quality of teaching 2. Types of patients seen 3. Residents' interpersonal skills 4. Time demands on residents	1. Residents, faculty 2. Records, charts 3. Residents, faculty, nurses, staff, patients 4. Residents, hospital administration
Product	1. Residents' knowledge 2. Residents' clinical performance 3. Perception of program nationally	1. In-training exam, certification exam 2. Supervisors, resident 3. Match applications

What Methods of Investigation Should Be Used?

Two generic approaches to investigation are frequently distinguished in the evaluation literature. These approaches may be differentiated along several dimensions: (a) philosophical assumptions, (b) types of questions addressed, (c) types of data collected, (d) methods of synthesizing data, and (e) techniques of data analysis. The first approach is measurement, numerical data, and statistical analysis. The other takes a holistic approach, relying more on verbal and textual data that are categorized and interpreted. These two approaches are frequently labeled "quantitative" and "qualitative" although the distinction is by no means clear-cut in all cases.

Quantitative Evaluation

Quantitative investigation includes experimental, quasi-experimental, and correlational techniques. Most clinicians are familiar with the experimental paradigm. The identifying features of experimental investigation are manipulation and measurement of the variables of interest, control of all other variables, and a heavy reliance on quantitative data. The experimenter seeks causal relationships among the variables of interest and attempts to remain as neutral as possible about the investigation and its outcome. This mode of inquiry is the preeminent method of investigation in clinical medicine and is one paradigm in evaluation.

Since the major objective of experimentation is the investigation of causal relationships, this method has been advocated in educational evaluation because of its potential to demonstrate that a particular educational program (or treatment, to use experimental terminology) "caused" specific outcomes. So, for example, through experimentation an evaluator might be able to establish a causal relationship between instruction given by a clinical pharmacologist in a residency program and subsequent prescription preferences of residents. Another intervention that might also affect the prescription preferences of residents is a computer-based record of individual prescription orders that each house officer reviews with a preceptor monthly. Certainly, the results of a properly conducted evaluation addressing these two interventions—one being made up of a direct instruction and the other a feedback and review system—would provide very useful information for program decision makers.

Unfortunately, experimentation in clinical settings is not as easy as it is in the laboratory (as anyone who has attempted a field experiment will readily attest). Although it may be possible to manipulate the variables of interest and measure their effects, control of extraneous influences is often difficult to achieve. For example, random assignment of learners to various conditions is often not feasible. Ethical considerations may preclude assigning some people to a control condition that is hypothesized to be less educationally effective than the experimental condition. For a comprehensive discourse on the challenges of field experiments see Cook and Campbell (18).

Experiments may be difficult to use in evaluating an entire program; however, there are times when a carefully controlled experiment is appropriate and feasible. It may be possible to isolate a specific aspect of an educational program for investigation using experimental methodology. In this context, the experimental design will be part of a formative attempt to study and improve the program rather than a summative assessment to determine accountability. In deciding what aspects of the program may be studied by the experimental method, the evaluator should consider the practicality of manipulating and controlling all the factors that may influence educational outcomes of interest.

Although it may be impractical to use experiments to evaluate a program, many sophisticated quantitative techniques exist that can be used to analyze numerical data generated by a program. These methods can assist in investigating the strength of associations among various elements of the program. Such techniques include simple correlational analysis as well as more complicated multivariate procedures including multiple regressions, causal modeling, and factor analysis (19–21).

Qualitative Evaluation

Other evaluators advocate more qualitative methodologies under a variety of different titles (9,11,22,23). Several features are common to these methods: emphasis on multiple perspectives of a program, use of a variety of sources of information, focus on decision-making needs, continuing interaction between the evaluator and program personnel, and explicit attention to diverse value orientation. Qualitative techniques are less standardized than some of the quantitative modes of investigation because they are more sensitive to the unique aspects of the program than to the requirements of the method.

Qualitative investigation is based on the premise that there is much more information available to the decision maker than that derived from merely counting things. The main components of any program are, of course, people. Clients, patients, residents, faculty, and staff all have needs that a program must address, expectations about what the program should accomplish, and feelings about how well the program is functioning. There may be problems or issues within the program that do not emerge until various perspectives are examined, compared, and analyzed. The investigator uses observations, interviews, organizational documents, and other sources of information to identify and verify important concerns. The information generated from qualitative methods is predominantly verbal, not numerical, thus posing a different set of requirements regarding analysis and interpretation. The evaluator seeks to categorize information into common themes and to verify these interpretations with multiple sources.

Several advantages result from this approach. First, qualitative methods can be incorporated easily into the ongoing process of program development and implementation, thus contributing to formative decision making. These methods provide a rich, explanatory context for quantitative data already being gathered such as standardized examination scores and performance ratings. Second, such information can change and improve the way in which decisions are made about the program. For example, many educational programs operate on a "red flag" approach to problem-solving in which nothing is done until a crisis may be avoided altogether. Evaluation becomes an organic part of the educational process when integrated into the day-to-day functioning of the program. Third, qualitative methods of evaluation pay attention to many types of information that may be misunderstood or ignored because of an overreliance on quantitative data. That is, qualitative approaches can use diverse bits of "informal" information (e.g., grapevine, observations, casual conversations) as legitimate observations.

There are, however, several issues involving the qualitative method that must be considered when developing an evaluation plan if potential problems are to be avoided. Guba (23) discusses three. First, as noted earlier in the discussion on the educational program as a system, the evaluator must establish boundaries on the scope of the inquiry. This involves setting limits to the inquiry and developing rules for inclusion, exclusion, and relevancy of information. Second, the evaluator needs to focus the inquiry in order to reduce the large volume of qualitative information into meaningful categories that can be interpreted for decision-making purposes. These meaningful categories can be thought of as specific questions that guide information collection and interpretation. Third, the evaluator must attend to the authenticity of the information through a systematic approach to gathering qualitative data. Professional colleagues must be convinced that the results represent more than the evaluator's personal beliefs about the program. Seeking evidence from multiple perspectives (e.g., residents, medical students, attendings, nursing staff, and patients) may develop themes that are corroborated from several sources.

The selection of a method will ultimately depend on a careful analysis of the decision-making needs, types of available information, and practical constraints. Each has strengths and weaknesses when applied to evaluation of departmental programs. Quantitative evaluation may be persuasive to decision makers but may focus on discrete and narrow facets of an educational program. For example, scores on in-training or board exams are often a major aspect of quantitative evaluation, but may provide little information as to the operational and educational processes in a program. Qualitative evaluation has the potential to capture the richness and depth of a program. For example, a knowledgeable observer may discern that the residents in a program are learning far more from each other about the nature and pitfalls of probabilistic judgments than they are from their interactions with faculty. If possible, the evaluator should integrate *both* quantitative and qualitative methods in the overall evaluation plan.

From Design to Data

Before the evaluator can begin to gather data about the components of the program, the key variables for each component under consideration must be identified. For example, if clinical instruction is a program component of interest, then key variables would include: teacher and resident roles, pacing of experiences, integration of didactic instruction and clinical practice, methods of assessment, and feedback rate. Next, the evaluator must operationalize the variables. This means specifying the indicators for each variable by making explicit how it will be counted, measured, or documented. Some variables are easily counted or measured, whereas others may require more descriptive documentation. Role of feedback, for example, can be counted by an observer or assessed by a questionnaire. In contrast, the mix of didactic instruction and clinical practice is not so easily reduced to numbers. In this case the evaluator must make explicit how he or she will document the nature of this integration. If qualitative procedures such as observation and interviews are to be used, then the evaluator must develop observation and interview schedules that guide the collection of information.

The credibility and persuasiveness of the evaluator's findings are enhanced when a variety of indicators are used for each variable. Since no single indicator completely reflects a given variable, the use of multiple indicators increases the likelihood of obtaining a comprehensive picture. Ideally, the information from multiple sources should yield similar results.

Analysis of Information

Once the information has been collected, the evaluator must analyze and intepret it in terms of the evaluation questions. Data analysis is a function of the method of inquiry and types of data collected. Analysis begins by reducing the data into meaningful categories. Quantitative data may be reduced to summary statistics and presented in tables and charts. Qualitative data can be refined into pertinent topical categories and presented with appropriate examples.

Analysis of quantitative data is accomplished by statistical procedures that reveal various patterns of relationships in the data. Depending upon the evaluator's methodological expertise, the assistance of a statistical consultant may be necessary. In contrast, the evaluator analyzes qualitative information gathered from interviews and observations by carefully examining transcripts and notes for recurring themes and issues.

The evaluator may find that the multiple sources of information fail to provide a consistent picture. More questions may surface that require additional information from other sources in order to make an interpretation. In some cases it may not be possible to resolve conflicting information and the evaluator will have to present the diverse findings. In other cases resolution of conflicts may lead to new insights. Generally, the results of the analysis should include answers to specific evaluation questions, but it may also contain other information that serves decision-making.

How Should the Findings Be Presented?

Reporting evaluation results should not be an isolated, single event. Audience commitment to the evaluation effort and its potential results are enhanced when the evaluator assumes a responsive and interactive role. The evaluator who takes on this role should keep the various program participants well informed throughout the course of the evaluation so that there are no surprises when the final report is presented.

A variety of report formats may be useful when presenting the findings of an evaluation. An oral, informal report could be made to a faculty meeting, program staff meeting, or other setting in which interested parties would be present. This may be an occasion for further discussion and an opportunity for differing interpretations to be shared in an open forum. A written final report should also be prepared that reflects the evaluator's analysis as well as any alternative views that have emerged. Although a consensus on the written report is not necessary, sharing the preliminary findings with others and seeking their comments usually increases acceptance. If the questions that are explored in the evaluation are of interest to other educational programs, the evaluation results may be published in the specialty journals or in the general medical education and evaluation literature.

The original assignment may not require more than the presentation of pertinent information concerning a particular issue. If so, the report should address only the information needs of the relevant decision makers. On the other hand, the evaluator may suggest actions based on answers obtained to evaluation questions. When this is the case, the final report should present evidence that supports a coherent and credible argument for the recommendations. If possible, as noted earlier, a continuing process of evaluation should be set up so that information that serves decision making is readily available.

Dr. Peffer Revisited

When last seen, Dr. Peffer was slouching in his chair pondering the immensity of the task ahead of him, and wondering why he hadn't gone into private practice after his specialty training. How could he begin to plan an evaluation of the residency program?

This chapter has presented some considerations that will help the Dr. Peffers of academic medicine begin to plan a systematic program evaluation at the department level. To recapitulate; a hypothetical sequence of events is presented. Dr. Peffer's first task is to clarify the nature of the assignment. He meets with the department chairperson to identify the impetus for the evaluation request, the purposes of the evaluation, and the underlying issues. His discussion with the chairperson should also include the expected result or product and the resources and constraints that will affect the evaluation. It may be appropriate at this time for Dr. Peffer to negotiate a release from other responsibilities in the department if he is to carry out the evaluation alone. The result of this meeting is a written agreement that lays out the nature of the assignment and the type and amount of departmental support that Dr. Peffer can expect.

Dr. Peffer's next task is to describe the program. By viewing the program as an open system, Dr. Peffer identifies the components of the program that interact with other systems in the program's environment. He relies on organizational documents, accreditation reports, specialty board requirements, and interviews to describe the various elements of the program. The results of this process is a written description of the program to be included in the final report.

The initial discussion with the department chairperson revealed several purposes of the evaluation. Dr. Peffer learned that the chairperson considered this a formative evaluation and that he wanted to make decisions that would improve the program. Although the initial concern expressed by faculty members in the education committee meeting focused on goals of the training program, the chairperson wants Dr. Peffer to consider other outcomes as well. Having ascertained the purposes of the evaluation, Dr. Peffer can now begin to formulate specific questions that will be addressed in the evaluation process.

The written agreement stipulated that Dr. Peffer will keep the education committee informed on the interim findings of the evaluation by reporting at the monthly meetings. At these meetings he will receive ongoing direction regarding criteria and standards to be applied in assessing his findings.

Using the CIPP model as an outline, Dr. Peffer decides to emphasize program processes and products in this evaluation. In addition to resident performance, Dr. Peffer looks at faculty teaching, training experiences, resources, and types of patient encounters. For a broad evaluation effort such as this, he uses a number of investigation methods, including both quantitative and qualitative techniques. He examines documents and examination scores; interviews faculty, residents, and staff; and distributes questionnaires to graduates and their supervisors.

Finally, Dr. Peffer summarizes his findings and presents the preliminary report to the education committee for their review and comment. He makes recommendations of his own and synthesizes the consensus of the committee as part of the final evaluation report submitted to the department chairperson.

The Future and the Evaluation of Educational Programs

Two factors are likely to affect future efforts in medical program evaluation. First, the need for accountability to specialty boards, external funding agencies, regulatory agencies, employers of graduates, and the public is likely to increase. The need for evaluation efforts will increase in kind.

Second, greater use of and dependence on advanced information technology for administration and teaching will lead to increased use of this technology for evaluation. For example, computerized patient records will make it easier to document residents' prescription patterns and other diagnostic and management decisions. Computerized tutoring systems will help track the developing competencies of residents. A departmental information management system will help keep track of patient loads and types of patients seen. The information needs of future evaluations should be kept in mind as these elements of information technology are incorporated into programs.

Conclusions

Program evaluation is appropriate and possible at the department level. This chapter has provided the academic physician a practical framework for organizing evaluation efforts. Many technical considerations in evaluation have not been addressed; however, the reader may consult the references for more in-depth study. Technical competence is of little value, however, if the evaluator is not genuinely interested in studying and improving the educational process. Effective evaluators must be sensitive to a variety of values and concerns, willing to alter their opinion based on accumulated evidence, and aware that useful results require hard work and tolerance. This chapter should provide the academic physician with the concepts and confidence necessary to undertake a program evaluation. The value of initiating an evaluation and using the results lies in its positive effect on educational programs and on the competence of both teachers and learners involved in those programs.

References

1. Cronbach LJ. Course improvement through evaluation. *Teach Coll Rec 64*:672–683, 1963.
2. Alkin MC. Evaluation theory development. *Evaluat Comm 2*:2–7, 1969.
3. Stufflebeam DL. Alternative approaches to educational evaluation: A self-study guide for educators. In *Evaluation in Education*. W J Popham (Ed) Berkeley, California: McCutchan, 1974.
4. Scriven M. The methodology of evaluation. In *AERA Monograph Series on Curriculum Evaluation, No. 1*. RE Stake (Ed). Chicago: Rand McNally, 1967.
5. Tyler RW. *Basic Principles of Curriculum and Instruction*. Chicago: University of Chicago Press, 1950.
6. Scriven M. Evaluation perspectives and procedures. In *Evaluation in Education*. W J Popham (Ed) Berkeley, California: McCutchan, 1974.
7. Nevo D. The conceptualization of educational evaluation: an analytical review of the literature. *Rev Educ Res 53*:117–128, 1983.
8. Guba EG, and Lincoln YS. *Effective Evaluation*. San Francisco: Jossey-Bass, 1981.
9. Stake, RE. (Ed) *Evaluating the Arts in Education: A Responsive Approach*. Columbus, Ohio: Merrill, 1975.
10. Stake, RE. The case study method in social inquiry. *Educ Res, 7*:5–8, 1978.
11. MacDonald B. *Evaluation and the Control of Education*. Norwich, England: Centre for Applied Research in Education, 1974.
12. Parlett M, Hamilton D. Evaluation as illumination: a new approach to the study of innovatory programmes. In *Beyond the Numbers Game*. D Hamilton et al. (Eds) Berkeley, California: McCutchan, 1977.
13. Patton MQ. *Qualitative Evaluation Methods*. Beverly Hills, California: Sage Publications, 1980.
14. Thomas L. *The Medusa and the Snail*. New York: Viking Press, 1979.
15. Engel, GL. The need for a new medical model: A challenge for biomedicine. *Science 196*:129–135, 1977.
16. House ER. *Evaluating with Validity*. Beverly Hills, California: Sage, 1980.
17. Eisner EW. On the use of educational connoisseurship and (educational) criticism for evaluating classroom life. *Teach Coll Rec 78*:345–358, 1977.
18. Cook TD, Campbell DT. *Quasi-Experimentation: Design and Analysis Issues for Field Settings*. Boston: Houghton-Mifflin, 1979.
19. Pedhazur EJ. *Multiple Regression in Behavioral Research*. New York: Holt, Rinehart and Winston, 1982.
20. Duncan OD. *Introduction to Structural Equation Models*. New York: Academic Press, 1975.
21. Kerlinger FN. *Foundations of Behavioral Research*. New York: Holt, Rinehart and Winston, 1973.
22. Patton MQ. *Utilization-Focused Evaluation*. Beverly Hills, California: Sage Publications, 1978.
23. Guba EG. *Toward a Methodology of Naturalistic Inquiry in Educational Evaluation*. Monograph Series, No. 8. Los Angeles: Center for the Study of Evaluation, UCLA, 1978.

Suggested Readings on Medical Education

Cox K R, Ewan C E (Eds) *The Medical Teacher.* New York: Churchill Livingstone Publishers, 1983.

This systematic description of the viewpoints of several Australian medical educators provides an excellent guideline for clinical instructors in any country. The topics are learners, the instructor, various instructional methods, instructional resources such as slides, simulations, and computer assisted instruction, and assessing learner performance. Each chapter in this book is brief, self-contained, and can act as a concise reference for medical teachers.

Cronbach LJ. *Designing Evaluations of Educational and Social Programs.* San Francisco, California: Jossey-Bass, 1982.

Program evaluation is addressed in this book as a scholarly enterprise, an activity that not only uses the best available tools of science but that is also sensitive to political realities. Readers will find this book to be a rich source of insight and ideas about program evaluation, treating the subject in breadth and depth.

Hubbard JP. *Measuring Medical Education: The Tests and the Experience of the National Board of Medical Examiners* (2nd ed.). Philadelphia, Pennsylvania: Lea & Febiger, 1978.

This book offers a detailed description of the evaluation practices of the National Board including the formation of test committees, preparation of test materials, psychometric issues, development of qualifying and specialty examinations, and research and development activities. Given the influence of the National Board on the evaluation of medical learners nationally and at individual medical schools, academic physicians will want to become familiar with its policies and practices.

Johnson DG. *Physicians in the Making.* San Francisco, California: Jossey-Bass, 1983.

This volume presents a detailed profile of, "Personal, academic, and socioeconomic characteristics of medical students from 1950 to 2000." The book is rich in data, drawing heavily from the research files of the Association of American Medical Colleges. In addition to presenting data about medical student characteristics, the book covers other topics including events and trends influencing medical students, and services and activities affecting medical student progress.

McGaghie WC, Miller GE, Sajid AW, Telder TV Competency-based curriculum development in medical education: an introduction. *Public Health Paper No. 68*. Geneva, Switzerland: World Health Organization, 1978.

Approaches to developing medical curricula that are closely matched to the health care needs of the population are the principal message of this slim volume. The book is a very basic primer on medical curriculum development that can serve as a point of departure for more advanced reading about the topic.

Miller GE (Ed) *Teaching and Learning in Medical School.* Cambridge, Massachusetts: Harvard University Press, 1961.

Four major sections of this book cover the medical student, process of learning, tools of instruction, and evaluation of learning. While it is dated, the book remains a nearly classic text about medical education.

Popham WJ. *Modern Educational Measurement.* Englewood Cliffs, New Jersey: Prentice-Hall, 1981.

This is a well-written basic textbook on educational testing that assumes the reader has no prior knowledge of the subject. It is a highly practical book that clearly shows how to plan tests and other devices (checklists, rating scales) used for academic evaluation, write questions, interpret test scores, set performance standards, and establish grading procedures. The no-nonsense approach to educational evaluation that is embodied in this book makes it a fine primer for academic physicians.

Stritter FT, Flair MD. *Effective Clinical Teaching.* Bethesda, Maryland: National Library of Medicine, 1980.

This monograph is a valuable practical guide for individuals involved in clinical instruction. Included in the monograph are chapters on establishing goals, motivating learners, stimulating memory, directing attention, communicating with learners, providing practice, evaluating and reinforcing, and integrating one's teaching. Also included are examples of each one of the points from a variety of health professions instructors.

III Clinical Research

Research productivity is a key feature of an academic medical career. In many settings, research and writing is necessary for academic advancement, for promotion to a tenured position, and to receive the status given to such an achievement. Research is also a source of pride and genuine professional satisfaction for physicians who report their scholarly work in journals and other outlets. Clinical research is not done exclusively by physicians who work in academic settings because community practitioners, government laboratories, and private industry also contribute. However, academic physicians usually perform the lion's share of clinical research and eagerly accept the obligation to publish their findings for colleagues and for posterity.

Busy physicians in any setting want answers to practical, patient care problems. Good research serves that need while simultaneously encouraging investigators and practitioners to pose even more probing questions for study. Thus from the standpoint of investigators and clinicians, attention to research promotes a scholarly, thoughtful approach to patients and their problems.

Untrained physicians often believe that clinical research skills are esoteric, difficult to acquire, and unrelated to patient care. Those beliefs are false. Similar to other academic skills, clinical research skills can be learned by any physician with enough interest and desire. The skills also improve with practice, becoming a set of professional tools that grow sharper and more efficient with increased use.

Chapter 10 introduces the clinical research enterprise by discussing its contribution to primary care medicine. We learn that historically, as today, good clinical research has its roots in patient care problems. Physicians who see patients and conduct clinical investigations quickly discover that the two activities are complementary.

Chapter 11 presents a research case study in enough detail to illustrate that projects rarely operate exactly as planned, yet with commitment and hard work, can still lead to a successful outcome. The authors show that project management and attention to the human side of clinical research are vital to ensure that data are collected at all, much less subjected to analysis. The chapter is also instructive because the project it describes originated from a casual clinical observation, which underscores the realism of the case.

Chapter 12 addresses the most fundamental, and the most important, research skills that an academic physician can acquire. Basic steps in the research process are identified and discussed. The steps include stating a research objective, searching the literature, formulating operational hypotheses, and choosing a study strategy. Experience teaches that the skills needed to work through the steps of research planning are the most difficult for inexperienced academic physicians to learn and use with confidence. Such persons should find many valuable insights in these pages.

Concepts and suggestions introduced in Chapters 11 and 12 are amplified in Chapter 13, Conducting a Research Study. Here, special attention is given to such issues as developing a project workplan and timetable, staff supervision, and data collection procedures. It is shown that valuable studies can be conducted on a modest budget. The chapter also contains a section on the measurement of clinical variables.

Chapter 14 describes resources for clinical research in terms of people, organizations, and money. By design, the chapter has a slightly administrative tone, suggesting that one's managerial skills grow in importance when—usually with experience and success—one's projects grow in size and complexity. Sources of local research support that are available to most investigators, often just by asking, are identified. Basic information is also presented about how to obtain research funds from federal agencies and private foundations.

This section of the *Handbook* concludes with Chapter 15, on research data management. It describes how to design and manage systems that permit effective collection, coding, and storage of research data. The chapter is a testimonial to microcomputer technology, while acknowledging that older technologies are still useful in many research situations. Chapter 15 does not, however, cover statistics and data analysis. These are concepts and procedures that are beyond the scope of this book.

Experienced investigators will note that this six chapter treatment of clinical research is incomplete and perhaps superficial. We acknowledge this issue while pointing out that the intent is to expose readers to breadth rather than depth. We urge those who see this section as limited to view it as an opportunity to pursue topics further, either in the cited sources or in the suggested readings.

10
The Role of Research in Primary Care Medicine

PETER CURTIS

Many scholarly books have been written about research in the biological sciences and on research methods in disciplines ranging from genetics to medical anthropology. In this book it would be impossible to address comprehensively the research needs and interests of all academic physicians, particularly those who have had specialized research training. Our aim is to help generalists, particularly in primary care, to learn some basic research principles and to guide them to more detailed readings and sources. Throughout the discussion our approach is pragmatic and basic. We choose to present fundamentals without apology because few physicians, in our experience, have received an introduction to the conduct of clinical investigation.

Research: (1) careful or diligent search, (2) studious inquiry or examination; *esp*: investigation or experimentation aimed at the discovery and interpretation of facts, revision of accepted theories or laws in the light of new facts, or practical application of such new or revised theories or laws. (*Webster's New Collegiate Dictionary*, 1977.)

Historical Review

Use of the word "research" became common at the beginning of the 19th century, when the era of specialization in medicine was beginning and enormous strides were being made in the basic sciences of chemistry, physics, physiology, and anatomy. At the same time, dispensaries and hospitals were being established as a result of the increasing prosperity of the Western European countries. With the coming of the Enlightenment with its nascent social conscience as well as the accruing wealth accompanying the industrial revolution, the number of hospitals expanded and medical practice became financially rewarding for most physicians (1). The clustering of patients in institutions, the creation of research laboratories, and the rapid growth of medical specialization produced a distinct trend toward investigation in hospital settings. In the 19th century this was particularly apparent in France and Germany.

At the beginning of the 20th century the practice of medicine and medical education in the United States were in disarray. Taking as a model the German concept of institutes for medical research with staff freed of clinical responsibilities, Abraham Flexner influenced both private foundations and the government in the United States toward investing in "science" and the development of full-time academicians. Consequently, research became established in the minds of the majority of primary care physicians as a "laboratory" activity. This is reflected in a statement by Sir James MacKenzie, a noted British general practitioner, who undertook studies of cardiac disease in his own practice.

About 1883, I resolved to do a series of careful observations, entirely for my own improvement, never dreaming of research, for I was under the prevalent belief that medical research could only be undertaken in a laboratory or. . .in a hospital. (2)

Biomedical research continued to expand after World War II following, as it had done for approximately 100 years, the reductionist model that seeks a cellular explanation for disease. In conjunction with this highly successful approach to medical and biological research, clinicians became ever more specialized and primary care and hospital based medicine diverged even further. However, during this period not only did society change its perspective of the medical professional but the prevalence and pattern of disease also changed significantly (3). Physicians, previously confronted with serious acute infectious diseases, were now dealing with chronic multifactorial disorders and the effects of the postindustrial society on health. It also became apparent that little was known about the early symptoms of disease or about how to improve the work of primary care physicians.

Table 10.1 summarizes some trends in medical research and patient care in light of social events that occurred in the U.S. and Western Europe during the 18th, 19th, and 20th centuries.

Pioneer studies in England, mainly descriptive in nature, were directed toward studying the content of primary care practice (4,5). An important American article by White et al. in 1961 underscored the point that over a 1-month period, out of 1000 adults 750 would experience an episode of illness, 250 of these would consult a physician, nine would be hospitalized, and one would be admitted to a university medical center (6). The authors made a plea for an emphasis on health services research that would equal the established research programs on the biomedical causes of disease. Since that time there has been groundswell of opinion supporting research in primary care (7). In 1978, Thomas R. Dawber commented:

The tremendous growth of medical research. . . has created an echelon of professional researchers concentrated in academic institutions. . . it has had little effect on solutions of many of the problems seen in medical practice. . . . There is almost no opportunity for physicians in practice to participate in the decision-making process [about what should be a researchable question]. (8)

Philosophy and Purpose of Primary Care Research

The major health problems of modern society are closely related to the environment and human behavior. Personal and social factors including obesity, lack of exercise, smoking, alcoholism, exposure to toxic agents, industrial and automobile accidents, and family and social stress are all sources of ill health. Each falls within the purview of the generalist who functions as the mainstay of the health care system. Consequently, research in primary care encompasses a broad range of disciplines including sociology, anthropology, psychology, epidemiology, clinical medicine, and health services research. The methods of study in primary care may be either quantitative, using precise units of measurement (as in blood pressure readings), or qualitative and descriptive, as found in the humanities. This contrasts a holistic research approach with the reductionism of normal biomedical science. Such an approach can cover a

TABLE 10.1. Trends in primary care research and patient care.

	18th century	19th century	20th century
Research	Theories of medicine Development of systems and classifications Bedside clinicians Early chemistry, anatomy, and physiology First medical libraries	Expansion of basic sciences Invention of instruments/x-ray films First research laboratories Descriptive clinical studies (signs, symptoms)	Reform of medical education National support for research Basic science expansion Pharmacology/drug trials Epidemiological studies
Patient care	First dispensaries Early hospitals Lack of care for the poor Botanic pharmacopoieas	Heroic therapeutics (bleeding, etc.) Expansion of hospitals Specialization Epidemics	Public health services Initially low medical standards of care Development of medical care systems in primary care
Social	Political revolutions in France and America	Industrial and social democratic movement Industrial poverty and disease Alcoholism	World wars Health care "as a right" Raised standards of living
Dominant medical countries	France Holland England	Germany France England	USA England Germany

wide range of activities that, because of their variety, are both bewildering and attractive to the primary care researcher. There is a temptation for the generalist to pursue several areas of interest at once, sometimes superficially. Unfortunately, failure to concentrate one's research interests may prevent the development of a scholarly focus or "theme" that matures and expands as an investigator gains experience.

The paradox of the generalist undertaking in-depth research in one small area of primary care is similar to that of the teacher who is expected, in the academic setting, to have an in-depth knowledge of a specific area of medicine, while still modeling a generalist's clinical approach for students and residents. This paradox must be accepted, however, to achieve success and recognition in academic circles. The generalist must focus his or her efforts on a specific area of study.

The broad areas of interest to the primary care researcher are shown in Table 10.2. A study of these reveals that most are not suited to closed-system investigation where variables can be rigorously controlled.

TABLE 10.2. Areas of interest for primary care research.

Epidemiological and clinical
- Practice studies
  - Case content
  - Consultation rates
- Single-problem studies
  - Morbidity
  - Natural history
  - Preventive
  - Early diagnosis
  - Management
- Clinical decision-making
  - Theory/models
  - Specific problems
- Geriatric Problems

Health services research
- Consumers
  - Patterns of use
  - Needs/demands
  - Participation in care
  - Patient education
  - Compliance
- Providers
  - Manpower studies
  - M.D. performance
  - Referral patterns
  - Record keeping
  - Models of primary care

Interdisciplinary
- Preventive medicine
- Community health services
- Patient outcomes
- Cost/benefit studies
- Continuity of care
- Comprehensive care
- Hospital-based studies

Pharmaceutical
- Adverse effects
- Iatrogenesis
- Clinical trials

Behavioral and family research
- Doctor/patient relationship
- Communications in health
- Illness behavior
- Family studies: morbidity, prevention, intervention
- Family behavior: normal/abnormal, changing pattern, life cycle problems

Community studies
- Environmental problems
- Ethnic studies
- Health needs
- Interventions

Educational research
- Evaluation of program/aids
- Documentation of experience
- Continuing education outcomes and costs

Adapted from Shank.[10]

Instead, descriptive studies, clinical trials (controlled and uncontrolled), cross-sectional, and case-control studies have been used in most fields of clinical medicine. These methods are subject to the well-known problems of bias, chance, and confounding factors. In fact, the frequency of studies with weak research designs has increased over the years (9).

Although poorly designed studies are common, there is support from medical leaders for increased research effort in primary care. Alvan Feinstein has noted that the direction of clinical investigation is changing from establishing diagnoses to studying prognosis, from disease to disability, from cure to relief and comfort, from therapy to care (11). Feinstein believes that the major problems for primary care research, whether behavioral or clinical, are due to the lack of research models, measurement techniques, and an underdeveloped intellectual discipline. Feinstein suggests the major goals for clinical research should be prediction of events, development of clinimetric methods, a return to clinical observation, and the improved training of clinical researchers. The growth and success of primary care research will come with better ways of measuring physician and patient behavior, characteristics, and clinical phenomena. Clinical phenomena include such items as decision-making, the chronology of illness, the reasons for functional limitations, the effects of the family and environment on illness, and the relationship of symptoms and compliance with therapy (11).

Getting Started on Research

The reader may now be thinking: This background information on research is all very interesting, but how does one realistically get started? This is a difficult question to answer, since the development of a research project in its early stages depends on the local clinical setting and the physician's attitude and thought processes. The major obstacles are a reluctance to critically review one's own work or the work of others (be it clinical practice or the medical literature), the tendency to become set in routine clinical behaviors, and ultimately the inertia that prevents one from addressing a problem that has been identified. Perhaps the best way to illustrate how to get started is to review a personal case.

In 1974 I was working as a general practitioner in a rural community in England. I became interested in audit as a method of improving the quality of care for my patients. At this time, clinical audit was a relatively new idea in the medical literature. I felt, however, that audit should be undertaken by the practicing physicians themselves and not by external assessors. One particular clinical problem presented itself for audit because it involved only a few patients who could be easily identified and because the management of these patients did not seem to correlate with reported hospital practice. The "investigator" (for that was what I had suddenly become) selected all patients in the practice with heart disease on long-term digoxin therapy—there were 42 in all. A literature search revealed a large number of articles on digoxin toxicity and its high prevalence in hospital settings, yet almost nothing about the long-term management of chronic heart disease in primary care. Medical textbooks were even less helpful. In contrast to the literature reports, I could not recall any clinical cases of digoxin toxicity in my practice over the previous 6 years. I reasoned that if there were no established guidelines for general practitioners about managing these patients, other doctors had probably developed their own methods and perceptions about patient management. But did they in fact adhere to their own "innate" protocols in actual practice? I now had several questions that needed answers.

1. Did general practitioners use some "innate" (i.e., informal) protocol for following patients with heart disease who were taking digoxin?
2. Did general practitioners adhere to their protocol? If they did not, did this represent poor care for the patient?
3. How did hospital-based internists manage these patients on an outpatient basis? Was it different from the management by general practitioners?
4. Were patients with heart disease (and on digoxin) different in some way from hospital patients with the same disorder (thus accounting for the apparent lack of digoxin toxicity in general practice)?

The study that followed included a survey of a sample of hospital-based internists and general practitioners to establish the nature of their informal protocols (audit criteria), a subsequent analysis of the charts from 42 selected patients using the audit criteria, serum digoxin studies, and a health status assessment of the patients.

Several outcomes and conclusions were derived from this study.

1. Some of the patients were originally misdiagnosed or had not been accurately diagnosed as having heart disease in the first place and did not need to be on digoxin.
2. Most of the patients had single-system disease and were different from hospitalized patients who had multiple problems.
3. Digoxin levels, i.e., therapeutic levels, seemed to bear little relationship to the patient's cardiac status and feeling of well-being.
4. According to medical records, the doctors in the investigator's practice did not manage patients the way they reported when asked to state their protocol. However, this was just as likely due to a recording bias as to poor care.
5. Steps were taken in the practice to modify the physicians' approach to such patients.
6. Clinical audit should be used as an educational tool. Audit should not be used to measure quality of care because there are so many opportunities for error and bias in the audit process.

7. Primary care clinical practice should create its own literature and academic base and not necessarily look to hospital-based medicine for the "gold standard" of care.

This project was undertaken in a busy practice without outside funding at a cost of approximately $100. An article describing the results was published in the *British Medical Journal* one year after the study was launched (12). The successful conclusion to the study, the first I attempted, can be ascribed to several factors.

1. The research idea—clinical audit—was one that at the time was receiving considerable attention in medical journals.
2. The objectives were limited to a small area of clinical practice.
3. The study population was small.
4. The initial inertia (problems of setting up the project, disturbances of normal practice routines, etc.) was small.

Expectations for Primary Care Physicians in Academic Settings

One of the greatest problems for the academic primary care physician is to stay informed about his or her specialty at a level equivalent to that of colleagues in other disciplines. The problem is based on the breadth of knowledge encompassed in primary care and the perceived need to "keep up" with information from a multitude of sources—the "jack of all trades and master of none" syndrome. The solution is to accept the fact that maintaining currency in all areas is impossible and to select specific areas of interest to pursue in depth while relying on colleagues to fill the gaps when needed. The academic physician will therefore develop a personal data bank in areas of special interest, but should also be adept at information retrieval. Apart from colleagues who can often point out useful information faster than a computer, the researcher must use textbooks, journals, review articles, government data sources, and abstracts from meetings. Efficient use of these and other information sources requires skill in using a medical library.

Telephone contacts are a rarely used, yet valuable way to obtain advice and information from "experts" in one's field of inquiry. This technique, though requiring some courage, usually gets results "from the horse's mouth" and has potential for building collegial relationships.

Another important trait for the academician is skepticism, both in clinical practice and about medical information. Practicing physicians are often far too accepting of medical articles and other resource reports, particularly when they are published by eminent colleagues in prestigious journals. Yet there is considerable evidence that many research reports have faulty methodologies and are published on grounds of the author's reputation rather than merit (9, 13, 14). Reading the medical literature critically takes skill and time. It can be facilitated by journal clubs, discussion groups, or by critiquing drafts of papers by colleagues. Guidance to critical reading and assessment of the literature can be gained from an excellent text by Gehlbach (15).

With skepticism should come a questioning attitude and a persistent curiosity. Most physicians are faced with a continuous stream of questions from patients and colleagues. Some can be answered "off the cuff," whereas responses to others require reference to the literature. In a few cases there is no easy response or answer; the effort to dig it out seems too great. Persistence with this type of question can pay off and research ideas can grow from the effort. In addition to this "organized curiosity," a term coined by Eimerl, the academic physician should develop skill at making connections between clinical observations, questions, ideas, and the ability to stimulate others to do the same (16).

Research is, of course, just one part of the academic physician's activities, which also include teaching, patient care, administration, and role modeling for learners. One of the common problems for physician faculty is reaching a decision about the proportion of their time and effort to allocate to research work. This depends on interest, capability, funding, and support in terms of time and staff. Tolly has reported that approximately 35% of M.D. faculty (without Ph.D.) spend less than 10% of their time on research (17). These data are derived from specialist and subspecialist medical school faculty where generalists are in the minority. Yet despite the tension that results from the numerous, and often conflicting, demands on the academic physician's time, research productivity is still expected for academic advancement. To illustrate, in a study of perceptions about promotion criteria held by faculty members and departmental chairmen, Gjerde and Colombo identified the following as the most important academic activities (18).

Rank	Faculty perceptions	Rank	Chairperson perceptions
1	Be primary author of a journal article	1	Be primary author of a journal article
2	Publish articles in refereed journals	2	Publish articles in refereed journals
3	Publish articles that make a contribution	3	Be respected by departmental members
4	Publish a certain number of articles each year	4	Publish articles that make a contribution to the field
5	Present papers at professional meetings	5	Present papers at professional meetings
6	Achieve national recognition for research	6	Accept responsibilities delegated by the chairperson

It follows that research and its sequelae (publication) are vital to the career of the academic physician. Respect for clinical skills and teaching are usually inadequate criteria for institutional advancement (see Chapter 3). Although promotion is a major reward for research activities, other equally satisfying outcomes are the respect that one receives from colleagues and the pleasure of seeing one's work in print.

References

1. Garrison FH. *An Introduction to the History of Medicine.* Philadelphia: W.B. Saunders, 1984.
2. Mackenzie Sir J. *Symptoms and Their Interpretations,* ed 4. London: Shaw & Sons, 1921.
3. McKeown T. *The Role of Medicine: Dream, Mirage or Nemesis?* London: Nuffield Provincial Hospital Trust, 1976.
4. Pickles WN. *Epidemiology in Country Practice.* Baltimore: Williams & Wilkins, 1939.
5. Hodgkin K. *Towards Earlier Diagnosis.* London: Churchill and Livingston, 1973.
6. White KL, Williams TF, Greenberg BG. The ecology of medical care. *N Engl J Med 265*:885–892, 1961.
7. Mumford E. Selective inattention to some subjects in medical research. *Man Medi 2*:65–70,1965.
8. Dawber TR. Annual discourse—unproved hypotheses. *N Engl J Med 299*:452–458, 1978.
9. Fletcher RH, Fletcher SW. Clinical research in general medicine journals. *N Engl J Med 103*:180–183, 1979.
10. Shank JC. A taxonomy for research. *Fam Med Teach 12*:22–23, 1980.
11. Feinstein A. Invited Lecture. Robert Wood Johnson Foundation Annual Meeting, Princeton, New Jersey, February, 1983.
12. Curtis P. Long-term digoxin treatment in general practice. *Brit Med J 4*:747–749, 1975.
13. Peters DP, Ceci S. Peer review practices of psychological journals: the fate of published articles, submitted again. *Behav Brain Sci 5*:187–255, 1982.
14. McGaghie WC. Medical problem solving: a reanalysis. *J Med Educ 55*:912–921, 1980.
15. Gehlbach SH. *Interpreting the Medical Literature.* Lexington, Massachusetts: Collamore Press, 1982.
16. Eimerl TA. Organized curiosity. *J Royal Coll Gen Pract 3*:246–252, 1960.
17. Tolly P. Physician faculty involvement in research. Datagram. *J Med Educ 58*:73–76, 1983.
18. Gjerde CL, Colombo SE. Promotion criteria: perceptions of faculty members and departmental chairmen. *J Med Educ 57*:157–162, 1982.

11
A Research Case Study

PETER CURTIS AND JACQUELINE RESNICK

Clinical research is an interactive enterprise. Research progress stems from a constant ebb and flow between research questions, methods, and results as data accumulate and as findings are incorporated into clinical practice. There is, however, no completely standardized protocol to shape the conduct of a study. Projects twist and turn in direction as they run their course as a result of serendipity, preliminary results, logistical hurdles, and most importantly, the conduct of the investigators themselves.

Despite the absence of a completely standardized protocol for clinical investigation, most research studies employ a fairly uniform set of procedures. Many texts that describe clinical research procedures are available (1–6). They generally emphasize the scientific method and the methodological tasks that investigators face. However, textbooks rarely present practical tips about techniques and procedures that experienced researchers use in their work. Books usually fail to convey "inside" information picked up "on the job" by investigators, or learned during research training or scientific fellowships.

The purpose of this chapter is to convey some of the practical and useful aspects of how a study is conducted by presenting a research "case." The intent is to make the research process more relevant and immediate, although no claim is made that the approach described is airtight or the only way to perform clinical investigation. The purpose is to demonstrate that as a project grows in complexity, it usually requires additional financial support and increases the need for careful organization and discipline in managing data collection and analysis. The reader should develop an understanding of the broad range of needs in the conduct of a study and adapt or extract what may be most appropriate to his or her own project.

The research "case" used in this chapter describes a recent clinical trial. The pilot stage was undertaken with no outside or private financial support apart from the cost of computerized literature searches. The major study that ensued was funded to support two half-time research assistants and to pay for computer programming and data analysis. This does not constitute a major clinical research project, but still provides a contrast to the "bargain basement" digoxin study presented in the preceding chapter.

Background

A Research Idea

Dr. Michaels, a family physician and residency faculty member with a special interest in obstetrics, had been moderately involved in the movement toward greater humanization of hospital deliveries. He philosophically supported home-style birthing rooms in hospitals, greater participation by families in the birthing process, and believed that women should have a say in the control of their own deliveries. At one delivery he attended, the woman, who had a delayed first stage of labor, was given breast stimulation by her husband based on the notion that this would produce uterine contractions similar to those induced by suckling. This indeed seemed to be effective and the delivery proceeded smoothly. Some weeks later another woman was admitted to the hospital with premature rupture of the membranes. She was anxious for a natural childbirth and rather than begin a pitocin augmentation, Dr. Michaels, impressed by the previous experience, decided to try the breast stimulation technique. Within 20 minutes of her massaging one breast she started a strong labor pattern and proceeded rapidly to a normal delivery.

The technique of breast stimulation thus appeared to enhance uterine contractions. Dr. Michaels wondered whether breast stimulation could be useful in augmenting labor without the potential dangers of the intravenous use of pitocin and the complexities of supervising oxytocin infusion. In addition, the stimulation technique could give women a more participative role in their own labor. Dr. Michaels wondered whether this clinical phenomenon might be worth investigating, since he was not aware of any published data on breast stimulation. However, before he started to collect data, Dr. Michaels realized it was important to find out for certain whether breast stimulation for augmenting labor had been investigated before or had been used for similar clinical purposes.

Literature Search

Although he used the technique of computerized medical literature searching (Medline) by submitting a request at his local library, Dr. Michaels was disappointed with the results. The literature search yielded only six reported studies on breast stimulation over the previous 40 years, one each in Russia, Hungary, Israel and Britain; two in Germany. None was designed as a controlled clinical trial; their data were subject to bias. All reported success rates in response to stimulation in the region of 70–80%. Study of the Surgeon-General's *Index-Catalogue* (7) for some anthropological literature revealed that breast stimulation had been and still is used for labor augmentation in tribal societies in South America, Africa, and the Far East.

He read from the extensive literature on oxytocin, the hormone involved in the pituitary/breast/uterine reflex of lactation and found that although a great deal was known about oxytocin levels in the blood before and after delivery, almost nothing was published on oxytocin blood levels during delivery. Further, the role of oxytocin in the process of initiation and maintenance of parturition was still obscure.

Dr. Michaels did not, however, follow up intensively on the literature search. For example, he did not contact the authors of the published articles by letter or telephone, or spend much time talking to expert colleagues in the field. This was in part a consequence of his initial explorations which revealed an almost total lack of information on the subject and his belief that further pursuit would be of little value. He also did not use the available resources to identify current research in progress. Had Dr. Michaels done so, he would have found several studies in progress on breast stimulation to induce uterine contractions, as well as physiological studies of oxytocin in labor. He did not realize that published papers often represent at least 3 years of prior clinical investigation and therefore do not always represent the state-of-the-art. Current work in progress can be researched in several sources:

1. *Grants and Awards:* Covers broad and detailed subject areas. Published annually by the National Science Foundation.
2. *Research Awards Index:* Published in two volumes annually by the National Institutes of Health. Volume 1: Subject list. Volume 2: Project titles and investigators.
3. *Coresearch* :Listing of grants awarded by Foundations; covers science, education, and the humanities.
4. *Foundations Grants Index:* Published by the Foundation Center.
5. *Government Reports Announcements Index:* Published by the National Technical Information Service.

There are several useful things to remember about researching the literature when a University or Medical Center library is locally available:

- Become acquainted with the librarian and discuss your search face-to-face at least once.
- Spend some time working out the key words you will use to search the literature, so librarians will access the references you need. This will save a lot of time and money.
- Start out with a reference card system and annotate articles.
- Don't assume references are accurate; at least 10% have errors, so read all cited articles.

- Don't always accept other people's interpretation of reference articles because they may be inaccurate or lack real understanding.
- Review reference articles and cards regularly. This will spark new ideas or open new approaches to the problem you are investigating.

Hypothesis

Dr. Michaels concluded that although there was no direct physiological evidence that breast stimulation in labor caused oxytocin release and thereby uterine contractions, it could be reasonably assumed. He also decided that a controlled clinical trial of the breast stimulation-uterine contraction hypothesis would be worth undertaking. Expressed in null form his hypothesis was:

Ho: *Breast Stimulation has no effect on uterine contractions in women who are experiencing inactive labor.*

Dr. Michaels reasoned that empirical data would likely reject or falsify the null hypothesis, i.e., show that there *is* a relationship between breast stimulation and uterine contractions. In terms of formal logic, rejection of the null hypothesis must necessarily lead to acceptance of one alternative hypothesis from two possibilities.

Ha: *Breast stimulation facilitates uterine contractions in women who are experiencing inactive labor.*

Ha′: *Breast stimulation inhibits uterine contractions in women who are experiencing inactive labor.*

Clearly, the most likely conclusion from a well designed clinical trial would be acceptance of the first alternative. In fact, given the armchair evidence in the literature and from clinical observations, Dr. Michaels would be quite surprised if either the null hypothesis or the second alternative hypothesis were supported by data. He still felt, however, that it was important to write down the null hypothesis in clear-cut terms to give focus to the study and to provide a target to "shoot down."

Stating one's research hypothesis in operational terms is one of the most important steps in clinical investigation. Unfortunately, many studies launched by inexperienced investigators are not based on the solid foundation of cerebral effort that is needed to write out one's research hypothesis. Many of these studies fail or drift off course due to a lack of focus. In Chapter 12, a slightly different approach is used to formulate Dr. Michaels' research hypothesis—that of stating the research objective. The words may differ but the message is the same. It is crucial for the clinical investigator to "think through" and write down the research hypothesis or objective *before* collecting a shred of data.

Pilot Study

Before attempting to design a clinical trial to investigate the breast stimulation hypothesis, Dr. Michaels decided to try the method out on a small number of patients and use the results as a pilot study.

As the months slipped by, he developed a simple data collection instrument to use when a woman was admitted for delivery who seemed suitable for the stimulation technique.He included only women at low obstetric risk, with premature rupture of the membranes, who were pregnant for 36 weeks or more, as the sample population. Eventually he acquired 12 "cases" over an 18-month period. From this experience and by jotting down clinical observations on cards, he was able to assess how the technique might be used on a busy labor ward, how acceptable it might be, and document some of the difficulties involved in measuring outcomes in a meaningful way. The apparent success rate as measured by an increase in uterine contractions after stimulation remained impressive, in the region of 70%, and 60% of the women proceeded to uneventful deliveries. Michaels concluded that the method seemed promising and decided to go ahead with a controlled clinical trial.

Dr. Michaels' decision to undertake a small pilot study was an important and correct step. Clinical ideas and hunches need supporting evidence before resources and personal energy are committed to a major study. A pilot study may be used at different stages of an investigation depending on what information is needed. It can provide data on feasibility, potential units of measurement, ways to increase efficiency, and costs. One of the major questions regarding a pilot study is, "How extensive should it be?" In general, a pilot is used for the reasons stated in Table 11.1.

TABLE 11.1. Purposes of a pilot study.

Purpose	Case example
To test untried procedures (This is what Dr. Michaels did)	The technique of breast stimulation during labor
To identify characteristics of the study population	Low-risk pregnant women at 36 weeks with premature rupture of the membranes
To check data sources	Obstetric medical record; fetal monitor strip
To clarify and identify the resources needed to do main study	Funds, fetal monitor machines, computer capability; nursing staff; faculty time
To develop and test data collection instrument(s)	Outpatient protocol; inpatient protocol
To identify and train personnel	Research person, nursing staff, and physician training and secretarial support

Comment

A number of points in the research process deserve reiteration given Dr. Michaels' decision to initiate a full-scale randomized clinical trial. His idea had been sparked by an unexpected clinical observation, a common occurrence for most physicians. Following up on ideas can be time-consuming and one often forgets them, an important reason to carry a small notebook to jot down thoughts. Intuitive flashes often come at inconvenient times and it is wise to be prepared to make them part of a permanent record.

Ideas for clinical investigations typically come from several sources.

- Simple observation of events such as interesting cases, an unusual response to therapy, etc.
- Discussion of clinical and behavioral events with friends or colleagues to obtain corroboration or to argue fine points
- Reading journals and books to learn about other scholars' theories, thoughts, and observations
- Questioning why an event occurred
- Asking, "If an intervention were made, what would happen?"

An investigator should not be discouraged at the early stages of a research project if answers do not surface immediately. One should also not be discouraged by the lukewarm reaction a study will inevitably receive from some colleagues whose interests are not vested in its success. Clinical investigators should be confident, yet not headstrong. They should also acknowledge that good research takes time, skill, hard work, organization, and at least a touch of compulsiveness.

The Clinical Trial

The outcome of Dr. Michael's pilot study led to a study design that was operationalized into two phases: (a) a 1-year outpatient study to evaluate the most acceptable and effective method of breast stimulation in producing uterine contractions,and (b) a 2-year inpatient study using the "selected" method on a group of women who either had premature rupture of the membranes and inactive labor or had inadequate labor patterns in the first stage of labor. The inpatient study would compare women receiving breast stimulation with those receiving pitocin for augmentation in a random fashion. It was hoped that this method would identify the degree to which breast stimulation could be used as an alternative to pitocin or as an initial intervention in augmenting uterine contractions.

In neither the outpatient nor the inpatient study could the patients or the investigator be "blinded." There was no way of hiding which intervention was being used to stimulate uterine contractions. There were, therefore, opportunities for bias resulting from both the patients' attitudes toward the intervention being tested (which might inhibit or promote the uterine response) or the investigator's conscious or unconscious desire to establish the value of breast stimulation as a labor-enhancing technique. The potential biases could be reduced somewhat by (1) assessing patient attitudes (using a questionnaire) and adjusting the outcome data statistically, and (2) removing the investigator from direct interaction with patients and from a knowledge of the assignment of cases into the study.

Phase 1: Outpatient

The outpatient phase occurred in an obstetric clinic, a setting where breast stimulation was already being used as a prenatal test for fetal well-being (contraction stress testing). Consequently, the research project was readily accepted by the clinical staff.

One nurse managed the study as a routine part of her work at running the contraction stress tests. A protocol was developed to test the efficacy of three different breast stimulation methods in producing adequate uterine contractions. Adequate was defined as three contractions in 10 minutes. A placebo technique was also assessed. The methods of stimulation were: manual self-stimulation through a gown by the patient, a breast pump, heating pad on the breast, and an unplugged heating pad (placebo) placed on the abdomen (Fig. 11–1).

(Dependent variable)

Outcome measures

1. No. contractions/5 min
2. Percent women with adequate contractions

Stimulation method
(Independent variable) . . . . . . >Monitoring of uterine contractions. . . . . >

	Baseline	Intervention	Resting
Manual	10 min resting	15 min stimulation	5 min resting
Pump	Resting	Stimulation	Resting
Heating pad	Resting	Monitoring activity Stimulation	Resting
Placebo	Resting	Monitoring activity 15 min placebo / manual stimulation Monitoring activity	Resting

FIGURE 11.1. Phase I. Protocol design (outpatient).

Dr. Michaels wisely obtained a 1-hour consultation with an epidemiologist colleague. Together they studied data from the published literature and the pilot study. They subsequently decided to select a sample size of 106 patients per group to test rigorously each breast stimulation technique. The placebo was the only exception because it might require fewer cases if it proved ineffective in producing contractions early in Phase 1 (8). The sample size was also based on an anticipated difference of 20% in the uterine response rates between the four groups. A total sample of 360 patients was finally determined to allow for attrition due to technical failures, patient refusals, and other problems. The patients would be randomly assigned in the study, and the comparison would be based on the number of uterine contractions produced by each breast stimulation method over 5-minute intervals recorded by an external uterine monitor (Fig. 11.1).

Implementation of the outpatient phase required the completion of a number of important tasks. For example, a separate protocol was developed for each of the stimulation methods and the placebo. These took 5 months to develop and refine based on continual field testing (Table 11.2). A package of informational materials about the project was produced to orient the nursing and medical staff, and Dr. Michaels made several presentations to both groups. There was also consultation with the obstetricians in charge of patient care and further discussion with the epidemiologist regarding research design issues. An "on-site" logbook was developed to record all cases and all completed protocols were photocopied as a back-up. A patient consent form was developed and the project was approved by the Committee for the Protection of the Rights of Human Subjects (Table 11–3). Weekly staff meetings were held with the research personnel to review progress. Data analyses were done in stages to examine trends and institute necessary changes in the study. Finally, the work was carefully monitored to ensure the project was proceeding according to a defined time schedule.

Phase 1 was completed successfully for a number of reasons. A close relationship developed between the research staff and the nurses who had responsibility for using the stimulation methods. The relationship was enhanced by frequent visits and telephone conversations and full discussion of problems that arose. The data collection instruments were refined until everyone involved felt the forms were clinically relevant and unambiguous. However, one major problem emerged. There were serious delays in obtaining the patients' medical records, a "glitch" that could have been foreseen. It took about 6-weeks to complete data collection on each patient and this threw the timeline of the project off course.

Phase 1 took 6-months longer to complete than expected. The results showed that manual breast stimulation was the most effective technique for producing uterine contractions and was quite acceptable to the patients. The careful recording of many aspects of contraction stress test-

TABLE 11.2. Antepartum SNILP protocol

(Keypunch Codes)

| A | 1 | serial | 1 2 3

1. Patient chart no.________
2. Protocol code___ __ __ (6 11 12 18)
   (case no.) (p) (n)
   (P = protocol initial; n = nurse's initials)
3. Date of test__ __ __ (19 24)
   mo. day yr.
4. Test no. __1 or __ __ (FILL IN ACTUAL TEST NUMBER) (25)
   (if OTHER than "1" is checked please answer question 5)
5. Repeat test information (please check one)
   a. __Tested within 24 hours
   b. __Tested within a week
   c. __Tested within 2 weeks
   d. __Other (Please specify)
   ______________________________ (27)
6. Birthdate __ __ __ (28 33)
   mo. day yr.
7. Gestational age of current pregnancy __ __weeks (34)
8. Reason for test (please check)
   a. ___ Routine surveillance
   Placental insufficiency: (36 41)
   b. ___ IUGR
   c. ___ Post maturity (42 weeks)
   d. ___ Diabetes
   e. ___ Gestational diabetes
   f. ___ Preeclampsia
   g. ___ Hypertension (chronic)
   h. ___ Adolescent
   i. ___ Other medical conditions
9. Parity (complete a, b, and c)
   a. Term pregnancies __ __
   b. Premature __ __ (42)
   c. Abortion: spontaneous __ __
   d. Abortion: induced __ __
10. Non-stress test (NST) (Please check one) (48)
    a. ___ Cannot be interpreted
    b. ___ Nonreactive (50)
    c. ___ Reactive (if "c" is checked, please answer question 11)
11. Disposition if NST is reactive (please check one)
    a. ___ Contraction stress test
    b. ___ No further tests done (51)
12. Breast stimulated (please check one)
    a. ___ Left breast OR b. ___ right breast
    c. ___ Both breast ALTERNATING (52)
    d. ___ Both breasts SIMULTANEOUSLY
13. Contraction stress test (CST) (15-minute stimulation plus 5-minute poststimulation resting strip) (please check one)
    a. ___ Cannot be interpreted
    b. ___ Nonreactive
    c. ___ Reactive

TABLE 11.2. *Continued*

(Keypunch Codes)

14. CST results (please check applicable items) |_|_| serial|_|_|_
    a. ___ Negative |_|_|_|_| 14 . 17
    b. ___ Positive
    c. ___ Fewer than three contractions in 10 minutes
    d. ___ Cannot be interpreted
    e. ___ Equivocal
    f. ___ Hyperstimulation (longlasting contraction greater than 90 seconds or 5 contractions in 10 minutes)
    g. ___ Other (please specify)______
15. Disposition if CST is UNSATISFACTORY (please check one)
    a. ___ Repeat NST-CST within 24 hours |_| 18
    b. ___ OCT (challenge test)
    c. ___ Nothing further done (reason) ______
16. Additional comments or observations (Y/N)|_| 19

____

____

Tracing interpretation

Procedure			Keypunch Codes
Procedure: (NST)	A.	Ten-minute resting monitor strip. Completed Yes___ No___	(Y/N)|_|
		Number of contractions in first 5 (1–5) minutes ___ ___	|_|20|
		Number of contractions in next 5 (6–10) minutes ___ ___	|_|_|
	B.	Fifteen-minute manual stimulation. Completed Yes___ No___	(Y/N)|_|
		Number of contractions in first 5 (1–5) minutes ___ ___	|_|25|
(CST)		Number of contractions in next 5 (6–10) minutes ___ ___	|_|_|
		Number of contractions in next 5 (11–15) minutes ___ ___	|_|_|
	C.	Five-minute resting poststimulation strip. Completed Yes___ No___	30 (Y/N)|_|
		Number of contractions in 5 minutes ___ ___	|_|_| 33 34

SNILP, Stimulation of nipple in Labor Project.

ing led to other areas of inquiry that were of interest; for instance, the characteristics of uterine hyperstimulation produced by the manual technique (9).

Phase 2: Inpatient

As Phase 1 approached completion, preliminary data showed that the manual (self-applied) method was the statistically superior method for producing uterine contractions. In addition, Dr. Michaels was pleased to discover that other studies on breast stimulation currently underway confirmed the effectiveness of the manual technique (10).

At this stage preparations began for the implementation of Phase 2 (Fig. 11.2). The objective of the second stage was to test breast stimulation in women admitted to the obstetric floor with two labor "delay" problems: (a) premature rupture of the membranes with no active labor,

TABLE 11.3. Patient consent form: Breast stimulation and the induction of labor.

Inside the mother, a baby is surrounded by a bag of water. Before delivery, the bag breaks and the mother usually begins labor. Sometimes, after the bag of water has broken and the baby is no longer protected from the outside environment, the mother does not go into labor. Since most doctors feel it is important that the baby should be delivered within 24 hours of the water breaking to prevent infection, sometimes labor has to be started by the doctor. This is usually done by giving a hormone called pitocin through an intravenous line (i.v.) into the mother's arm. Although this usual procedure is safe, it does require close medical supervision and does not allow the mother complete control over her own labor.

STUDY PURPOSE

There is strong scientific information that nipple stimulation causes the womb to contract. Nipple stimulation can be performed by the mother herself, her support person, or by a breast pump during labor. *The purpose of this study is to test breast stimulation as a safe alternative method of making the womb contract when the bag of water has broken and labor does not start or if labor has started but is not proceeding quickly enough because of irregular or weak contractions.* We hope to find out if the use of nipple stimulation will avoid the need for i.v. pitocin and strict medical supervision so that YOU, THE MOTHER, WILL HAVE MORE CONTROL OVER YOUR OWN LABOR. About 300 mothers are being asked to participate in this project.

PROCEDURE

If you and your doctor decide to take part in this study, you will be asked to participate in ONE of the following three ways:

Group 1. Use breast stimulation throughout your labor

or

Group 2. Use breast stimulation for a 2-hour period at the start of your labor

or

Group 3. Have routine care using pitocin.

If you decide to take part in this study and are in group 1 or 2, the following procedures will take place.

1. The nurse will explain how the stimulation will be done. She will stay with you until you are sure you understand the method and will supervise your labor carefully.
2. A monitor will be placed either on your belly or inside your uterus to measure the baby's heart rate. THIS IS A ROUTINE PROCEDURE.
3. Stimulation will start. You can do it yourself, have a friend or husband do it, or ask the nurse to show you how to use the breast pump.
4. If your labor does not progress in the normal matter, if complications arise, or if you want to DISCONTINUE THE STIMULATION AT ANY TIME, THE STUDY WILL BE STOPPED, AND YOU WILL BE CARED FOR IN THE USUAL ROUTINE MANNER.
5. The day following your baby's birth a questionnaire will be given to you or mailed to you to ask you your opinion about this method of study.

RISK/COSTS

As far as is known there is no scientific information of danger or side effects using the nipple stimulation method. There are NO EXTRA COSTS for taking part in this study.

ALL RESULTS OF THIS TEST WILL BE KEPT CONFIDENTIAL.

We thank you for your time and cooperation. If you wish more information at any time, you can contact Dr. Michaels at the University of North Carolina. Dr. Michaels is

TABLE 11.3. *Continued*

the principal investigator for this study. His phone number is (919) 966-2091. If you feel any infringement on your rights, you may contact John Q. Mensch, M.D., Chairman of the Committee on the Protection of the Rights of Human Subjects at the University of North Carolina at Chapel Hill (919) 966-3641.

Project participation consent

I, ______________________________, UNDERSTAND THE PURPOSE OF THIS PROJECT as stated above and I am fully aware of the procedures to be followed. I understand that in the event of physical injury directly resulting from the research procedures, financial compensation cannot be provided. However, every effort will be made to make available to me the facilities and professional skills of Memorial Hospital and Medical Center. I understand I am free to end my involvement in this study at any time.

______________________	______________________	
WITNESS	PATIENT SIGNATURE	DATE
	______________________	
	PARENT OR GUARDIAN (if patient is under the age of consent)	

requiring augmentation, and (b) inadequate uterine contractions during the first stage of labor. After informed consent was obtained, breast stimulation would be used as an intervention at the same stage at which intravenous pitocin would otherwise be used by a physician. Patients would be divided sequentially into three groups: (a) intermittent breast stimulation throughout labor; (b) intermittent breast stimulation for 2 hours only, later reverting to orthodox management; and (c) orthodox management with pitocin (control group). Patients would be given the option of using either the manual or breast pump stimulation method depending on which method was the most comfortable for them. They could also switch methods if the one being used became too uncomfortable during active labor.

The outcome of labor in the three groups was compared. Outcome measures included:

1. Uterine contraction patterns (frequency, length, height of contraction)
2. Time intervals (cervical dilation, delivery)
3. Interventions (forceps, Cesarean section, etc.)
4. Analgesia used
5. Apgar scores for the babies.

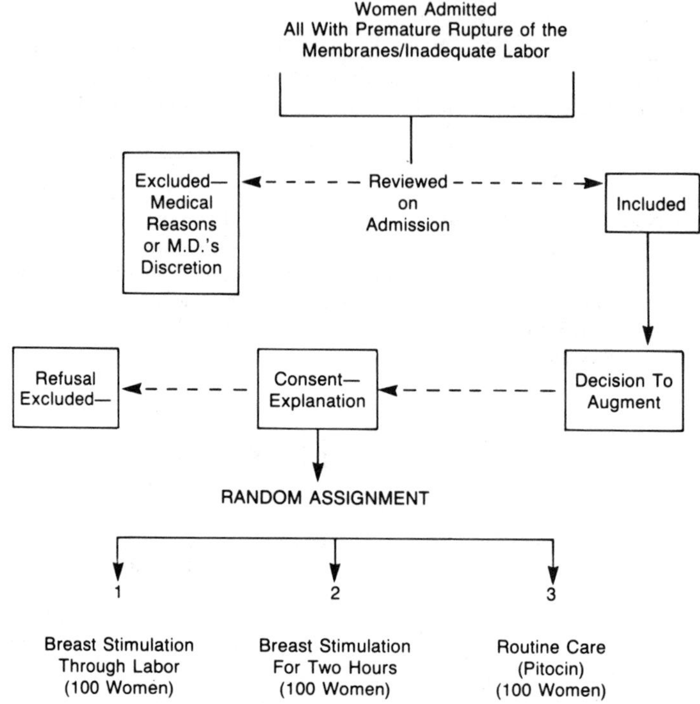

FIGURE 11.2. Phase II. Study design for an inpatient randomized controlled trial using manual or breast pump stimulation.

Another consult with the epidemiologist was arranged to reach a decision about sample size for the Phase 2 clinical trial. It was determined that each study group should contain 100 women. Attention was also given to the source of subjects. On the average, 10% of the women admitted to the labor ward of the hospital serving as the research site would have premature rupture of the membranes. Consequently, Dr. Michaels felt that the clinical trial could probably be completed in 18 months.

The inpatient phase was much more complex, logistically, than Phase 1. It took place on a busy obstetric ward and involved carefully defining and making operational all the protocol terminology. For example, what is meant by "active labor," "premature rupture of the membranes," and "full dilation of the cervix?" Inclusion and exclusion criteria for the enrollment of subjects were developed in consultation with attending obstetricians (Table 11.4). A patient consent form and an inpatient data collection protocol were designed and tested in conjunction with the obstetric ward staff. Ward procedures were observed so that the sequence of events required to implement the clinical trial would not obstruct the staff in their routine duties or cause severe disruption (Fig. 11.3). The

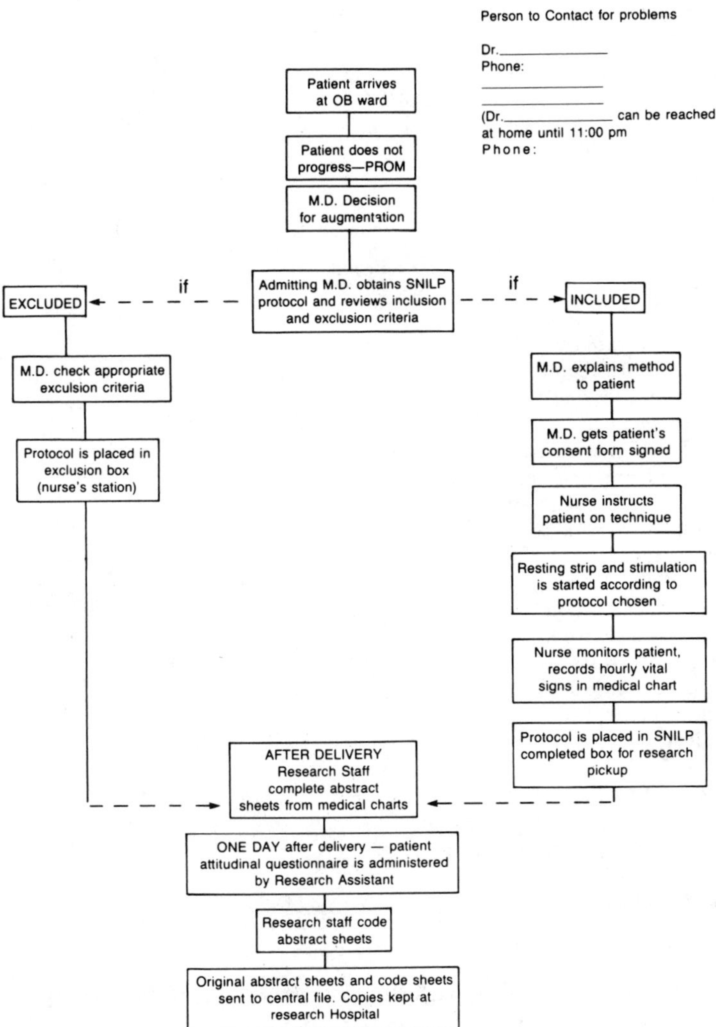

FIGURE 11.3. SNILP sequence of events (inpatient).

nurses and physicians needed to be oriented to the methods and sequencing of the breast stimulation method which was then pilot-tested on the obstetric ward. Cooperation was enhanced by giving the involved staff members feedback about the results of Phase 1 and preliminary results of Phase 2.

A separate data collection form was required for each of the interventions being tested. Consequently, each study package contained common materials (inclusion-exclusion criteria, patient consent form, guidelines for the physician, procedures for the nurse, abstracting form for the

TABLE 11.4. Inclusion/exclusion sheet: SNILP

Case No. ___

ADMITTING OB PHYSICIAN
PLEASE COMPLETE THIS PAGE.

Admitting OB physician ______________________________
Name

Time of evaluation ______________________________

Estimated gestational age: (circle 33 34 35 36 37 38 39 40 41 42 43

I. Exclusion factors

Check where appropriate:

___ Membrane rupture > 10 hours previous
___ Preeclampsia (requiring)
___ MgSulph: Diastolic > 90
___ Uterine anomaly
___ Bleeding: undiagnosed in third trimester
___ Twins, breech, or C-section planned
___ Gestational age of 35 weeks or less
___ Less than 16 years old
___ Diabetes
___ Hydramnios
___ Pelvic disproportion
___ Patient is in active labor
___ History of serious mental illness
___ Substance abuse-active, including alcohol, excluding tobacco
___ Patient refuses
___ Evidence of chorioamnioitis

DISCRETIONARY EXCLUSION CRITERIA
___ Previous fetal anomaly
___ Amniocentesis in this pregnancy
___ Previous C-section
___ Abnormal lie

research assistant) together with the instruction sheet dealing with one of the three specific interventions being tested.

Like the outpatient phase, refinement of the new protocols took approximately 6 months. An early discovery was that busy nurses and physicians balk at having to collect extra data. Thus the protocols were simplified to require minimal recording (a checklist for physicians on a single page) and to fit in with the established nursing routines. Despite these efforts, it took the staff several months to overcome their reaction to what seemed a thick bundle of paper for each patient. The protocol package contained:

- One page outline of the project
- Inclusion/exclusion criteria sheet—to be completed by the physician
- Patient consent form
- Instructions to nursing staff on breast stimulation techniques
- Three data abstract forms for research staff

IF EXCLUDED, PLACE IN BOX IN NURSING SECTION AND DO NOT GO ON TO SECTION II

Check where appropriate:

II. Inclusion factors

Premature rupture of membranes: ___ No rupture ___ Rupture (complete A and B)

A. Documentation of rupture: _______ Date of rupture _______ Time of rupture

1. Fluid seeping from os ___ Yes ___ No ___ Clear ___ Bloody ___ Meconium
2. Nitrazine test ___ + ___ − ___ Not done
3. Fern test ___ + ___ − ___ Not done

B. Inactive labor over a 30-minute time period. Contractions less frequent than 1 in 5 minutes: ___ Irregular contractions ___ Inactive or ___ Active labor

(Please check which applies)

Lack of progress of labor

C. ___ Cervix: Lack of descent or dilation without contraindications to augmentation

D. ___ Uterus: Irregular and/or infrequent/inadequate contractions

IF PROM IS ESTABLISHED AND LABOR IS INACTIVE, AND NO EXCLUSION FACTORS ARE CHECKED, INCLUDE IN STUDY—GIVE THIS SHEET TO OB NURSE.

IF BOTH C AND D ARE PRESENT AND NO EXCLUSION FACTORS ARE CHECKED, INCLUDE IN STUDY—GIVE THIS SHEET TO OB NURSE.

PATIENT STAMP

SNILP, Stimulation of Nipple for Induction of Labor Project.

It took about 5 months for the clinical trial to get underway. The time was needed for staff to learn how the trial fitted into their normal clinical activities and to adjust to the protocol and sequence of the intervention.

Comment

The attitudes of the physicians and nursing staff were one of the most important factors involved in implementing this phase of the study. The physicians resented having to record more data (although this was minimal) and disliked having their routines disturbed. The nurses, by contrast, saw a positive outcome as saving them a lot of work supervising pitocin infusions. It was therefore crucial for Dr. Michaels to spend time on the obstetric floor, explaining the study and working with the staff, and to make sure that the randomization of interventions was occurring correctly to eliminate bias. Unfortunately, he had not taken into account the changeover of interns on the obstetric floor every 3 months, so that every quarter a new group of physicians became involved in the study without adequate orientation. This resulted in reduced patient recruitment and Dr. Michaels had to schedule training sessions with each set of new interns.

Another problem was the tendency for some nurses and physicians to want to use the breast stimulation technique because they felt it was effective. This meant that when the randomization process indicated that a suitable patient should be assigned to the control (i.e., non-breast-stimulation) group, the staff were reluctant to do so and wanted all suitable patients to receive the new intervention. Obviously, this behavior threatened the design of the study. Others tended not to use it because they preferred the orthodox way of patient management. Thus, control of the study occasionally became an important issue. It was handled by educating the staff about basic research principles and about the need to carry out the study completely before reaching conclusions.

As patients became enrolled in the project Dr. Michaels and his part-time staff tested the ease with which data could be abstracted from the medical record onto the data collection forms. This turned out to be difficult and time-consuming because there were problems deciphering handwriting and data were sometimes missing or inadequate. This important exercise led to drastic changes in the data collection forms to streamline the process of abstracting patient data. However, the process still took about 30 minutes per record, the equivalent of $3\frac{1}{2}$ weeks of continuous work.

As the study continued, Dr. Michaels regularly visited the obstetric floor, appeared at work rounds, and "talked up" the project among the nursing staff and physicians. Telephone contact was maintained with the ward staff, and this form of communication and feedback proved to be a key element in the continued motivation of nurses and physicians to recruit patients and maintain the project. Continuous personal contact is an essential element of a collaborative project.

At the time of writing this chapter the clinical trial is still underway but nearing completion. Unfortunately, patient recruitment is behind schedule and research funds may dry up before the target number of subjects is reached. This may threaten the successful conclusion of the study unless strenuous efforts are made to increase patient recruitment or find extra support.

In the years that it took to transform a clinical observation into a clinical trial, many new ideas had grown out of the study, along with unexpected approaches to analyzing the collected data. Dr. Michaels had become less of a novice in conducting a study but no less of an enthusiast in the search for new knowledge. He had learned that careful development of a clearly defined and operational hypothesis was an essential prerequisite to setting up a study—that thoughtful planning and, above all, communication were the foundation of success.

Implementation of the project required the following activities:

- Negotiation with the obstetric staff regarding logistics and staff support
- Development of an organizational plan for the entire project
- Hiring of research staff to assist with development of data collection instruments, training of clinical staff in stimulation methods, data col-

lection, facilitating data flow and organization, updating literature files, maintaining supervision of the two phases of the project, and data analysis and reporting
- Development of task assignments and a data management plan
- Obtaining approval from committees, agencies, and institutions regarding patient consent, access to patients and medical records
- Obtaining funds to support the study.

A Study Checklist

A number of basic elements must be considered whenever a major research study is designed and implemented. An inventory of these items is listed below. A smaller project will not require attention to all of the items identified (11, 12). Many of the points given in the checklist are amplified elsewhere in this volume, especially in Chapters 12–15.

Project organization plan:

- Identify the project leader
- Establish communication lines
- Establish individual supervision of research staff
- Clarify staff activities and roles
- Outline and budget financial costs data entry, copying, literature searches, telephone, supplies, travel, typing, salaries, computer resources
- Establish a projected time plan for project activities
- Establish a sequence of data flow and management.

Management issues:

- Obtain cooperation from data providers: time, costs, space, attitudes, commitment, compensation.
- Obtain adequate work space for project staff.
- Train staff in data abstracting methods.
- Estimate abstracting costs, and time taken to collect data.
- Follow hospital/clinic protocols to access medical records.
- Pilot abstracting procedures to (1) identify problem areas, and (2) develop quality control criteria.
- Maintain research logs and note problems.
- Set up protocols for data entry and coding.

Development of data collection instruments:

- Review raw data sources (i.e., sample medical records).
- Identify items to be included on the form.
- Identify data entry method, i.e., precoded, keypunch, optical scanning.

- Review format of form for (1) clinician, (2) research staff, (3) keypuncher; format should be visually acceptable.
- Decide on coding methods.
- Reduce data collection errors—standard coding, exhaustive coding, data entry audits, quality control.
- Submit the data collection instrument to peer review.
- Field test data collection instrument (i.e., several people abstract one medical record).
- Evaluate logistics of data collection on site (i.e., sequence of data collection, confidentiality, storage).
- Document and store all drafts of the data collection instruments.

Acceptance of the study:

- Involve project staff in planning and implementation.
- Prepare timely submission of protocols and patient consent forms to the Human Subjects/Ethics Committee.
- Obtain approval of the study through the appropriate Research Committees.
- Sequence approvals through committees—in general, Human Rights approval is a prerequisite for other committees.
- Prepare press release with care and communicate with colleagues and administrative staff in the academic institution first.
- Maintain contact with participant clinicians to ensure support for the study.

Quality control:

- Observe initial data collection process.
- Verify the accuracy of abstracting the data—sampling 10% of the data. Use of collection forms and checking with the raw data sources are usually adequate.
- Review each data form for completeness and errors.
- Ensure that data collection forms are ready for data entry.

Considerations for research staff:

- Qualifications and experience
- Work environment
- Time availability (day/night)
- Hours of work
- Travel (if applicable)
- Training—staff manual, training sessions, role plays
- Weekly meetings
- Schedule of tasks
- Performance review
- Regular reports

Study documentation:

- Proposal
- Study design
- Data collection instruments
- Progress reports
- Coding manual
- Location of computer files/programs
- Publications, budget reports
- Literature files/bibliography

Finally, in addition to the items listed above as key elements in designing a research project, several other recommendations not generally found in textbooks on research methods may be valuable for young academics. These and other "pearls" are from a book on clinical research by British surgeon James Calnan (13).

Here are ten [nevers] that may help in research:

- Never criticize another researcher, only his work. He may have more friends in high places than you.
- Never show anybody a half-worked idea; finish the thinking and write it out in precise language first.
- Never talk about brilliant projects until you have the work in hand, for another may beat you to the post and publish first.
- Never undertake to do work for a commercial company unless you are fairly certain of a result either way.
- Never forget to acknowledge the finance from a grant-giving body and those who helped materially (the chief, medical artist, technicians, nurses); better still, put their names on the paper for publication.
- Never acknowledge help that was not given, even if you think it will please (it will, of course, but you may bitterly regret it later).
- Never forget that you *might* be wrong, even if you name is Wright.
- Never promise to deliver work by a certain date, unless you have already completed it.
- Never pursue a lost cause. If others have been unable to solve a problem and you clearly cannot, then move on to something else, while you are still healthy.
- Never give up.

References

1. Hamilton M. *Lectures on the Methodology of Clinical Research*, ed 2. Edinburgh, England: Churchill Livingstone, 1974.
2. Chalmers TC. The clinical trial. *Mil Mem Fund Q 59*:324–339, 1981.
3. SchlesselmanJJ. *Case-Control Studies: Design, Conduct, Analysis.* New York: Oxford University Press, 1982.

4. Fletcher RH, Fletcher SW, Wagner EH. *Clinical Epidemiology: The Essentials.* Baltimore: Williams and Wilkins, 1982.
5. Cartwright A. *Health Surveys in Practice and in Potential: A Critical Review of Their Scope and Methods.* London: King Edward's Hospital Fund for London, 1983. (Distributed by Oxford University Press.)
6. Gordon MJ. Research workbook: a guide for initial planning of clinical, social and behavioral research projects. *J Fam Pract 7*:145–160, 1978.
7. *Index-Catalogue of the Library of the Surgeon-General's Office.* United States Army. Washington, DC: U.S. Government Printing Office. Series 1, Vol. 1, 1880.
8. Young MJ, Bresnitz EA, Strom BL. Sample size nomograms for interpreting negative clinical studies. *Ann Intern Med 99*:248–251, 1983.
9. Druzin ML, Paul RH, Gratacos J. Current status of the contraction stress test. *J Rep Med 23*:222–226, 1979.
10. Oki EY. A protocol for the nipple stimulation CST. *Contemp Obstet Gynecol 22*:157–159, 1983.
11. *Quality Control Manual for Epidemiologic Studies.* WESTAT, Environmental Studies Section, National Cancer Institute, National Institutes of Health, Bethesda, Maryland, September 1982.
12. Babbie ER. *The Practice of Social Research,* ed 2. Belmont, California: Wadsworth Publishing, 1979.
13. Calnan J. *One Way to Do Research: The A-Z for Those Who Must.* London: William Heinemann Medical Books, 1976.

12
Planning a Research Study

CARL M. SHY AND WILLIAM C. MCGAGHIE

To organize and conduct his study of breast stimulation among women in inactive labor, Dr. Michaels followed a sequence of steps that are worth retracing in some detail, since they are common to all well-designed clinical research projects. First, let us identify each step and then track Dr. Michaels' thinking about the issues central to that step.

Step One: Have an Intense Desire to Know

Research should be initiated because the researcher is keenly interested in gaining new knowledge, information that will help solve an important problem, fill a crucial gap in the literature, or simply satisfy the researcher's innate desire to know why things are the way they are. The key element in this first step is the investigator's intense desire to know. Without this sense of enthusiasm for solving a problem or gaining new knowledge, research can easily become a drudgery or a waste of time.

The idea for Dr. Michaels' study came from his observation that one of his obstetrical patients who had a delayed first stage of labor apparently developed effective uterine contractions on stimulating her breast. This observation would not have engendered a research project unless Dr. Michaels had first been concerned with greater humanization of hospital deliveries. He was disturbed by the use of intravenous pitocin to augment labor, and he saw in breast stimulation the opportunity to avoid pitocin use and involve women more directly in the progress of their labor. Dr. Michaels first tried the procedure with another of his patients, found it to be successful, and then sought to put it to a rigorous test, one that could withstand critical review by his peers.

Ideas for research can originate from many sources: observations, experiences, reading the literature, conversations, reflections. In Dr. Michaels' case, he not only observed for the first time the effect of breast stimulation on uterine contractions, but he reflected on this observation because of his abiding concern for greater participation of families in the birthing process. Reflection and keen interest led to the desire to gain further knowledge about the initial observation. Similarly, potential research ideas abound in our experience, but these seeds must fall on the fertile soil of a reflective, interested, and enthusiastic mind. The desire to know, to solve problems, and to evaluate is the crucial element necessary to convert ideas into research. This process requires reflection on one's experiences, a mindset oriented to acquisition of new knowledge, and a sense of excitement about gaining this knowledge.

Step Two: State the Research Objective Clearly

When Dr. Michaels reflected on his observations and experiences in the delivery room, many possibilities for research were considered, if only fleetingly. What is an appropriate and efficient method to stimulate the breast while a woman is in labor? When and how long should the breast be stimulated? What is the physiological mechanism by which breast stimulation enhances uterine contractions? Has breast stimulation been tried by others? What is known about blood oxytocin levels during labor? Are there modifying circumstances that alter the effectiveness of breast stimulation? A stream of ideas of this sort is not unusual, once a problem is identified. At this point, it was important for Dr. Michaels to focus his thoughts and state exactly what he wanted to learn, realizing that he could not address all possible questions that already had presented themselves to him. Like a military commander in the field, Dr. Michaels had to stand back, examine his objectives, and concentrate his thinking on those objectives he could reasonably achieve. For the researcher, this requires a clear and explicit statement of a research objective. Taking his various ideas and possible avenues of research into account, what exactly does Dr. Michaels want to accomplish?

In thinking about his goals, Dr. Michaels realized that he is a clinician, not a physiologist, biochemist, or psychologist. His desire to know is motivated by his desire to apply his knowledge to a clinical situation. Hence he is more inclined initially to learn about observable factors that directly produce a useful clinical outcome than to study physiological mechanisms or behavioral determinants of this outcome. Further, his competence lies in his contact with patients and in making observations on the clinical course of his patients, less so in exploring physiological pathways or measuring psychological aspects of human behavior. Dr. Michaels needs to take his resources and abilities into consideration in developing his research objective. The important step is for the investigator to consider these various aspects of the problem he faces, realize what is feasible for him to accomplish, and state explicitly what he wants to do. Without an explicit statement of objectives, he will pursue too many aspects of the problem, diffuse his resources, and end up with an ill-defined research project. Even if approved or initiated without external funding, a research project that begins with diffuse, uncertain goals is virtually doomed to failure.

What are the features of a clear and sharply defined objective? Dr. Michaels could state his in a simple declarative sentence: "Breast stimulation will produce uterine contractions in women who are experiencing inactive labor." Note first that the objective is formulated as a grammatically complete sentence, with subject, verb, and object. In contrast, the objective could have been stated as: "Breast stimulation and labor" or "Noninvasive methods to promote uterine contractions in labor." These incomplete sentences may be appropriate as project titles, but a research objective is more clearly and forcefully presented in a sentence tying the "experimental" factor to the outcome of interest by means of a transitive verb, implying a cause and effect relationship. Dr. Michaels included each of these elements in the statement of his research objective:

1. An experimental or *study factor of primary interest*
2. The *outcome* expected from use of, or exposure to, the study factor
3. The anticipated *nature of the relationship* between study factor and outcome
4. The *study population*: persons who will be studied.

In Dr. Michaels' study, breast stimulation was the study factor of primary interest, uterine contractions were the outcome, the relationship between the two was a causal one as expressed by the verb "will produce," and women in inactive labor were the primary persons studied. Dr. Michaels' statement identified each of these elements and expressed their relationship in a specified population.

The importance of a concise and explicit statement of a research objective deserves great emphasis. Dr. Michaels had to determine what he was studying (the study factor of primary interest), what outcome he expected, and in whom this outcome should be observed (the study population). All subsequent planning, literature search, selection of a study strategy, recruitment of subjects, and development of data collection forms and study protocols will be guided by the contents of the research objective. These components of a research project will require the expenditure of considerable energy and attention to details, and these efforts will be diffused if they are not sharply contained within boundaries clearly established by the research objective.

The approach to stating a research objective described here is similar, but not identical, to the method of writing hypotheses in null and alternative forms that was given in Chapter 11. Either approach can be used to accomplish the same important end: to direct the investigator's attention toward a sharply defined research problem and to keep the problem in focus.

Step Three: Search the Literature

Avoid reinventing the wheel. If others have researched the problem you are facing, it would be foolish to proceed in ignorance of their work. You might replicate their study to confirm independently what they report, but you can do this much more effectively if you are aware of their methods and results. Dr. Michaels realized early that it was important to learn whether anyone had reported the use of breast stimulation for augmenting labor. He found case reports and anthropological observations that breast stimulation appeared to be a successful practice, but there were no systematic attempts to study the effectiveness of the procedure under scientifically controlled conditions. This reinforced his feeling that breast stimulation should be put to a rigorous test.

Even if the literature on the topic were more abundant, its review would still have helped Dr. Michaels direct his research objective to fill important gaps in the state of knowledge. Every report of a research project has deficiencies or limitations when applied under circumstances different from those that occurred during the study. For instance, Dr. Michaels might have found a paper from Britain reporting that 70% of 30 women in one obstetrical clinic successfully responded to breast stimulation after labor became inactive. There were no controls in this study, such as women in inactive labor given no further treatment or a sham treatment for a specified interval of observation. On reviewing this study, Dr. Michaels should be unsure whether labor would spontaneously have become active again in many of the treated cases, whether the women responded to the psychological effect of something being done to them (the "Hawthorne Effect"), or whether breast stimulation per se was the effective agent. Even if these questions could be resolved, he should be concerned that something about the British style of obstetrical care might interact with the breast stimulation procedure to yield an outcome that could not be reproduced elsewhere.

Often, a thorough search of the literature helps the researcher to refine and sharpen the original statement of objective. Dr. Michaels may have found reports on different methods to stimulate the breast, such as a breast pump, manual nipple massage, and a mechanical vibrator. Some of these methods may be easier to use, may be more acceptable, or more successful in producing uterine contractions. Dr. Michaels could elect to study the method that was most acceptable to the potential study subjects or an alternate method that appeared most successful but was more complicated to use. Likewise he might have found evidence that breast stimulation was not effective until it had been tried for at least 30 minutes of inactive labor. He might have discovered reports that the procedure worked far better in multiparous than in primiparous women.

With this information in hand, Dr. Michaels could considerably enrich his initial research plan. Note that his original research objective does not have to be thrown out. He is still basically interested in the effectiveness of breast stimulation in augmenting labor. Now he has more evidence that the procedure may be more effective under certain circumstances determined by the type and duration of the stimulation and the parity of the pregnant woman. Without this information, Dr. Michaels might not have thought about these modifying circumstances, would not have measured them, or explicitly incorporated them into his research plan. If Dr. Michaels began his study by using only 5 minutes of stimulation with a breast pump, and if 80% of his subjects were primiparous, he would have likely concluded that breast stimulation would not work, when in fact nipple massage for 15–20 minutes would be effective, particularly amoung multiparous women.

The search of the literature, then, is an extremely important step after the initial research objective has been stated. The literature can affirm the researcher in his resolve to proceed with a research plan, can identify shortcomings or limitations that should be avoided, and can suggest modifying factors that should be incorporated in the subsequent design of the study. If the literature on the subject is reasonably abundant, the review will be likely to identify a number of important factors that the researcher will want to measure or otherwise control in his study. The researcher might decide that some small-scale preliminary investigation will be needed. In Dr. Michaels' case, he chose to perform a pilot study to evaluate procedures and then to divide the study itself into two phases. The first phase tested different methods of breast stimulation in the outpatient clinic prior to onset of labor. The second and definitive phase involved a clinical trial in an inpatient setting.

Eight to twelve hours of literature review may save eight to twelve months of new research that simply rediscovers modifying factors already reported. It is far better to develop a research project on a firm base of knowledge than to replow already tilled ground that others have worked.

Step Four: Formulate Operational Hypotheses

The purpose of this step is to convert the general, nonmeasureable concepts of the research objective into specific, operational terms that define exactly the who, what, and how of the study. Dr. Michaels' research objective states: Breast stimulation will produce uterine contractions in women who are experiencing inactive labor. Although a clear statement, the individual words and phrases of the objective are not expressed as scientifically measurable variables. The study factor, breast stimulation, is itself a concept rather than a specified, measureable variable. Dr. Michaels needs to describe the method for breast stimulation in sufficient detail for readers or reviewers to judge whether the method is reasonable and reproducible. Research objectives are expressed conceptually, whereas operational hypotheses are stated in specific, measurable terms. To make this translation, Dr. Michaels must choose methods to stimulate the breast and to measure uterine contractions among a group of women defined as to their age, geographical location, and possibly other factors such as parity, race, and calendar interval of entry into the study. The results of this translation might appear as follows:

Concepts in Research Objective	Terms in Operational Hypothesis
Breast stimulation	Self-administered manual massage alternating between right and left nipple, 5 minutes each for a total of 20 minutes, repeated at 20-minute intervals over 2 hours.
Will produce uterine contractions	Uterine contractions will recur within the 2-hour interval and will progress through the third stage of labor.
In women who are in inactive labor	Study populationwill be limited to obstetrical patients admitted to the research hospital between July 1, 19XX and June 30, 19XY. Women must be between the ages of 18 and 34 years and have have completed at least 36 weeks gestation, have no preexisting major obstetrical complications, and be in inactive labor for a minimum of 2 hours after rupture of membranes or after at least 2 hours of normal uterine contractions.

The operational hypothesis converts the research objective into measurable and replicable terms specified by time, person, place, and methods of measurement. This conversion cannot be made in a vacuum. Dr. Michaels conducted a pilot study in which he gained valuable experience with different women's reactions to breast stimulation and with different methods of stimulation. He developed a sense for the length of time breast stimulation could be appropriately used on each nipple, how soon uterine contractions would begin if the woman was going to respond to the stimulation, and how long to continue with stimulation if no response was forthcoming. He conducted interviews with some of the subjects in the pilot study to gain further perspective on their tolerance of the procedure and to obtain suggestions on how to make the maneuver more acceptable or comfortable. Reflecting on these experiences, Dr. Michaels was able to specify with greater precision the terms of the operational hypothesis.

At this point in the research plan, Dr. Michaels knows whom he wants to study and how he will measure both the study factor and the outcome. Through his pilot study, he has gained experience with the procedures, the operating conditions, and the reactions of his subjects to the test situation.

Step Five: Select a Study Strategy

The study strategy functions like a blueprint for building a home. The strategy will determine how the "building materials," as specified by the operational hypothesis, are to be assembled and organized. For his study, Dr. Michaels selected a clinical trial. In general, clinicians can choose from one of four strategies for making observations and drawing inferences regarding associations between a study factor and some clinical outcome. These strategies are: clinical trial, cohort (synonyms: follow-up or longitudinal), case-control, and cross-sectional study. Most research objectives can be pursued with any one of these study strategies but one may be particularly appropriate depending on the availability of clinical material, the stage of knowledge about this material, and the type of study factors and clinical outcomes being investigated. In brief, each strategy can be characterized as follows:

Strategy	Characteristics
Clinical trial	The investigator experimentally assigns study subjects into a treated and a control, or placebo, group. After assignment, subjects are followed over time to the end of the study, and rates of disease or response to treatment are compared in the two groups.
Cohort study	The investigator selects for study two similar groups, or populations, one exposed to a suspected hazard or possessing an indigenous risk factor for disease, the other not so exposed. Both groups are free of the outcome of interest at the beginning of the study and are followed over time. At the end of the study, rates of disease or development of the outcome are compared in the two groups.

Case-control study	The investigator selects for study a group of patients with a selected clinical disease or cause of death. Similar persons free of the disease or dying of other causes are chosen as controls. The frequency of past exposure to a suspected risk factor is then compared in the case and control groups.
Cross-sectional study	The investigator surveys a population at one point in time and compares the prevalence of disease or of some other clinical outcome in two groups of persons, one of whom is exposed to an external hazard or to an indigenous risk factor and the other group is not so exposed.

Clinical Trial

Since Dr. Michaels was primarily interested in evaluating the effectiveness of a therapeutic regimen, the clinical trial was clearly the method of choice for his study. The clinical trial is the only clinical study strategy that enables the investigator to simulate the experimental conditions of a laboratory investigation. Generally, study subjects are randomly allocated to treatment or control groups. Random allocation is simply a means to let chance distribute other factors that may affect the clinical outcome equally between treated and control groups. In Dr. Michaels' case, he might be concerned that parity, age, and possibly race could affect progression of labor in women whose labor became inactive. Random assignment of all eligible women in inactive labor to receive or not receive breast stimulation would tend to distribute these other factors equally if the sample size of the two groups is reasonable. This distribution of factors could be evaluated once the two groups are formed.

The clinical trial is the most convincing and straightforward approach to evaluating the effectiveness of drugs, therapies, or preventive measures. However, the use of this strategy is limited by many practical and ethical considerations. For example, we could not evaluate a risk factor, such as smoking or alcohol, among pregnant women by randomly allocating women to smoke or drink during pregnancy. It would also be unfeasible to allocate randomly pregnant women to quit or not quit smoking. The inference we can draw from clinical trials is often limited by the relatively small proportion of eligible subjects who will actually participate in a trial. Participants are usually more health-conscious and compliant than nonparticipants, and the effectiveness of a therapeutic regimen in the former group may not be predictive of what will happen in the general population.

Cohort Study

A cohort study is similar to a clinical trial in most respects but one, allocation of persons to exposed and control groups. When a cohort study starts, the investigator can merely assemble groups of people who are already exposed or not exposed; no experimental control can be exerted over who becomes exposed to a suspected hazard or risk factor. Because of this lack of experimental control, the investigator must deal with the possibility that exposed and nonexposed groups will differ not only in exposure status but also in a number of other characteristics associated with exposure. For example, smokers may differ from nonsmokers, as a group, in education, occupation, alcohol consumption, and other lifestyle characteristics. Each of these may be independent risk factors for the clinical outcome of interest, e.g., heart disease. In contrast to a clinical trial, these "extraneous" risk factors or confounders cannot be equalized between the exposed and control groups by the process of randomization. However, methods are available to exert some degree of control over confounding factors. These methods are: matching exposed and nonexposed persons on confounders during the selection process, and stratifying exposed and nonexposed groups into identical categories of each confounder in the analysis phase of the study. For example, in a study of heart disease risk due to smoking, an investigator might compare smokers with nonsmokers, both of whom are blue collar workers with less than college education who consume less than a specified amount of alcoholic drinks per week. To make valid comparisons of exposed and nonexposed groups in a cohort study, it is essential to identify potential confounders beforehand, measure their distribution in the study groups, and control for their effects in the final analysis.

After the clinical trial, the cohort study is one of the most convincing methods of associating risk with clinical outcome. Since exposed and nonexposed groups are chosen to be free of the outcome at the start of the study, the antecedent-consequent nature of the relationship between risk and disease outcome is not in doubt. Risk of disease due to exposure can be directly measured from the incidence rate in exposed compared with the incidence rate in nonexposed persons. Incidence rates themselves are the proper and direct measure of disease risk. Further, since the investigator selects the study population and observes this population over a follow-up interval, measurements can be made of risk factor or exposure before disease outcomes occur. Once disease occurs, a person may alter his habits or exposures accordingly, and the antecedent-consequent relationship will no longer obtain.

The major limitations of cohort studies are logistical: they require large sample sizes and long periods (often several years to decades) of follow-up. If a disease is relatively rare, i.e., an incidence rate of less than 1 per 100 per year, large populations of exposed and nonexposed persons must be studied to detect a doubling or tripling of incidence rates in exposed persons. Cohort studies typically require several hundred to several thousand study subjects even for research on the more common diseases. For more rare events such as breast or lung cancer, several thousand exposed and nonexposed persons must be followed over 10–20 years. The stronger the effect of the risk factor, the smaller the sample size can be. Yet even strong risk factors (such as smoking as a risk for lung cancer) require large sample sizes to study diseases where the annual incidence rate is less than 1 per 1000, as is true for most cancers. By extending the length of follow-up from one to many years, a smaller sample size can be studied, because the number of incident cases in a cohort study is determined by the population size, times the annual indicence rate, times the number of years of follow-up.

Case-Control Study

This approach to the study of disease risk is foreign to many clinicians, even though they have direct access to the subject material required for such studies. In the case-control study, the investigator argues from the fact of an effect to a likely cause, a backwards logic that is common in clinical experience. The objective of this approach is to compare prior exposures to a suspected risk factor among cases of a specific disease and among controls free of that disease but otherwise representing the population from which cases arose. Unlike the clinical trial and cohort study, the starting point of the case-control study is the clinical outcome. There is no prolonged follow-up period during which the investigator must observe sizeable groups of exposed and nonexposed, or treated and control, persons. Consequently, case-control studies require considerably smaller sample sizes than do cohort studies, and are therefore particularly efficient for studying disease of low frequency as is true for any organ-specific cancer.

Two major obstacles stand in the way of an investigator planning a case-control study: (a) obtaining accurate information on prior exposures of cases and controls, and (b) selecting controls representative of the source population from which cases arose. If exposure to the risk factor of interest has been documented in some record, such as a medical chart, the first of these obstacles can be coped with. Medical records often contain considerable information on prior risk factors, such as drug use, smoking and alcohol habits, family history of disease, and blood pressure. If medical records are the source of prior exposure data, controls will most likely be selected from a hospital or clinic population and this selection can lead to a biased representation of the risk factor experience of the population from which cases are derived. In general, if controls are drawn from a medically based source, selection biases can be minimized by obtaining controls from the broadest clinical base possible. Hence the investigator should avoid choosing controls from only one clinical service or from one set of clinicians, since the prior exposure histories of such controls is less likely to be representative of that in the general population than is true for a more diverse group.

Among the four strategies for investigating disease risk, clinical trials and case-control studies are the most feasible for clinicians. The former is particularly useful for evaluating therapeutic effectiveness whereas the latter is desirable for etiological research. Participants in a case-control study will generally be willing to respond to interviews or questionnaires regarding prior exposures, whereas clinical trials require subjects to receive blindly a drug or other treatment regimen versus a placebo, and this "experimentation" is usually met with more resistance. In some case-control studies, it is not even necessary to contact subjects, as medical records can provide the necessary data both on clinical outcome and prior exposures. A recent publication by James J. Schlesselman provides an in-depth discussion of this study strategy and is recommended reading for the serious clinical investigator (1).

Cross-Sectional Study

Whereas the three methods of clinical investigation described above are applicable for evaluating therapies or studying disease etiologies, the cross-sectional, or prevalence study is useful to describe the existing, or prevailing, state of affairs regarding the coexistence of risk factor (or of some other study factor of interest) and a clinical outcome. Like the cohort study, the cross-sectional study conceptually begins with groups of persons exposed and not exposed to a factor of interest, e.g., persons with high blood cholesterol versus persons with normal cholesterol, and proceeds to compare the simultaneous occurrence of the clinical outcome in the two groups, e.g., abnormal stress electrocardiograms. However, being cross-sectional, there is no follow-up interval over which persons initially free of disease are observed for the incidence of new clinical events. As a result, the investigator cannot be certain that the risk factor preceded the clinical outcome nor can the investigator directly measure the risk of disease in persons exposed to the risk factor. The number of cases observed by an investigator conducting a cross-sectional study is the sum of recent or "incident" cases and surviving "prevalent" cases. This sum is a measure of disease prevalence, which is determined not only by incidence rates but also by the duration of disease in affected persons. Consequently, a higher prevalence of disease in exposed persons, or persons with the risk factor of interest, may occur because of a higher incidence among exposed, because disease lasts longer in exposed than nonexposed persons, or because cure rates are greater in nonexposed persons. The prevalence of disease in a group, therefore, is not a direct reflection of risk for that disease.

Although cross-sectional studies present difficulties in assessing antecedent–consequent relationships and in interpreting the association between risk and clinical outcome, they are very useful for obtaining preliminary evidence for subsequent etiological studies. The investigator can obtain the required data at a single point in time. These studies are also valuable to obtain descriptive information on the proportion of persons having a risk factor of interest such as high blood pressure, smoking, and obesity, or on the prevailing use of a clinical service or of a drug such as the proportion of pregnant women examined within the first 8 weeks of gestation or taking vitamins during the first trimester.

Conclusions

This chapter has identified five basic steps that need to be accomplished while planning a research study. The investigator must (a) have an intense desire to know, (b) state the research objective clearly, (c) search the literature, (d) formulate operational hypotheses, and (e) select a study strategy. The aim has been to demonstrate that clinical research is highly deliberate, cannot be done casually, and requires discipline and meticulous attention to detail.

Experienced clinical investigators know that the hallmark of good research is cerebral effort, not numbers or swanky statistics. Our aim has been met if readers understand that the most important phase of a clinical research project is its design, where early decisions dictate the form and quality of data that are collected later. Time spent planning a study, and getting early consultation from methodological experts, is never wasted. Academic physicians are urged to take time to "think through" their research ideas *before* collecting data, perhaps with the aid of step-by-step guidelines (2). The conduct, and certainly the outcomes of clinical research will be improved by such a scholarly approach.

References

1. Schlesselman JJ. *Case-Control Studies: Design Conduct Analysis.* New York: Oxford University Press, 1982.
2. Gordon MJ. Research workbook: a guide for initial planning of clinical, social, and behavioral research projects. *J Fam Pract* 7:145–160, 1978.

13 Conducting A Research Study

William C. McGaghie

Earlier chapters in this part of the *Handbook* have discussed the contribution of clinical research to primary care, presented an example of a research project, and introduced the basic steps involved in planning a study. This chapter extends the previous writing by offering some practical suggestions about how to conduct a study once it is planned. The reason for including this chapter is to convey the idea that data are precious and hard to obtain. Their acquisition, analysis, and presentation to the scientific community stem not only from research that is thoughtfully designed, but also from projects that are carefully managed.

The chapter has three sections. The first section deals with project organization and management, including the development of timetables, personnel supervision, financial matters, and other administrative duties. Second, attention is turned to measurement of clinical variables. Issues of data quality are covered along with several different measurement strategies that are available to clinical investigators. The short third section covers practical approaches to data collection and offers tips about how to avoid problems like missing data and messy files.

Project Organization and Management

Clinical research needs to be conducted in an orderly fashion. There needs to be a carefully developed workplan and timetable coupled with clearly stated research procedures, personnel duties and responsibilities, and a plan for writing progress reports and journal manuscripts. All this requires the clinical investigator to give time and attention to project organization well before data are collected, especially if other persons such as research assistants, medical students, or nursing staff have a role in the work. Projects quickly deteriorate into chaos unless a clear-cut plan of action is established and followed.

Workplan

A project workplan lists in detail the many discrete steps that need to be accomplished for a study to reach a successful conclusion. An illustration is given in Table 13.1, based on the breast stimulation for the induction of labor study described in Chapter 11. The workplan has been highly condensed for presentation in this volume. Its true form would be much more precise identifying, for example, the specific data collection instruments to be developed in Year One and the individual pilot tests that would be undertaken. Some investigators develop workplans in such detail as to identify a member of the research staff who is responsible for each project task and an expected date of task completion. In particular, large research projects involving many people often require highly detailed workplans. The size and scope of a study will dictate the form and level of precision of the workplan needed to move a project from the planning phase through data collection, analysis, and report writing. Such

TABLE 13.1. Condensed workplan: Breast stimulation for the induction of labor study.

Year 1: Planning

The first phase of the project will involve the following development tasks.

1. Hiring staff to implement the project
2. Development of the project design using biostatistical and obstetrical consultation
3. Development of inclusion and exclusion criteria
4. Development of data collection instruments
5. Development of attitudinal questionnaires; patient consent forms; protocols for staff; etc.
6. Computer programming
7. Purchase and testing of equipment
8. Training of research assistant and nursing staff
9. Pilot projects to pretest methodology, equipment, and protocol
10. Preparation of reports

Year 1: Clinical implementation

Initiation of the second phase of the project—the study comparing breast pump versus nipple stimulation, involving 360 randomly allocated women. The tasks will include:

1. Supervision of the project—continued training, data collection, site visits, administration of attitude questionnaires
2. Data analysis and interpretation
3. Computer programming
4. Preparation of reports
5. Planning for second-year inpatient study

Year 2: Inpatient study

The third phase will include the implementation of the radomized trial of breast stimulation compared to orthodox therapy (300 patients). Tasks will include:

1. Supervision of project—continued training, data collection, administration of attitude questionnaires
2. Data analysis and interpretation
3. Programming
4. Preparation of reports

a workplan needs to be prepared in writing before a clinical investigation is started so that tasks, responsibilities, and expectations are known to everyone involved.

Timetable

A timetable complements a project workplan by showing the sequence and duration of the separate tasks embedded in a clinical investigation. A timetable can help the researcher visualize the work that needs to be done at different phases of a project and to plan ahead for busy and slack periods. Figure 13.1 presents a timetable that codifies the research tasks set forth in the previous workplan.

The timetable has several noteworthy features. It shows, for example, that some tasks such as hiring and training staff and pilot tests of equipment and procedures are expected to be episodic, covering fixed and clear-cut periods of time. By contrast, other research tasks such as report preparation, computer programming, and data analysis are cyclic. They recur on predictable occasions during the course of a study. Project supervision, the day-to-day management of research details and personnel, is also different because supervision must be continuous. Experienced investigators know that close daily attention to the planned (e.g., write progress reports) and unplanned (e.g., staff turnover) events that occur

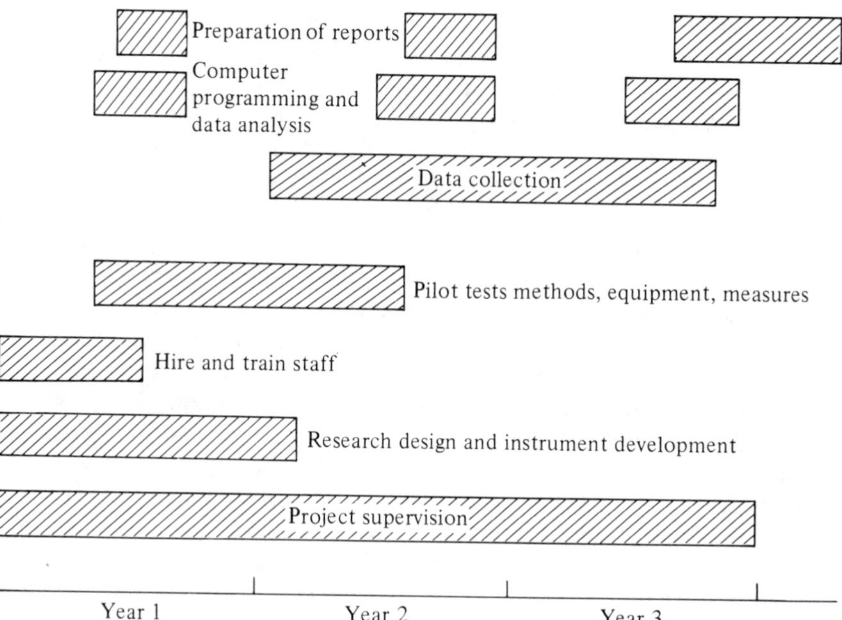

Figure 13.1 Research project timetable.

during a research project is essential for the study to be finished successfully.

Inexperienced investigators frequently fail to give enough time and attention to identifying the specific research tasks and the organization and management of project work that is reflected in a detailed workplan and timetable. Yet no clinical investigation will "run by itself" and unless it is carefully planned and managed, even the most elegantly designed scientific study may drift off course.

Personnel Supervision

People are a key ingredient in a clinical research project, a resource that may require as much of an investigator's attention as the technical details of a study. Academic physicians rarely work in solitude, neither in the clinic seeing patients, when teaching, nor when engaged in scholarship. Instead, individuals including clerical and nursing staff, research assistants, medical students, subspecialty fellows, and colleagues who are coinvestigators may be involved in a clinical study. A research team, where members rely on each other to accomplish mutual scholarly goals, has the potential to be far more productive than any individual investigator working alone. Conversely, a research team can be inefficient, unproductive, and full of friction without careful supervision of its personnel.

Similar to managing a clinical team, the academic physician who supervises research personnel needs to keep a number of simple rules in mind. First, productive research can occur only when research goals are established, put in writing, and endorsed by team members. Everyone needs to have a common understanding of the project's final destination. Second, a realistic workplan and timetable need to be prepared that takes account of team members' interests, skills, and ambitions. Although the principal investigator (PI) needs to be flexible enough to revise the workplan and timetable to accomodate unexpected problems (which occur inevitably), an unambiguous list of tasks and expected dates of completion should be on paper. Third, expectations about the work and workload for each team member (including the PI) should be made clear. The reward system also needs to be explained, usually in the form of salary for clerical staff and research assistants, and publication credit for students, fellows, and colleagues. Because publication credit often represents "academic currency," and can be a source of project discord, it is important for team members to agree about authorship early in a project's life. Fourth, a system of performance appraisal should be established similar to the procedures described in Chapter 3. No matter if the research team is large or just two or three people, all members deserve to know "how they are doing" in a fair, reliable, and regular manner.

Following these few rules will not guarantee success at clinical investigation but should enhance staff efficiency and morale. Morale is important and should not be taken lightly. A happy staff is far more likely to be productive than one where rivalries, bickering, and "hidden agendas" are the norm. Evenhanded personnel supervision is the best way to ensure that teamwork prevails, not conflict.

A final word about supervision of research personnel who work in clinical settings warrants explicit attention. Simply, it is imperative to educate the research staff about the importance of maintaining the confidentiality of research data, just as they would for clinical data that are used for patient care. The privacy of research subjects, many of whom are also patients, needs to be carefully protected. The consent of individual subjects to participate in a program of clinical research should be returned with a guarantee that the data they provide in confidence will not be made public in other than aggregate form. Research staff members need to remember to keep their files locked and their mouths closed.

Budget

Clinical research not only takes time, it also costs money. Sometimes the money is specifically appropriated to support a research project, usually in the form of a grant or contract. Frequently, however, clinical research is done without formal financial backing as when a community practitioner does a study to satisfy personal curiosity (see Chapter 10), or when an academic physician does an investigation "on a shoestring," piggybacking the work onto routine clinical duties. For example, Fletcher and Hamann (1), retrospectively evaluated the care given to 169 patients who presented to an emergency room (ER) with sore throat. Their data indicate that ER housestaff managed the patients properly. However, physician effectiveness was reduced by a slow bacteriology laboratory and inefficient hospital administrative practices, health care delivery problems that have straightforward solutions. To paraphrase one of these investigators, the only cost involved in this highly useful clinical study was "shoeleather" (S. Fletcher, personal communication).

Similar to designing the scientific features of a research project, the budget needed to conduct a study—however modest—should be planned carefully. Whether or not a project needs a formal budget clearly depends on its size, which usually means the number of professional, technical, and clerical people involved. For work supported by a grant or contract, the lion's share of research money is used to pay staff salaries and fringe benefits. Other customary budget items include research materials or apparatus, telecommunications, mainframe computer time, printing, and travel to professional or scientific meetings (see Chapter 14). The exact breakdown of expenditures for a research project will, of course, stem from the scope and ambition of its workplan and timetable. Small, yet highly significant, clinical studies can be done at very low cost (e.g., duplicating a questionnaire) by physicians working in a community hospital or office practice setting. By contrast, controlled multicenter clinical trials that are needed to test the efficacy of proposed cancer chemotherapeutic agents frequently have multimillion dollar budgets.

Inexperienced academic physicians who anticipate the need to obtain information about setting up and managing a research project budget can turn to a number of sources for help. The first source is experienced colleagues, other investigators who "know the local system" and can give good advice about what costs to expect and how to plan a budget that takes the costs into account. Most university departments, and administrative units in community hosptials, have an accountant or bookkeeper who is responsible for financial management. Such a person can be a valuable source of advice during budget preparation and as money is spent while a study is underway. Finally, the use of spreadsheets in microcomputer software packages can simplify project planning and budgeting. Academic physicans who become familiar with this new technology will be able to plan and conduct clinical investigations with greatly increased efficiency. (See Chapter 15 for a more thorough discussion.)

Writing Reports

As shown in Figure 13.1, report writing occurs throughout a research project, not just at its conclusion. Reports of project activities are needed for scientific and administrative purposes, and sometimes for public relations. In any case, just like the other managerial chores that have been discussed, report writing is best done according to a carefully prepared plan.

Seasoned academic investigators know that research project reports may be catagorized in at least three ways: required, optional, and expected. *Required reports* are those that need to be submitted if the researcher wants to stay in operation. Examples include progress reports to external funding agencies; for one's dean, department head, or other top administrator; and for the local committee that has jurisdiction over research on human subjects. In addition, a final project report is usually required for studies supported by a foundation or a government agency. *Optional reports* are written at the investigator's discretion. They include technical memoranda, notes "for the file," results from pilot studies, and other types of documentation about project activities. Preparing and filing such optional reports usually makes writing required reports much easier later on. *Expected reports* are manuscripts that are written to accomplish scientific or professional goals, rather than to fulfill adminsitrative purposes. Papers and abstracts for presentation at professional meetings, journal articles, and scholarly monographs comprise most of this category. As noted in Chapter 3, regular production and publication of "expected" reports is closely linked to career advancement in many academic medical settings.

The surest way to promote productivity in report writing is to start with a plan or agenda that includes deadlines. Table 13.2 illustrates such

TABLE 13.2. Research project writing agenda.

Reports	Audience	Eligible outlets	Submission deadline
1. Project final report	Funding agency	—	June 1986
2. Annual report about protecting subjects' rights	Human Subjects Committee	—	June, 1986
3. Presentation of clinical trial data	Obstetricians and gynecologists, primary care clinicians	*Am. J. Obstet. Gynecol.* *Br. J. Obstet. Gynecol.*	November, 1986
4. Historical review about approaches to manual induction of labor	Medical historians	*Medical History*	March, 1987
5. Use of oxytocin based on literature review and study data	Primary care physicians	*JAMA* *J. Fam. Pract.*	June, 1987
6. Prenatal hyperstimulation of uterus: first year data	Obstetricians and gynecologists	*Am. J. Obstet. Gynecol.*	July, 1987
7. Manual alternatives to biochemical induction of labor	Laywomen of childbearing age	Women's magazines, e.g., *Family Circle* *Women's Day*	September, 1987

a writing agenda, again based on the breast stimulation for the induction of labor study described in Chapter 11. Note that each report is written for a particular audience. Also observe that the audiences are served by different outlets (professional journals and lay magazines).This means that primary care physicians are more likely to read *JAMA*, the *New England Journal of Medicine*, or the *Journal of Family Practice* than, for example, the *Journal of Nervous and Mental Disease.* Articles intended for primary care physicians should, therefore, be carefully tailored for journals that they read (see also Chapter 17). Finally, the writing agenda gives a deadline for each report. Deadlines for the reports to the Funding Agency and the Human Rights Committee are imposed and cannot be changed; others are established at the discretion of the investigator.

Scheduled report preparation can also be aided by several activities that if done routinely can make large writing tasks less onerous. Clinical investigators are urged to keep up-to-date computer or paper files of project notes and memos. Keeping a daily logbook of project events (e.g., visitors, articles read, consultations) helps, in the long run, to place seemingly isolated workdays in a larger perspective. Careful documentation of all computer files and maintenance of a valid codebook of research variables are also essential. The bottom line message is that efficient report writing, as all other steps involved in clinical research, requires the investigator to plan carefully and thoughtfully manage the enterprise.

Measurement

Variables and Measures

Sound research depends upon an investigator's ability to measure key variables and to use the results of those measurements —data—to reach conclusions about the research question. The measures used in clinical research represent variables of interest. Measures are an operational expression of clinical variables that are frequently intangible. They may be numerical or categorical: plasma glucose for diabetes, intraocular pressure for glaucoma, WAIS for intelligence, enlarged liver for alcoholism. Some variables can be measured by different methods. Depression, for example, can be measured using psychological instruments including self-report scales, ratings by interviewers, and behavioral observations (2). Depression can also be measured biochemically, using the Dexamethasone Suppression Test (3). Different measures of the same variable should agree before clinical investigators place much stock in either the variable or its methods of assessment.

All of the variables that are identified in a research protocol need to be measured carefully. This holds for the outcome or *dependent* variable in which change or association is being studied and for the *independent* variables that are suspected of producing the change or association. Richard Riegelman points out that four criteria must be satisfied to obtain valid measures of a research outcome (4).

1. The investigator must use a measure of outcome that is *appropriate* to the question to be answered.
2. The measurement of outcome must be *precise.*
3. The measurement of outcome must be *complete.*
4. The outcome of the study must not be influenced by the *process of observation.*

These criteria should also be met by measures of the independent variables used in clinical research.

Table 13.3 identifies an illustrative set of five clinical research variables and sources of data, typical measurements, and an index of data quality for each measure. Careful investigators take time to think through the measurement issues involved in their research to be certain the measures employed are the best ones available.

Scales

Measurement in clinical research is done with different degrees of precision. Variables such as temperature, blood sodium, thyroid uptake, and grip strength can be measured with great precision, on a graduated scale, with equivalent and uniform intervals at all scale levels. This is termed *continuous* or interval-level measurement. It represents the investigator's measurement ideal on grounds of precision and for the sophisticated statistical analyses that can be performed on continuous data.

However, many important clinical variables defy expression on a continuous measurement scale. This situation does not reduce the clinical or scholarly importance of such variables, it only shows that "hard" science has yet to encompass fully clinical research. Here, the investigator must resort to measures of variables that are either *ordinal,* which involve data expressed as ranks, or *nominal,* which involve categorization without numerical order. Examples are shown in Table 13.4.

Much valuable clinical research involves the use of data measured at the nominal and ordinal levels of measurement. For example, Nancy Nelson and her colleagues conducted a randomized clinical trial comparing the Leboyer approach to childbirth with the conventional method of delivery (5). Among an array of biomedical and psychological measures used in this study were several nominal measures of maternal morbidity. Here, Leboyer and control group mothers were categorized according to the presence or absence of five factors: postpartum hemorrhage, third-degree extension of episiotomy, infected episiotomy, endometritis, and urinary-tract infection. The groups were then statistically compared on these categorical measures. The analysis did not reveal a statistically (or

Table 13.3. Measurement issues in clinical research.

Variable of interest	Source of clinical data	Typical measurements	Index of data quality
Cancer	Clinical impressions	Presence of enlarged liver	How careful and thorough is the examiner?
		20 lb. weight loss	Is the finding reproducible?
Myocardial infarction	Symptoms	History taking	Carefulness of examiner
	Lab tests	SGOT, LDH, EKG	Lab precision
Emphysema	Pulmonary function tests	FEV, RV, FVC	Lab precision
Alzheimer's disease	Psychological tests	Activities of daily living	Reliability and validity
Lung cancer	Medical chart	Biopsy evidence	Care in recording data
		Radiographic evidence	Definiteness of work-up
		Smoking history	

TABLE 13.4. Measurement scales.

Scale type	Examples
Continuous (interval)	Blood sugar, blood pressure, age, weight, height, intelligence
Ordinal (ranks)	Frequency of smoking, intensity of heart murmur, stage of tumor
Nominal (categories)	Diagnostic groups, race, sex, blood groups

clinically) significant difference between the two groups for any of the morbidity categories. However, the analysis does illustrate how clinically meaningful data measured on a nominal scale can be incorporated into a research study. Moses and colleagues make a similar argument for clinical research that involves measurement of ordinal variables (6).

Measures of Effect

Measures of effect are another way that results of a study can be expressed, especially when measurement is done at the nominal level. These indexes describe the degree of association between an independent variable, often a risk factor such as cigarette smoking, and a clinically important outcome (dependent variable) such as the presence or absence of lung cancer. Familiar indexes including *attributable risk* and *relative risk* are often reported for prospective investigations such as cohort studies and clinical trials. Attributable risk answers the question, "What is the additional risk of developing disease (incidence) following exposure to a risk factor, over and above that experienced by people who are not exposed?" Relative risk responds to a slightly different question, "How many times more likely are exposed persons to get the disease relative to nonexposed" (7)? A measure of effect similar to relative risk, the *odds ratio* , is ordinarily reported for retrospective, case-control studies.

Table 13.5 illustrates the calculation of attributable risk and relative risk for a disease or other outcome of clinical interest (e.g., disability) in prospective studies. The intent is to show that useful research—in this case, linking exposure to a risk factor with later manifestation of disease—can be done using such common measurements as clinical observations. Clinical investigators need only keep a systematic "boxscore"of patients exposed to a risk factor who later develop *and* fail to develop a disease entity to complete a useful prospective investigation.

Reporting research results as measures of effect makes sense clinically because they describe findings in a way that can be applied directly to patient care. The measures help the clinician/investigator to assess rates of new disease that are due to measureable risks that patients experience. As a result, indexes such as attributable risk and relative risk are more informative to the clinician than the probabilistic results of statistical

TABLE 13.5. Two measures of effect for prospective studies.

		Disease or outcome	
		Present	Absent
Risk Factor Exposure	Present	A	B
	Absent	C	D

$$\text{Attributable risk} = I_E - I_{\bar{E}} = \frac{A}{A+B} - \frac{C}{C+D}$$

$$\text{Relative risk} = \frac{I_E}{I_{\bar{E}}} = \frac{\frac{A}{A+B}}{\frac{C}{C+D}}$$

Where I_E = Incidence of the outcome in the exposed group
$I_{\bar{E}}$ = Incidence of the outcome in the nonexposed group

tests that journals usually report. Measures of effect address clinical *applications,* statistical tests suggest medical *implications.* Both methods of reporting research results have an important role in clinical investigation yet it is important to recognize their differences.

Interested readers are urged to consult other sources for a more detailed treatment of measures of effect than can be presented here . Several basic textbooks on clinical epidemiology are particularly recommended because they are written by and for active clinicians (4, 7). A more exhaustive discussion about measures of effect is available in an excellent epidemiology textbook authored by Kleinbaum et al., (8).

Reliability

An important index of the quality of measures taken for clinical or research purposes is their reliability. An investigator needs to establish the reliability of the measures used in a study to make a convincing argument that the study findings are trustworthy.

Reliability of measurement can have several slightly different meanings depending on the type of measurement apparatus being used (e.g., spirometer versus questionnaire), the scale of measurement involved (nominal, ordinal, continuous), and the design of the study in which the measures are used (e.g., cross-sectional versus longitudinal). In general, however, reliability refers to the precision, consistency, or reproducibility of data derived from a measurement procedure. Reliability coefficients are customarily expressed on a scale from zero to one with higher values denoting more trustworty data. A convenient way to think about reliability of measurement is as a signal-to-noise ratio. Reliable data send a clear, static-free signal. As the amount of static (measurement error due to random or systematic events) in the signal grows, the "message" contained in one's data becomes increasingly difficult to unscramble.

Serious investigators worry about reliability of measurement and take steps to ensure that their research data are trustworthy. Laboratory apparatus is calibrated frequently, observers are trained to interpret and record events (e.g., doctor–patient interactions) in the same way, and questionnaires are pilot tested before being used in a full-scale study. The use of procedures such as these, along with routine estimation of reliability coefficients for data derived from research measures, all increase the odds that high-quality data will be acquired. There is no quantitative procedure that will remove the error from ("clean up") a set of data once the measurements have been made.

Validity

Statements about the validity of data derived from a measurement procedure refer to the way in which the data are used or interpreted for a specific purpose. Thus the validity of a measurement is to a certain extent context bound, and data acquired for one purpose are unlikely to be useful for another. For example, data from a CBC can be used with validity to diagnose anemia, just as abnormal values for BUN and serum creatinine are valid signs of end stage renal disease. However, none of these laboratory tests is a valid measure of migraine headache, COPD, or depression because valid diagnosis of these clinical entities relies on different indicators.

For clinical and research applications, the reliability of a measure needs to be established before it can be stated that the measure is valid for a particular purpose. Clinical investigators need to show that their data convey a clear signal; only after reliability is shown can data be used to address a specific clinical or research problem. This is analogous to establishing the predictive value of a diagnostic test. Before a diagnostic test's predictive value can be determined, a researcher needs to demonstrate that data from the test have acceptable sensitivity and specificity, i.e., that the test correctly rules in or rules out disease according to a known "gold standard." Once the test data are shown to distinguish clearly the normal from the abnormal, their predictive validity is greatly enhanced (4,7).

The validation of a measure—actually, of the data that a measure yields—is therefore a judgment call. The investigator needs to take account of the reliability of the measure, the clinical or research problem at hand, and other factors such as the cutoff between normal and abnormal when reaching a diagnosis. These and many other elements can influence the proper interpretation and use of clinical or research measures, which is the essence of the measurement validity issue.

To illustrate, consider the presumed clinical correlation between patient reports of symptoms and the presence or absence of pathology as determined by objective tests. The presumed relationship is often justified, as in the association of angina and serum enzymes in diagnosing myocardial infarction. However, symptoms are not always valid indicators of the presence or absence of disease. For example, Berry et al. (9) present data that show no relationship exists between symptoms of byssinosis (chest tightness, difficulty breathing on Mondays) and annual declines in objectively measured lung function (FEV) in a large sample of British textile workers. These data indicate that byssinosis symptoms are not necessarily a valid measure of progressive lung disease and treating them as such would be a misinterpretation of the evidence.

Measures and Statistics

Although it is beyond the scope of this book to give statistics more than passing mention it is useful to point out the interplay between measures used for clinical research and the statistical analyses they allow. Table 13.6 illustrates the relationship between levels of measurement for clinical data and the analyses that can be done on measures having varied precision. At all three measurement levels shown, the research purpose is to assess the relationship between cigarette smoking and emphysema.

The study suggested in Table 13.6 could be either prospective, concurrent, or retrospective (depending,of course, on the availability of accurate records). Its design is left to the investigator's discretion. However, the way in which the variables of interest—smoking and emphysema—are conceptualized and measured will determine the statistical analysis the investigator can use. Getting precise, continuous measures of smoking (urine cotinene) and emphysema (pulmonary function tests) would generally be preferred to measuring variables at lower levels because the Pearson correlation is a slightly more rigorous statistic than its counterparts for ordinal and nominal data. Although the results of the study would be very similar if it were replicated across the three mesurement levels, most investigators would choose to perform the research using continuous data.

The link between measurement levels of research variables and statistical analysis of resulting data exists not only in studies where relationships are sought, but also in studies that assess group differences. Thus the commonly used t statistic is appropriate to determine if two groups differ, based on a variable measured on a continuous scale. Other statis-

TABLE 13.6. Assessing the relationship between cigarette smoking and emphysema using measures at different levels.

<table>
<tr><th>Measurement level</th><th colspan="2">Measures</th><th>Statistic</th></tr>
<tr><td>Continuous</td><td colspan="2">Urine cotinene (smoking), pulmonary function tests (emphysema)</td><td>Pearson correlation</td></tr>
<tr><td>Ordinal</td><td>Intensity of smoking

High
Moderate
Low</td><td>Degree of wheezing

Grade III
Grade II
Grade I
None</td><td>Spearman or Kendall Rank-order correlation</td></tr>
<tr><td>Nominal</td><td colspan="2">Smoking</td><td rowspan="4">Phi coefficient, relative risk, attributable risk, or odds ratio (if study is retrospective)</td></tr>
<tr><td>EMPHYSEMA</td><td>Present</td><td>Absent</td></tr>
<tr><td>Present</td><td></td><td></td></tr>
<tr><td>Absent</td><td></td><td></td></tr>
</table>

tics are used to assess group differences when ordinal and nominal measures are employed. The point to remember is that the statistical analysis used in a clinical investigation is a direct expression of the study's research design *and* measurement procedures.

There are many textbooks about statistics that are available for use by academic physicians. Two of the best, and most readable, are volumes by Fleiss (10) and Swinscow (11). In addition to using these or other written sources, inexperienced researchers should contact a seasoned colleague or statistical consultant for advice about measurement and statistical analysis of data. Such methodological advice should be secured early, when a study is being planned, to be of maximum benefit.

Data Collection

Once a study is planned including formulation of objectives or hypotheses, development of the research design, identification of measures and data analysis procedures, and preparation of a workplan, the investigator is ready to begin collecting data. This seemingly simple act involves taking and carefully recording measures of one's research variables, usually under uniform conditions. To illustrate, clinical investigations that include measurements of blood gases need to set up a standardized procedure for drawing blood, getting the samples to the laboratory for analysis, receiving the lab reports, recording the data in the research file, and being certain that each subject's blood gas data are correctly entered in the file, i.e., merged with other research data about the same person. Studies involving other types of measures such as questionnaires or data extracted from medical records should follow similar procedures. In data collection, as in all other phases of the research process, compulsiveness is a virtue. Productive researchers collect and file their data with great care and attention to detail, and see to it that assistants and coinvestigators share the obsession.

Data collection needs to be a deliberate, planned activity. It should be carefully thought out, economical of time and energy and, in particular, efficient from the standpoint of research subjects. For example, researchers should never ask patients who consent to be subjects to make unnecessary clinic visits just to serve research goals. Not only are such tactics questionable on ethical grounds, but they also place the burden of research on patients and increase the odds that subjects will actively or passively withdraw from the study.

Good investigators stay "close" to their data. They never hesitate to gain first-hand experience with measurement chores such as distributing questionnaires, interviewing patients, drawing blood, or obtaining sputum samples. Not all data collection tasks, even the most mundane, are delegated. A useful rule-of-thumb about data collection in clinical research is that, if possible, investigators should never take a measure on research subjects that they have not experienced personally. One's respect and gratitude toward subjects are quickly reinforced by the experience of filling out long questionnaires seeking sensitive personal information or by experiencing repeated "sticks" for venous blood samples.

Data organization and management are covered at length in Chapter 15. It is enough to mention here that researchers should design their data abstracting and coding forms far in advance of gathering any research data. Completing this step requires the investigator to decide exactly which data will be collected, which potential research data will be ignored, whether fresh responses are needed from subjects or if archived data (e.g., from charts) can be used, and how to physically gather research information and organize it into files.

In conclusion, clinical investigators need to remember that a source of data is not a measuring instrument (12). Investigators must understand that data from different sources vary widely in quality (validity), especially if the data are gathered for purposes other than research (13). Thus one's early decisions about study variables, measures of the variables, and data collection procedures will together affect the scientific integrity and usefulness of a research project. Such decisions should be calculated carefully.

References

1. Fletcher SW, Hamann C. Emergency room management of patients with sore throats in a teaching hospital: influence of non-physician factors. *J Commun Health 1*:196–204, 1976.
2. Moran PW, Lambert MJ. A review of current assessment tools for monitoring changes in depression. In *The Assessment of Psychotherapy Outcomes*. Lambert ML, Christensen ER, DeJulio S. (Eds). New York: John Wiley, 1983.
3. Shapiro MF, Lehman, AF. The diagnosis of depression in different clinical settings: an analysis of the literature on the dexamethasone suppression test. *J Nerv Ment Dis 171*:714–720, 1983.
4. Riegelman, RK. *Studying a Study and Testing a Test*. Boston: Little, Brown, 1981.
5. Nelson NM, Enkin MW, Saigal S, Bennett KJ, Milner R, Sackett DL. A randomized clinical trial of the Leboyer approach to childbirth. *N Engl J Med 302*:655–660, 1980.
6. Moses LE, Emerson JD, Hosseini H. Analyzing data from ordered categories. *N Engl J Med 311*:442–448, 1984.
7. Fletcher, RH, Fletcher SW, Wagner EH. *Clinical Epidemiology— The Essentials*. Baltimore: Williams & Wilkins, 1982.
8. Kleinbaum, DG, Kupper LL, Morgenstern H. *Epidemiologic Research: Principles and Quantitative Methods*. Belmont, California: Wadsworth, 1982.
9. Berry G, McKerrow CB, Molyneux MKB, Rossiter CE, Tombleson JBL. A study of the acute and chronic changes in ventilatory capacity of workers in Lancashire cotton mills. *Br J Industr Med 30*:25–36,1973.
10. Fleiss JL. *Statistical Methods for Rates and Proportions*, ed 2. New York: Wiley, 1981.
11. Swinscow TDV. *Statistics at Square One*, ed 5. London: British Medical Association, 1979.
12. Fiske DW. A source of data is not a measuring instrument. *Psychol Bull 84*:20–23, 1975.
13. Romm FJ, Putnam SM. The validity of the medical record. *Med Care 19*:310–315, 1981.

14 Resources for Clinical Research

JANE E. ARNDT

Research productivity is the outcome of a wide range of individual and organizational factors. Prerequisites for an effective research career in a clinical setting, as in other academic settings, include the individual investigator's talent, ideas, and initiative, an organizational unit in which research is an acknowledged priority, and access to adequate resources.

This chapter focuses primarily on the fourth prerequisite, resources that facilitate research in clinical settings and strategies for obtaining necessary resources. The first section discusses the importance of colleagues as resources for research—as consultants, as collaborators, and as peer reviewers. The second section examines research support that may be available within departments and institutions. The third section focuses on external funding of research. The fourth section, Research Environment, summarizes organizational factors that help to promote research and reflect the priority the organization gives to research.

Resources such as those noted above are required throughout the research process. At different points in this process, different types of resources are of particular value. Table 14.1 summarizes key requirements during three phases of the research process: project development, implementation of the study, and dissemination of research results through publication or presentation at professional meetings.

Colleagues as Research Resources

Although discussions of resources for research usually focus on funds and facilities, colleagues with research expertise are an equally valuable resource. Experienced colleagues can help shape a study so that its impact and its chances for successful completion are enhanced. Their contributions may range from advice about local resources or strategies for obtaining external funding to extensive participation throughout the entire research process. Although academic research is based on a norm of collegiality, many institutions are characterized by internal competi-

TABLE 14.1. Resources for research: An overview.

Phases of a research project		
Development	Implementation	Dissemination
Colleagues:		
Consultants re: research question, study design and procedures, data analysis plans	Consultants re: study procedures, data analyses	Consultants re: reporting of methods and findings
Coinvestigators (involvement in all phases of research project)		
Peer review re: research question, project proposal		Peer review re: presentation of study findings
Institutional resources:		
Library (especially computerized bibliographic searches to identify relevant literature and possible funding sources)		Library for relevant literature published since original search and to identify appropriate audience for papers
Human Subjects Committee re: project approval for protection of human rights and safety	Departmental staff and equipment to assist in carrying out the study	
	Data processing facilities	
Students to assist with review of literature, proposal preparation	Students to assist with data collection and analysis	Students to assist with preparation of articles/ presentations
Research funding:		
Identification of one or more appropriate funding sources; application for funding	Funding for project	Final report to funding agency

tion for scarce resources. The following discussion assumes that a traditional collegial model can have mutual benefits for both novice and experienced investigators.

Consultants

A consultant may serve any of a wide range of functions during planning and implementation of a research project. Appropriate roles for a consultant vary with the nature of the project, the skills of the principal investigator, the breadth of expertise of the proposed consultant, and the funds available. A consultant's time commitment can range from 1 or 2 days of technical assistance on a limited question related to study design or statistical analysis to continuing regular participation during all phases of the project.

Consultants offer resources of several types. A research consultant can complement the principal investigator in training, perspective, or skills.

For example, a behavioral scientist consultant may offer expertise in attitude measurement that complements a physician researcher's clinical knowledge. A consultant within the same field can provide a complementary area of expertise, such as in-depth knowledge of a specific topic or of principles of research methodology. Consultants are often selected for their ability to supplement areas in which the primary researcher needs additional support. For example, a consultant with an established research reputation may provide greater credibility for a study proposed by a new investigator. A consultant who has extensive experience in a specialized data collection technique that is relatively new to the principal investigator can provide essential guidance on its applications.

Although the involvement of a consultant in clinical research can be rewarding both to the principal investigator and to the consultant, clear negotiation and agreement about their respective roles and rewards are required. It is particularly important to identify the specific tasks or areas of responsibility for which a consultant is being sought during this negotiation. Lack of initial clarity is a frequent source of later problems that may become serious enough to jeopardize the project. In general, although consultants can offer advice on technical issues or on the general direction of the study, they do not carry out the study or make all the major decisions. If this level of involvement is necessary, the role of coinvestigator is more appropriate.

If external funding is being sought for a project, a letter of agreement from the prospective consultant will usually be needed. Such a letter for inclusion in a grant proposal confirms the consultant's interest in the project but ordinarily does not include all aspects of the working relationship between consultant and principal investigator. Whether or not external funding is involved, a detailed letter of understanding between the principal investigator and the consultant is useful. Such a letter might outline: (a) the extent of the consultant's expected participation stated in terms of time and/or effort; (b) payment (if any) and the basis for payment; e.g., at regular specified intervals or contingent on satisfactory completion of a defined task such as a draft questionnaire or a report summarizing specified statistical analyses; (c) agreements on access to data during and after the study period; and (d) publication guidelines regarding issues such as authorship or prior approval of manuscripts before submission for publication. For a project involving one or more coinvestigators, all investigators should concur with this letter of understanding.

Although the need for outside consultants may become apparent only after a study is underway, it is especially valuable to bring in consultants during the planning phase. Statistical consultation *before* data collection will help ensure that the desired analyses can be performed. Many statistical consultants are reluctant to become involved in a study that has progressed beyond the initial design stage.

Careful selection of consultants is a prerequisite for their effective involvement. Several strategies are useful for identifying suitable candidates. Regular contacts with other departments or nearby institutions (for example, by attending seminars or receiving their newsletters) provide an avenue for identifying people with similar research interests or appropriate skills and for establishing collegial relationships. If these outside contacts have not yet been established, departmental colleagues may be able to suggest local faculty with shared interests. Likely sources of consultants include: (a) other medical school departments, especially for a clinical study; (b) behavioral science departments (sociology, anthropology, psychology) at a nearby university, college, or community college; (c) other health professional schools or divisions such as nursing, pharmacy, public health; and (d) specialized research institutes, such as health services research centers or institutes specializing in the study of specific diseases.

For many research projects, interested and able consultants can be found in the principal investigator's home institution or in nearby universities or colleges. If local resources are sparse or the research is highly specialized, attendance at professional meetings or a review of the pertinent literature to identify those working actively in a topic area will help to locate likely consultants outside the principal investigator's locale. Under these circumstances, consultation takes place primarily by telephone and correspondence unless a project has access to funds for consultant travel.

Consultants have special value for new investigators. If external funding is being sought for the project, a consultant with a recognized reputation in the field may make the difference between a grant and no grant by assuring the funding agency that the inexperienced investigator will not be conducting the project in isolation.

Collaborative Research

The earlier discussion focused on a situation in which the principal investigator for a study engages one or more consultants to provide expertise in certain areas. In a collaborative research situation, commitments and responsibilities on the part of each investigator are equal. This may involve collaboration within a department or between departments or institutions. The general principles for establishing working relationships for consultation and collaboration are similar, although the mechanisms will differ and the ratio of costs to benefits may not be the same.

In considering the benefits and costs of collaborative research, the value of collaboration must be weighed in terms of both the individuals and their departments or organizations. Effective collaboration can yield the advantages noted earlier for consultants, with the added benefits of closer involvement throughout the study period and greater commitment to a project's success. For example, collaboration between physicians and behavioral scientists can be productive and stimulating because they bring different perspectives to the research question and different conceptual frameworks with which to organize data. The benefits of collaborative research can accrue to organizations as well, building trust and organizational linkages that can transcend a single project. In addition, members of a department or unit who lack experience in administration of funded research may gain valuable insight through collaboration with more experienced administrators.

The costs of collaboration, like the benefits, may accrue either to the investigators involved or to their organizations. Partial loss of control over the study—either intellectual control or administrative control—may sometimes be seen as a disadvantage of collaboration. Timely progress of the research may be hindered by a need to schedule frequent meetings between the coinvestigators, while support staff are unable to proceed with the work of the project. Collaboration between individuals who are unable to agree on key aspects of the research or on their relative contributions may raise barriers between departments that continue after the end of the project. Another less direct cost of collaborative research is the dilution of such benefits as prestige, grant funds, or bargaining power within the larger institution.

Successful research collaboration may be facilitated or hindered by a number of factors. Effective prior working relationships between individuals and departments facilitate the process, as do congruent interests and priorities of the coinvestigators. A shared perception that collaboration will enhance access to scarce resources increases individual and organizational commitment. On the other hand, competition or conflict between departments may hinder successful collaboration even in the presence of cooperation between investigators. Organizational policies and bureaucratic alliances may make collaboration difficult to carry out, especially if external funds are involved.

Establishing a mutually rewarding research collaboration involves many of the same considerations that are important when engaging a consultant. Like other partnerships, research collaboration is ideally based on commensurate benefits for each participant. Although benefits do not have to be of the same type for each partner, the exchange must be perceived as equitable by both parties. For example, a department may offer access to a desirable patient population in order to secure a collaborator whose reputation in a specific clinical area will help win grant funds.

Clearly negotiated agreements on rights and responsibilities are particularly important when a coinvestigator has substantial involvement in a research project. Responsibility for timely completion of research tasks, procedures for making decisions that alter the previously established protocols, access to data during and after the project period, and publication rights (including prior review of manuscripts and questions of authorship) are areas in which misunderstandings and acrimony are especially common.

Peer Review

The process of peer review—an objective evaluation by fellow scientists—is the method that many funding agencies use to reach decisions about the allocation of money. Refereed journals use a similar procedure to select articles for publication. Some institutions have administrative mechanisms that offer a limited form of peer review of research. These include reviews for protection of human subjects or for approval of proposed studies using specialized facilities such as Clinical Research Units.

In addition to these formal mechanisms, the principles of peer review can be applied to an informal but still rigorous critical process among colleagues in a department. To provide valuable feedback, peer reviewers do not have to be experts in the investigator's field or possess in-depth knowledge about the specific research topic. They may instead be colleagues respected for inventive ideas, astute criticism, good judgment,or writing ability.

Informal, voluntary peer review is particularly valuable during research project planning and later during the preparation of reports. At any time in the research process, peer review can provide feedback about the study itself (for example, the research design, data collection techniques, or statistical analysis) and about the wider implications of the study (for example, the study's clinical significance or its value to the discipline). At certain times, some of these areas are of special importance. During proposal development, peer reviewers can focus particularly on the value of the research question in the context of existing work in the field and the soundness of the proposed research design and its planned implementation.They can also provide feedback about whether a proposal for outside funding is clear and well documented and whether the requested funds and time period appear reasonable for the scope of the project.

During preparation of final reports, papers, or presentations, peer reviewers can evaluate the soundness and appropriateness of the statistical analyses and can help assure clear and accurate reporting of study procedures (including any limitations on data quality or generalizability) and conclusions that are clearly drawn from the data at hand. Reviewers can also comment on a manuscript's suitability for its intended audience and perhaps suggest alternative means of disseminating research results (for example, in a roundtable discussion at a professional meeting instead of a journal article).

Departmental research committees offer one mechanism for organized peer review. In addition, other sources of peer review are typically available to an investigator, especially in a university setting. These include colleagues within one's department, elsewhere in the organization or the local area, and colleagues within a wider network established through contacts at national meetings. Academic custom encourages busy clinicians to put critical thought into a review (usually under time pressure) in exchange for future review of their own work.

Novice researchers are often reluctant to submit their work for peer review because of concern about criticism and about abuses of the process of collegial review (such as theft of ideas). Although abuses can occur, the process usually provides an objective assessment of the strong and weak points of a study under development and may yield suggestions for its improvement. Participation in peer review can serve as a learning experience for both the reviewer and the reviewed, helping sharpen the analytic skills of both.

Organizational Support of Research

In addition to collegial advice and perhaps external funding (discussed in the next section), other resources are needed for the development and implementation of research in clinical settings. These research support resources, which may be available within a department or organization, include staff, equipment, library resources, and data processing facilities.

Departmental Resources

For a pilot study or other project that does not require extensive data collection, resources within the investigator's department may be sufficient to carry out the entire project. Even if departmental resources are not formally designated for research support, there may be personnel and equipment that are available for research uses on a limited basis.

Personnel available in clinical departments typically include research staff, administrative and clerical staff, nursing staff, and others such as audiovisual specialists. Some staff members may have formal training or prior experience in research. Others are familiar with clinic procedures or medical records and can offer valuable suggestions for planning a study that is realistic in that setting.

A clear work plan that details tasks to be accomplished, the skills required to accomplish them, and time requirements for each phase of the study is helpful to identify optimal ways to use existing personnel in a proposed research project. This work plan can then be used as a basis for negotiating research support from departmental staff who are assigned to other responsibilities. Skills of available personnel can be compared with the skills needed to perform the necessary research tasks. If staff members do not already possess the required skills, additional time and supervision for a training period must be requested in addition to the time commitment required for the project.

In addition to technical skills, a key feature of such training would be to educate research assistants and associates about protecting the privacy and confidentiality of persons who consent to serve as research subjects. Because instrument development, data collection, data processing, data analysis, preparation of audiovisual materials, and drafting reports or articles require different capabilities, it is likely that staff members without prior research experience can be most effectively involved in a project episodically, during specific limited phases.

Equipment that is used primarily for routine department operations may be available on a shared basis if research tasks can be scheduled not to conflict with regular uses. Access to essential equipment is important for both funded and nonfunded projects, because funding agencies often place severe restrictions on purchase of general purpose equipment. Microcomputers or computer terminals, videotape or audiotape equipment, and laboratory facilities under departmental control are often accessible for research use with prior arrangement.

Institutional Resources

Institutional resources vary greatly beyond the departmental level. A broad range of services is typically available to researchers in university settings. Community-based residency programs may have formal links with local or regional universities or colleges that offer some or all of the same types of support. Libraries, data processing facilities, and research assistance provided by students are resources of particular importance in any setting.

Libraries and library personnel are used most intensively during the project planning and reporting phases. Reference librarians at specialized health science or medical libraries can direct researchers to current literature on topics of interest. They may be able to suggest valuable reference sources, such as specialized journals, which are outside the clinical literature and thus unfamiliar to many physician-investigators. General-purpose university libraries offer many of the same services, although their medical collections may be limited. Libraries that do not have an academic affiliation, such as public libraries in large urban areas, often have access to medical databases or medical reference collections at distant libraries.

Computerized bibliographic searches have become a vital tool for scholarly work. The usefulness of these searches, widely used for developing up-to-date literature reviews, depends on the searcher's knowledge of the field, judgment, and experience with a specific data base. Recent availability of consumer-oriented bibliographic data base systems has increased researchers' direct access to data bases such as the National Library of Medicine's MEDLINE. Rapid growth in use of microcomputers has produced an expanding market for a wide range of information services, including medical information. If direct use of a computerized data base is not feasible or desired, the quality of the search is improved if the researcher can discuss the project and the literature sought with the library search specialist, and (if library policy permits) be present while the search is being performed.

After the relevant literature has been identified through a computerized or manual search, some items may be unavailable in the holdings of the institution's library. Inter-library loan programs are almost universally available but often are not widely publicized. These programs, which are typically international in scope, allow access to virtually any published materials required through loan of the original document or a photocopy.

Data processing facilities available through a university or college often provide all the data processing services needed for research applications. For community-based residency programs, these services may be available through agreement with a local community college, a distant university, or a commercial data processing service. Unlike library services, which are available at minimal cost at most institutions, data processing costs and policies vary widely among organizations. Services may include: (a) data entry services (e.g., keypunch or optical card reader), (b) use of on-site computer facilities, (c) use of off-site computer facilities through telecommunications links, or (d) programming consultation. Guidelines for selection and use of these resources are discussed in Chapter 15. Like inter-library loan programs, the range of available computer services is often much greater than is generally publicized.

In a university-affiliated setting, students frequently seek opportunities to participate in research to fulfill course or practice requirements. Students who are interested in an apprenticeship experience can be a valuable source of volunteer research assistance when the project's needs are congruent with the student's learning needs. Medical students and graduate students in health-related or behavioral science fields often have specialized training that can contribute to a study's success. In a university setting, work-study students may be available for various types of work at little or no cost to the requesting department. When involving students in research on either a volunteer or paid basis, the guidelines for research collaboration discussed earlier are especially important in order to avoid perceptions of exploitation. Questions of authorship and access to data during and after the project period are frequent sources of misunderstanding.

Advantages of involving students in research include access to well-trained and often highly motivated research workers who may be familiar with the latest literature in the field, frequently at no cost or at lower cost than hiring a permanent staff member. Disadvantages may include students' inability to make a long-term commitment to a project and the need for school obligations and schedules to take top priority.

In addition to the libraries, data processing facilities, and students that are available at most universities and many other settings in academic medicine, many institutions offer a variety of other services that support research. These may include central laboratory facilities, specialized equipment, medical illustration and other audiovisual services, microfilming, as well as administrative services. Information about these services is frequently not well publicized and must be obtained through informal networks.

External Funding of Research

Even when departmental or institutional resources exist, many research projects cannot be carried out unless funds are obtained specifically for this purpose. Studies that require extensive data collection and analysis or investigations conducted at multiple sites almost always require outside funding. Research funds are available from a wide range of sources, depending on the nature of the project and the qualifications or institutional affiliation of the investigator. Because funding agency priorities and guidelines change frequently, the following discussion is designed to provide a framework in which to view research funding and strategies for obtaining current information rather than to provide procedural details that would soon be obsolete. Basic reference sources that are widely available in academic and large public libraries are listed at the end of this chapter.

Types of Funding for Research

Most research funding available to investigators in academic primary care is awarded in one of three forms: project grants, contracts, and training grants. Although terminology and guidelines differ from agency to agency, some distinctions among these funding mechanisms are generally applicable. Grants and contracts usually differ in (a) the source of the research idea and the study design, (b) the extent to which the funding agency monitors progress and outcomes, and (c) the degree of flexibility permitted in either intellectual or administrative management of the study. Regardless of the type of award, it is customarily awarded to an institution (such as a university or hospital) on behalf of the principal investigator rather than being awarded directly to the individual principal investigator. Although the individual investigator is responsible for the scientific aspects of the project, the institution frequently has responsibility for administering the award in accordance with agency fiscal guidelines. Because of the need for established accounting and auditing procedures, research grants and contracts are seldom made to individuals who do not have an institutional affiliation.

Research grants, the most common type of external funding, include pilot or seed grants for short-term studies of limited scope that are expected to form the basis of larger projects, as well as research project grants that support the costs associated with controlled trials that may require several years to complete. Small grants may also be made for methodology development, for feasibility studies of particularly difficult or high-risk designs, and occasionally for proposal development.

With grant funding, the principal investigator typically originates the research idea, develops the methods to be used, identifies the research population, and plans the proposed analyses. The funding agency expects the principal investigator to pursue the general goals and objectives outlined in the proposal, but does not preclude modifications of the study protocol if promising leads emerge during the course of the research. The principal investigator is expected to keep the funding agency informed about work on the project, typically through annual progress reports, and to follow the proposed time schedule closely enough so that project activities can be completed during the funding period. Regular financial reports (usually handled by a designated division of the institution such as Grants Administration or Accounting) are required, but granting agency policies typically allow some flexibility in use of grant funds if unanticipated needs arise.

Research contracts, by contrast, are issued by a funding agency as a mechanism for purchasing a specified product or service, just as contracts for purchase of equipment or laboratory services specify the product sought. The research idea and often the specific methods to be used are specified by the funding agency, which solicits proposals to carry out the desired study. Such a Request for Proposals (RFP) may spell out the agency's requirements in considerable detail. RFPs are issued for research studies, or less frequently for related work such as analytic literature reviews or state-of-the-art papers, and specific analyses of existing data bases (such as those from the continuing health surveys conducted by several Federal agencies).

Because the contract is the result of a funding agency initiative rather than an unsolicited proposal from the principal investigator, a contract officer from the agency usually maintains fairly close contact with the research to ensure timely progress toward the specified objectives. Progress reports may be required much more frequently than for grants, and agency approval may be required before moving to the next phase of a study. Typically, investigators are given less latitude in modifying the research objectives or methods. Guidelines regarding budget management and accountability are generally less flexible for research contracts than for grants, even from the same funding agency. Drug studies sponsored by pharmaceutical firms are a common example of contracted research done by clinical investigators in university settings.

Research project grants and contracts are typically awarded to support a specific study. Research training grants, although administratively similar to project grants, are awarded to support the development of research skills by an individual or a group of individuals. A research project may comprise part of the training and may be evaluated in reviewing proposals for training grants, but the grant support goes beyond the research project alone.

Some research training grants are awarded to an individual, usually for a prescribed period of supervised training and experience in research methods, which may focus on a specific topic area. Postdoctoral fellowships or mid-career awards for research training in conjunction with a sabbatical are typical of these individual training grants. Other types of training grants are awarded to a department or institution that offers a training program in a specific topic area to a group of trainees. Residency training grants or training grants in specialized areas such as cardiology are common examples. Training grants of either type may be awarded on the basis of an independently developed proposal or a response to an agency solicitation. The latter is more typical.

Sources of Research Funding

Academic physicians frequently think about funding for research only in terms of federal grants. Although federal agencies are an important source of external funding, other sources are often overlooked that may be more appropriate and accessible, especially for the neophyte researcher.

For university-based researchers, the applicant's own institution is frequently a valuable source of funds for pilot work or small research grants. Money may be available from departmental or school discretionary funds or from formal small grant programs (often targeted to junior faculty). In a typical situation, such funding opportunities are not widely publicized. They have few rigid application criteria and are quite flexible in terms of permissible uses. In some institutions research money is available for pilot projects or to bridge a funding gap before a larger approved project begins. Applications for institutional funds usually do not require an extensive proposal. Funds are available soon after the application is reviewed and approved. Institutional research grants are, however, usually small in amount (for example, $1500 maximum) and/or duration (1 year or even 1 semester) and may have other restrictions on how the funds may be spent (for example, travel and equipment may not be permissible expenses). Accordingly, these sources are seldom suitable for large-scale studies. For a beginning researcher, however, local small grants are an excellent source of support for pilot work that can be used to make a proposal for a subsequent larger study more competitive.

Private foundations have received increased attention during recent years as Federal funding has become more restricted and more highly competitive. Foundations range from locally oriented organizations with extremely limited funds to national organizations whose professional staffs and large grant budgets help to shape the direction of U.S. medical research. Although this wide range is reflected in an equally wide range of policies and interests, private foundations are typically more flexible than federal agencies in the types of projects they support, and often in their application and review procedures. Individual contact by letter, telephone, or in person before a formal application is submitted is very useful and indeed is often required. Many foundations, especially those that are small or have geographic restrictions on their giving, are not widely publicized. They may have a broad general mandate (such as improving health care) rather than a specific list of priority areas. Even where funding priorities are stated, these may not be inclusive. Foundations are able to take into account a broader range of factors in deciding which projects to fund and may be able to support projects that do not clearly fall into the categories of research or demonstration projects. In the past, foundations were perceived as less competitive than Federal agencies. However, with increased publicity about foundations and with added pressure to obtain research funding, this is no longer true.

Funding from federal agencies continues to be an important source of financial support for research and demonstration projects. However, most novice investigators will find it difficult to obtain federal funding because the competition for agency dollars is intense and one's scholarly "track record" is usually a key factor in the funding decision. The availability of funds varies widely across agencies and from year to year as a result of shifts in agency and Congressional priorities and budgets. Although eligibility criteria vary for different categories of federal grants and contracts, such awards are almost always available only to principal investigators who have a formal affiliation with, and apply through, an organization such as a university, residency training program, or hospital.

The process by which federal research grant applications are reviewed for funding has been widely used as a model by other funding sources. In most Public Health Service agencies (including the National Institutes of Health), final approval of a project first requires its approval on the basis of scientific merit and only then evaluates its congruence with the goals and priorities of the funding agency. Sound science is a prerequisite to approval. Proposals are reviewed for scientific merit by peer review groups composed of nongovernment scientists. In the NIH review system, these groups are called Study Sections. Although staff members from the relevant funding agencies attend Study Section meetings they normally are not voting members. After the Study Section has discussed each proposal, members vote to approve for funding, to defer for further consideration pending additional information, or to disapprove the application. Each approved proposal is assigned a numerical score based on its funding priority. Approved proposals are then examined by agency staff, who usually have scientific training, in terms of the current topic areas or approaches being emphasized by the agency. Funding decisions generally follow the priority rankings assigned by the Study Section, but may deviate from these if this is felt to be in the interest of the agency.

Corporate funding for biomedical research has traditionally come from drug or medical supply companies. Other corporate sources may provide money for medical research in a limited way. Such funding may include research project grants to individual investigators or contracts with an organization to conduct clinical trials of experimental drugs or devices. In a clinical drug trial, all details of the protocol are often worked out in advance by corporate personnel. Participating organizations may gain experience in collecting data and adhering to an established protocol but may not have input into the design of the trial. The availability of corporate biomedical research funds and procedures for negotiating such funding are in general less publicized and formalized than grant application procedures to Federal agencies or private foundations.

Sources of Information about Funding Opportunities

Background research on funding agencies is an essential part of developing a grant proposal because the types of funding available from various sources, their funding priorities, and guidelines for applications change frequently. A survey of publications from or about potential funding agencies should provide enough information to identify a number of likely sources. Several important reference works that are useful for this purpose are listed in Table 14.2. Agencies themselves publish newsletters, statements of policy or priority areas, or (especially in the case of private foundations) annual reports that summarize grantmaking activity during the past year and planned directions for the future. An initial list compiled from these and other sources can then be narrowed down through contacts (by letter, phone, or personal visit) with staff members of the most likely agencies.

Libraries at most research institutions and many smaller organizations offer basic reference materials on outside funding, such as those noted in Table 14.2, as well as some specialized information on federal and other agencies of particular interest. Many suggested references are revised annually; frequent changes in virtually all categories of information make it imperative to consult the most recent edition available. In addition to printed material, libraries may have access to computerized data bases with up-to-date information on public and private agencies and the grants they have awarded. Some data bases, such as the Smithsonian Science Information Exchange, include information on studies in progress. This is useful for supplementing a review of published literature with information about research that is too recent to have published results, as well as for suggesting funding sources.

Funding information is often available at an applicant's own institu-

TABLE 14.2. Sources of information on outside funding.*

Publication	Comments
The Foundation Directory (revised approximately annually)	Contains brief descriptions of over 3000 U.S. foundations, listed by state. Most useful for quick reference re: officers and mailing addresses, number and dollar range of grants awarded, geographic limitations, etc. Subject index is very general.
Foundation Center Source Book Profiles (published bimonthly)	More detailed descriptions of over 500 large U.S. foundations, with analyses of funding patterns by type of topic and recipient. Date of preparation of individual profiles varies; information on procedures for contacting agency is up to date.
Foundation Grants Index (revised annually)	Well-indexed listing of grants over $5000 awarded by more than 300 major U.S. foundations. Keyword and subject indexes are especially useful. Good source of information on current funding patterns of a limited number of foundations (but does not reflect planned changes in priorities)
NIH Guide for Grants and Contracts (published at irregular intervals, every 3–6 weeks)	Announcements of NIH agency policies, programs, and procedures, including Requests for Grant Applications (RFAs). Best single source of current NIH research priorities
Commerce Business Daily (published 5 days/week)	Announcements of Requests for Proposals (RFPs) issued by any federal agency and announcements of contracts awarded. No grant information
Catalog of Federal Domestic Assistance (revised annually with several interim updates)	Summaries of all federal aid programs (excluding international aid), with program objectives, eligibility, guidelines, restrictions, funding levels, etc. Extensively indexed (by subject, agency, type of applicant, etc.). Time-consuming to use because of format and comprehensiveness, but contains information unavailable elsewhere.

*Adequate information on grant and contract opportunities is basic to a successful search for funds. The books and periodicals listed are recommended as a starting point in this search. They are widely available in libraries and in university administrative offices. Complete citations are given in the list of Suggested Readings for Section III, which begins on p. 277.

tion through a central office with a title such as Office of Research Administration, Office of Sponsored Projects, or Institutional Development Office. In addition to maintaining a collection of publications from funding agencies, such offices may be able to suggest agencies that have previously supported research or other projects at the local institution and to direct applicants to key contact people. Such offices frequently keep files of successful proposals for inspection by inexperienced grant applicants. Modeling one's grant application—in form, not substance—after a proposal that was funded is a useful tactic that a novice investigator might adopt.

Staff of research support offices can provide information regarding institutional policies and practices that may restrict searches for outside funding or that may make certain agencies especially desirable. At some institutions, sponsored research offices offer assistance with proposal development; this may include preparation of budgets and similar materials or assistance in proposal writing. For researchers without a university affiliation, the research administration office at a nearby university may be able to provide access to its reference collection even though it may not be able to offer consultation or other services.

Colleagues (especially those with similar research interests) may be able to suggest appropriate funding sources based on their own experience as a grant recipient or as a peer review group member. Research colleagues can be especially valuable sources of contact people at potential funding agencies and of insights regarding working relationships between agency representatives and their grantees.

Research Environment

This chapter has focused to this point on research resources from the perspective of an individual investigator. Consultation from colleagues, research support from the department or institution, and external funding for specific studies or training programs have been discussed as important aids for a productive research career. Few academic physicians work in isolation. When resources such as those described earlier are most effective, they exist in an organizational environment that promotes research as a departmental priority and views research as an integral part of its operation.

Effective organizational support for research may take a variety of forms. Ideally, they help increase a departmental commitment to research, as well as reflect an existing commitment. The list given in Table 14.3 identifies a number of approaches for developing an organizational environment that fosters research. The items mentioned can be seen both as strategies for promoting research and as characteristics of a

TABLE 14.3. Strategies for promoting research.

1. Incorporate research topics into existing schedule of conferences or seminars for residents and faculty. This might include didactic or experiential sessions on research methods and presentation of research that has been conducted within the department.
2. Offer a separate research seminar series for faculty and residents that focuses on research (both within and outside the department) that is particularly relevant to department priorities and faculty interests.
3. Present a series of didactic seminars on research methodology and ethical issues in research (tailored to the level of experience of most faculty members).
4. Include research in the resident or undergraduate curriculum either as a required or an elective offering.
5. Provide a forum for presentation of research findings to interested colleagues outside the immediate department or division. For example, schedule an annual Conference Day at which residents and faculty give formal presentations to an audience of local researchers. A department-sponsored journal represents another forum for presentation of research findings.
6. Present a competitive research award for resident research or faculty work (especially junior faculty).
7. Develop a workable system for internal circulation of research publications including departmental subscriptions to research journals, articles, or books of special interest, and especially publications by departmental or institutional faculty.
8. Establish a departmental research committee to promote constructive peer review and scholarly exchange on both the process and products of research.
9. Encourage faculty representation on institutionwide research committees such as hospital or medical school committees that review research regarding access to patient populations, protection of rights of human subjects, or institutional funding.

productive research environment. This list is not intended as an inclusive catalog. The value of specific strategies depends in part on the level of research activity already existing within a department and the degree of research sophistication of the faculty, as well as on resources and traditions elsewhere in the institution.

Involvement in research—as an investigator and as a critical consumer of the research of others—is an essential part of the tradition of academic medicine. Direct involvement often begins on a limited scale with a pilot study of a problem suggested by a clinical observation or by a controversy in the existing literature. As investigators develop experience and a scholarly reputation, they may proceed to carry out studies that are larger in scope, with separate funding. Such scholarly productivity is enhanced by an environment in which colleagues are actively engaged in research and in which departmental and institutional resources are accessible.

15 Research Data Management

PAUL GILCHRIST

This chapter deals with the "nuts and bolts" of a research project, the management of research data. The main purpose is to introduce a number of concepts and principles, and to provide practical illustrations of some data management procedures. Although jargon will be avoided, some attention to terminology is necessary to clarify basic ideas.

The topic of research data management, in its general meaning, can refer to tasks and procedures at nearly every step of the research process—from searching and handling bibliographic references to printing or typesetting a final report. It is helpful here to break down the research process into a number of component steps (Table 15.1). The steps provide a framework for more detailed discussion later.

The breakdown given in Table 15.1 does not distinguish between a pilot study and a full-scale research project. If a project involves a pilot phase, the investigator may perform steps 1–12 in the pilot, and then repeat steps 6–13 in the full study. A pilot study can be thought of as a miniature version of the full study whose primary purpose is to test methodology rather than to produce conclusive statistical results.

Many investigators think about research data management narrowly, referring only to those stages of the project from data collection and analysis to preparation of graphics (Table 15.1, steps 7–11). This narrow view usually refers to data that are collected in the course of conducting the project, especially about data intended for quantitative summary and analysis. A broader view of data management includes numerical data *and* textual or verbal information, such as reference materials gathered early in the project or the written report resulting from the research.

This chapter will discuss research data management in its broad sense, including both numerical and textual information. It should be noted that the term "quantitative" does not mean only numerical data. Data measured at the nominal (categorical) level can also be analyzed statistically. Chapter 13 includes a discussion about levels of measurement for research data.

TABLE 15.1. Steps in the research process.

Development of a study:
1. Identify research idea; develop preliminary outline.
2. Conduct bibliographic search.
3. Compile personal reference files.
4. Formulate conceptual and theoretical context.
5. Formulate operational questions and hypotheses.
6. Develop research design, including analytic design.
7. Design or obtain data collection instruments (if any).

Implementation of study procedures:
8. Collect data.
9. Process data: coding, data entry, and editing.
10. Analyze data: summaries, tabulation, statistical analyses.

Communication of research findings:
11. Present results in graphic form.
12. Prepare written report.
13. Disseminate results by means of publication or presentation.

Automation

A discussion of data management inevitably involves the subject of computers, whether mainframe, mini, or microcomputers. In recent years, the cost and physical size of computers have been going down steadily, while capabilities, accessibility, and ease-of-use have improved. Data management programs are now among the most popular microcomputer software packages. Central questions facing the investigator planning a research project concern the role of automation in the project and the extent to which its data management should be "computerized."

Inexperienced investigators may feel apprehensive about automated data management, especially if they are not acquainted with computers. Yet learning how to use, say, a microcomputer for file management or word processing is easy and fun to do. Several hours of reading and practice is usually all it takes to "master the basics"; more advanced computing skills can be acquired PRN. The amount of time and energy devoted to developing computer skills will be shaped by each academic physician's professional goals and interests and the academic environment in which he or she works.

Questions about automation usually arise early in a project, during its early development, and in at least two ways. One set of questions involves bibliographic searching and project reference files, data management tasks that require working with textual information. The second set of questions is associated with research design. They include planning for data management during project implementation, especially in data collection, summary, and analysis. Decisions are needed about the use of computers in both these areas, the former being a more imminent task, since it occurs early in a project, and the latter concerning stages later in the research process.

The value of automated data management needs to be kept in proper perspective. One should not assume that automation is always essential or the most advantageous course. Computer use is a means rather than a research goal in itself. The computer is a tool that can increase research efficiency and productivity if used properly. It can also result in a good deal of difficulty, both in technical and human terms, if used improperly or ineptly.

Personal Goals

Beyond the scientific goals of a particular research project, the academic physician's personal goals are an important element in determining the method of data management. For the novice researcher, a project may have more value as a way to learn about research methods than as a means for creating new knowledge about the subject under investigation. This can also be the case for experienced investigators who lack first-hand experience with automated data management and computer use. Such an investigator may want to become closely involved in the data management process in order to learn about it. The academic physician who plans to conduct future projects might not always intend to have such a close contact with data management, but may want to establish the capability to understand, direct, and evaluate data management work by subordinates in the future.

In any of these situations, the physician-investigator is demonstrating an interest in learning about an unfamiliar area so that new knowledge can be applied in later research efforts. In fact, many clinical investigators note that as their experience grows with computers and data management, they develop a greater ability to discern and tackle research questions that would otherwise go unrecognized. As the investigator's degree and sense of control over the research increases, there should also be positive attitudinal effects in terms of satisfaction, achievement, and confidence.

In summary, an investigator's personal goals in conducting a research project can include learning about computerized data management. This might involve creation of computerized data files and computer data analysis even when these tasks could be done by less automated methods. If, however, the investigator just wants to get the project finished, and if data processing and analysis are uncomplicated enough to grant automation no advantage, then managing the data "by hand" is probably indicated. (One might argue that the ability to make this decision presupposes some prior knowledge of computerized data management.)

Factors External to the Investigator

Decisions about management of research data, and the role of computers in the process, are also influenced by factors external to the individual academic physician. How many others in the organization or department are engaged in research, and how do they manage their data? What are the department's goals and priorities regarding research? How important is clinical research among one's colleagues? How many projects are likely to be going on in 6 months or a year, both by the investigator and by colleagues?

As noted in Chapter 14, the variety of resources available locally is also important. The available facilities, equipment, personnel, and funding, even the organizational structure, will determine the methods used in data management. The factors external to the investigator vary among organizations, often according to whether departments are situated in university or community settings.

In situations where there is much interest in developing research competence, but where experience and skills are low (both among faculty and staff), a research project may serve as a vehicle for acquiring the desired proficiency. In other words, a project can become part of a plan for group or departmental development.

Data Management during Project Development

A list of steps in the research process was presented at the beginning of this chapter (Table 15.1). There it was noted that for studies that include a pilot phase, a number of the steps are repeated during the full-scale project. This is because the pilot study is a trial-and-error run. The difference between a pilot and the full-scale version of a study involves more than just the volume of data or number of cases collected; their purposes are different.

The pilot's main purposes are to test methods and data instruments, estimate costs, and often to provide preliminary statistics necessary for sample size determination for the full-scale study. This tryout of methods can include any aspect of the data management and analysis process. Hence, a pilot study as a dry run of a larger project offers an opportunity to develop, learn, and experiment with data management techniques. Different methods can be tried for speed, ease, accuracy, and cost. These trials are a sort of methodological hypothesis–testing process. One or more methods are proposed, tested, and the inferior ones rejected in favor of the superior. If an investigator is uncertain about the use of a computer, it can be tried during the pilot; experience and training can be acquired, and problems solved. In this way, the idea of a pilot study is conceptually similar to the idea of using a research project as a learning vehicle as part of a personal plan for professional development.

Bibliographic Searching

Searching the literature is an important activity as a research project develops. In compiling a bibliographic base, typical modes of searching the literature include "manual" searching and "online" computer searching of bibliographic data bases performed by a trained reference librarian or conducted by the researcher. A research effort may involve any or all of these modes. Although the large literature data bases are created, maintained, and managed by others, the researcher is personally involved in data management whenever the library or a user-friendly data base is used.

The bibliographic information derived from a search provides a theoretical context for the research, helps formulate hypotheses and questions, and provides information about research methods used by other investigators. Productive researchers are well-read; they are masters of the literature in their field and are aware of its areas of controversy and weakness. Such individuals read not only the medical literature, but also literature in related and peripheral fields (e.g., behavioral sciences) for information and methodological ideas.

Manual Searching

A library may be viewed as a large data base with information organized, filed, and available for use according to several classification systems. The researcher can conduct a manual search based on the classification schemes of, for example, Excerpta Medica or MeSH (Medical Subject Headings), and by referring to indexes such as Index Medicus, and Science Citation Index. Another common method involves using recent articles on a topic as a starting point, and searching backward through references cited in these and previous articles. The Science Citation Index allows a forward-looking search to be performed, searching for articles that cite one or more earlier articles. One should look especially for literature reviews that summarize information in an area of interest, and the review sections of individual articles.

Online Searching

There are many computerized bibliographic data bases that can be used for online literature searching. Expansion of the number and size of data bases and improvements in their accessibility have evolved in-kind with developments in hardware and software capabilities and telecommunications technology necessary for gaining access to them.

The data base that is most frequently used in medical research is MEDLINE, the online data base of Medical Literature Analysis and Retrieval Systems (MEDLARS) produced by the National Library of Medicine. MEDLINE is accessible directly by trained reference librarians; other users ordinarily gain access to the data base through commercial information services, such as Dialog or Bibliographic Retrieval Services (BRS/

After Dark). Both of these commercial services offer many other data bases as well, including data bases in fields peripheral to the health and medical sciences. The MeSH classification system is typically used in conducting a MEDLINE search. One advantage of online literature bases is that the references are sometimes more up-to-date than in published indexes, and may even precede publication of pertinent journal articles.

The online data base of Excerpta Medica (EMBASE) is another bibliographic resource covering a broad spectrum of medical and health literature. SCISEARCH is the online version of the Science Citation Index from the Institute for Scientific Information. A more recent medical information service called AMA/NET is provided by the GTE Telenet Medical Information Network. It offers information bases on drugs, diseases, procedure codes and nomenclatures, nonclinical aspects of health care, and an electronic mail service. Other data and services are planned. Online data bases for other academic fields include Psycinfo (Psychological Abstracts), Social Scisearch (Social Science Citation Index), Biosis (Biological Abstracts), and Chemical Abstracts.

The traditional method of online literature searching entails having the search performed by a trained professional, such as a reference librarian or search analyst specializing in health sciences. The researcher discusses the project with the search professional and provides key words, concepts, methods, specific references, and authors. Together they devise a search strategy. If possible, the investigator should be present when the search is conducted to supply the searcher with clarifications and decisions about questions that arise during the course of a search. For example, if an initial search results in a larger number of references than is manageable or necessary, the investigator can help narrow the focus by suggesting additional criteria or key words. Conversely, a very small result would mean the researcher should broaden the search criteria.

Self-conducted online searches using data bases accessed through commercial vendors are growing in popularity. This is a consequence of several factors. At the provider end, hardware and software systems have developed to the point where their operation has become more user-oriented. There has been an increase in the number of providers and data bases, and the competition for users has led to "friendlier" systems, better manuals and documentation, and more qualified instructors and training seminars. User commands have become simpler. Telecommunications and computer interfaces have improved greatly in terms of speed, reliability, and cost. Abstracts are often available in the data bases. For example, for the past several years, about 40% of MEDLINE citations include an abstract; about 25% of non-English articles have an English abstract.

At the user end, the increasing popularity and capabilities of microcomputers, storage media, telecommunications devices, and printers mean that more researchers have easy access to the equipment necessary for conducting online searches in their offices or at home. Searches can be conducted from home at night at reduced rates and greater speed than during the busy daylight hours. Reference librarians are serving increasingly in a teaching and consulting capacity, in addition to their work as searchers. Teaching the principles, concepts, and strategies of bibliographic searching to investigators will enable them to conduct their own searches with greater efficiency and effectiveness. It is necessary, however, for researchers to use their online searching skills frequently or to register periodically for refresher seminars, because techniques quickly get rusty from lack of practice.

Professional organizations such as the American Society of Information Science are in the forefront of disseminating information about research and development of bibliographic systems oriented toward academic users. Their professional journals are a valuable scholarly resource.

Project Reference Files

Handling bibliographic information obtained from a literature search is an important aspect of research data management. The way in which the information is managed determines its value to a project by providing theoretical and conceptual foundations, deriving questions and hypotheses, describing research methods, and revealing the state of knowledge in the area under investigation.

To enhance the utility of the bibliographic information, the researcher can organize it into reference files for personal and project use. When organizing reference files, great savings of time and effort can be gained by using information classification schemes that have already been developed, such as the Medical Subject Headings (MeSH) Index or Excerpta Medica. In addition, codes such as the International Classification of Health Problems in Primary Care (ICHPPC) and the International Classification of Diseases (ICD) can be incorporated into the filing scheme.

Some purposes served by the researcher's reference files are in preserving and maintaining the reference information in a convenient way, allowing additional references to be added in the appropriate places, allowing cross-classification of articles and chapters, permitting consistency with one or more classification schemes, and facilitating retrieval in a variety of ways and combinations. The files can serve both the current project and future investigations.

The researcher may choose to develop project reference files on index cards, in file folders (for reprints and other copies), or as computer files. In deciding whether or not to computerize one's reference files, the general considerations discussed earlier in this chapter are relevant. The amount of the information to be contained in the files should also be considered. In addition to citations, the researcher may wish to file annotations, abstracts, or copies of articles. These may require a system of files, rather than just one. For example, citations on index cards might be kept as one file, and abstracts and/or reprints in another.

Another consideration is the organization of the files, especially with regard to the capacity to file and retrieve references quickly and accurately, according to some classification scheme, and with convenient cross-classification. In an index card file, cross-classification can be accomplished by having multiple cards for references that need to be filed in more than one location. Multiple copies of the first page of an article can allow it to be filed in more than one place in a file of hard copies, with the complete article in only one location. The entry of the reference information onto the records in the files, and the extraction of that information when writing up an article or reference list, should also be considered on grounds of time, ease, and cost. A further consideration would involve the desirability of subdividing the files or combining them with other files, either during the course of the project or later.

If the investigator has easy access to a computer (whether micro, mini, or mainframe), it may be advantageous to use its capabilities to organize, maintain, and use reference files. For example, each time information from a literature search is manually entered into a reference file, costs are incurred in terms of both time and money. Typing up file cards and labels, including multiple ones for cross-classification, is laborious. By contrast, bibliographic information from an online search can easily be "downloaded" (i.e., transferred electronically over the telephone lines) from the data base source to the disk of the user's microcomputer. This includes citation, abstract, and even the entire article if it happens to be available. With minor alteration or editing, the information becomes part of the researcher's computerized reference files. The need to key (type) the references manually into the researcher's computer is averted; time and expense are saved and the potential for error is reduced. Other references that are found can be typed into the file from the keyboard.

When information is to be printed from the reference file, it can be selected and printed easily. For example, if a research article or literature review is being written with a word-processing program on the computer, the desired citations or any textual passages from the abstracts can be electronically extracted from the reference files and transferred into the article being written. Again, time and money are saved. Once information is in the computer, it should seldom have to be reentered.

In computerized reference files, organization according to indexing schemes, or cross-classification of references, can be accomplished by codes and keywords incorporated into the record, thus eliminating the need for multiple entries of an item. Computerized reference files enable the researcher to use simple and powerful data management procedures to subdivide the files or to join files quickly. Selecting, sorting, and editing are other functions that are easily accomplished. Many programs are available that enable one to handle personal bibliographic files on a computer.

Research Design

The relationship of data management to research design is somewhat abstract, yet very real. It was noted earlier that researchers should devote much thought to data management during the development of a research project. Beyond the bibliographic reference stage, this involves planning for data management in each of the subsequent project stages. The main point is that a researcher should make detailed plans for data management during project implementation and when communicating results before the investigation is started.

Inexperienced researchers commonly slight this principle, and rarely think beyond their data collection instruments and procedures. Data collection looms large early in the project and is fraught with uncertainty, making it difficult for new investigators to plan beyond that step. The danger is that the researcher may be inclined to operate on faith that solutions to analytical problems can be handled easily. This can lead to difficult situations, much anguish, strained relationships, and many regrets later in the project, not to mention inefficiencies and extra costs.

Attention to research design during project development should include data analysis, not just data collection. The researcher should decide early about how the data are to be analyzed, what types of summaries and tabulations will be done, and which statistical procedures will be performed to answer the questions or test the hypotheses the study is addressing. Ideally, the researcher should identify the equations and construct the outline of "dummy" tables and graphs that will be used to present data, lacking only the actual values that will ultimately be entered. For projects that involve study of a sample of subjects, the step of determining adequate sample size requires that the analysis be designed before data are collected.

The decision to use computers is affected by the volume of data (including width, length, and complexity of data sets), the complexity of the analysis and statistics employed, plans for use of graphics, and knowledge of the availability of computer software to do what is desired.

Data Management during Project Implementation

Data Entry, Editing, and Error Detection

These are the "input" steps in the data management process, in which the information generated during the project is prepared for summary and analysis. If a computer is being used for the data management and analysis, the objective is to get the data into the computer system with as little transfer error as possible, i.e., with as little discrepancy between the information that was gathered and the information as it ends up in the computer files ready for further processing. This does not address the issue of how accurate, valid, or reliable the information is to begin with; see Chapter 13 for discussion of those sources of error.

Data collection should always be carefully designed with an eye looking ahead to the data entry and analysis steps. This calls for thoughtful decisions about response codes and the layout of data collection instruments so that data entry will proceed smoothly and easily.

Data Management Systems

Many readers may not be familiar with data management systems, so it should be helpful to provide a brief explanation of what the term means. A data mangement system (DMS), sometimes called a database management system (DBMS), is a generic term for a category of computer application programs that store, manipulate, and retrieve different types of information without requiring the user to be a computer programmer. A DMS may be described as a group of modules that perform various procedures such as editing, sorting, and tabulating data, as well as printing reports. Other applications that are sometimes contained in the group of modules include programs for word-processing, financial "spreadsheets," communications, graphics, and statistics.

Data management systems have been available for about 20 years on mainframe computers, especially as integral parts of statistical packages. Lately, they have multiplied and are now widely used since the advent of microcomputers.

Data management systems do not require computer programming skills or knowledge of computer language in the traditional sense. In this regard, they may be conceptualized in terms of levels of computer "language," and the amount of programming effort required of the human user to accomplish a certain task or series of tasks. Figure 15.1 shows a simplified diagram representing "language level" on the vertical axis and "programming effort" required for a series of tasks along the horizontal axis.

In the years following World War II, as early computers became more widely available, programming was performed primarily at the machine or assembler language level. This is very laborious work directing the individual on-off switches of the machine at what is called the "bits and bytes" level. Over the years, as millions of programming hours accumulated on different machines in various locations, these low-level commands were combined into sequences that could be invoked by "macro" commands so that each sequence did not have to be rewritten each time it was needed. Such advancements helped "high-level" programming languages such as Fortran, COBOL, then BASIC and Pascal to emerge and improve. Introduction of "high-level" languages allowed programmers to work more efficiently and productively without having to use machine language directly. Devices inside the computer translated programmers' commands into machine language, i.e., for any given task, the computer would put in more effort while programmers put in less.

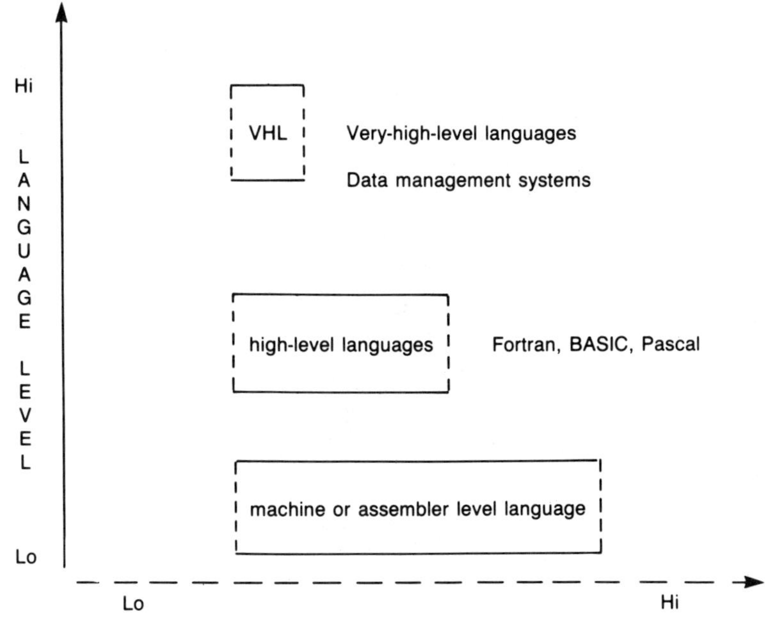

FIGURE 15.1. Levels of computer language.

Still, it requires much training and practice to program in high-level languages, and it is difficult enough for physicians to stay abreast in their specialties without having to develop and maintain programming skills. However, the "high-level" languages, in turn, have evolved into today's data management systems, which are even more user-efficient sets of "macros" that invoke combinations of instructions written in high-level languages. Thus, state-of-the-art data management systems represent "very-high-level" or VHL languages. Besides data management, some systems also contain modules for complex statistical procedures that may be invoked using simple commands without having to program the mathematical formulae. The point is that one does not need high-level programming skills to use a data management system. Data management systems, then, are called very-high-level "user-oriented" programs, in contrast with high-level "programmer-oriented" languages. The data management system used in the examples that follow is written in the Pascal langugage, but the user does not need to know Pascal to operate the programs.

Data management systems vary in the complexity of their VHL languages. Some require considerable study and training to be used effectively, and in this respect, they may actually seem close to the high-level languages. A well-known example of this type is *dBase II* by Ashton-Tate. Many other data management systems can be used more easily by the

TABLE 15.2. Antepartum SNILP protocol.

(Keypunch Codes)

|A|1| serial|_|_|_ (1, 2)

1. Patient chart no._______ |_|_|_|3|_|_
2. Protocol code___ __ __ |_|6|_|_|_|_|11| (12 — 18)
   (case no.) (p) (n)
   (P = protocol initial; n = nurse's initials)
3. Date of test__ __ __ |_|_|_|_|_|_| (19 — 24)
   mo. day yr.
4. Test no. _1 or __ (FILL IN ACTUAL TEST NUMBER) |_|_| (25)
   (if OTHER than "1" is checked please answer question 5)
5. Repeat test information (please check one)
   a. _Tested within 24 hours
   b. _Tested within a week
   c. _Tested within 2 weeks
   d. _Other (Please specify)

   ____________________ |_|
6. Birthdate __ __ __ |_|_|_|_|_|_|27| (28 — 33)
   mo. day yr.
7. Gestational age of current pregnancy __weeks |_|_| (34)
8. Reason for test (please check)
   a. ___ Routine surveillance
   Placental insufficiency: |_|_|_|_|_|_| (36 — 41)
   b. ___ IUGR
   c. ___ Post maturity (42 weeks)
   d. ___ Diabetes
   e. ___ Gestational diabetes
   f. ___ Preeclampsia
   g. ___ Hypertension (chronic)
   h. ___ Adolescent
   i. ___ Other medical conditions
9. Parity (complete a, b, and c)
   a. Term pregnancies __ |_|_|
   b. Premature __ |42|_|
   c. Abortion: spontaneous __ |_|_|
   d. Abortion: induced__ |_|_| (48)
10. Non-stress test (NST) (Please check one)
    a. ___ Cannot be interpreted |_| (50)
    b. ___ Nonreactive
    c. ___ Reactive (if "c" is checked, please answer question 11)
11. Disposition if NST is reactive (please check one)
    a. ___ Contraction stress test |_| (51)
    b. ___ No further tests done
12. Breast stimulated (please check one)
    a. ___ Left breast OR b. ___ right breast |_| (52)
    c. ___ Both breast ALTERNATING
    d. ___ Both breasts SIMULTANEOUSLY
13. Contraction stress test (CST) (15-minute stimulation plus 5-minute poststimulation resting strip) (please check one)
    a. ___ Cannot be interpreted |_| (13)
    b. ___ Nonreactive
    c. ___ Reactive
14. CST results (please check applicable items)
    a. ___ Negative |_|_|_|_| (14 — 17)
    b. ___ Positive

TABLE 15.2. (*Continued*)

c. ___ Fewer than three contractions in 10 minutes
d. ___ Cannot be interpreted
e. ___ Equivocal
f. ___ Hyperstimulation (longlasting contraction greater than 90 seconds or 5 contractions in 10 minutes)
g. ___ Other (please specify)___

15. Disposition if CST is UNSATISFACTORY (please check one)
a. ___ Repeat NST-CST within 24 hours |__| 18
b. ___ OCT (challenge test)
c. ___ Nothing further done (reason) ___

16. Additional comments or observations (Y/N)|__| 19

Tracing interpretation

Procedure: (NST)	A. Ten-minute resting monitor strip. Completed Yes___ No___	(Y/N)|__| 20	
	Number of contractions in first 5 (1–5) minutes ___ ___	|__|__|	
	Number of contractions in next 5 (6–10) minutes ___ ___	|__|__|	
	B. Fifteen-minute manual stimulation. Completed Yes___ No___	(Y/N)|__| 25	
	Number of contractions in first 5 (1–5) minutes ___ ___	|__|__|	
(CST)	Number of contractions in next 5 (6–10) minutes ___ ___	|__|__|	
	Number of contractions in next 5 (11–15) minutes ___ ___	|__|__|	
	C. Five-minute resting poststimulation strip. Completed Yes___ No___ 30	(Y/N)|__|	
	Number of contractions in 5 minutes ___ ___	|__|__| 33 34	

SNILP, Stimulation of nipple in Labor Project.

"average" research, administrative, or academic user with little training or experience. Fortunately, most DMSs are entirely satisfactory for nearly all research data management needs so that an investigator or research assistant need not consult a programmer. The fact that data management systems are available on mainframe, mini, and microcomputers means academic physicians are likely to have access to one or more machines that can be used for data management.

In conclusion, relatively simple, powerful, and user-oriented data management packages are readily available for virtually any brand of computer. They can easily handle most research data management without requiring much computer skill or experience. What follows is a discussion about data entry and editing, illustrated with an example. Construction of a data file will be demonstrated, and some common data management procedures will be performed. The illustrations are based on a data management package called PDMS (Pascal Data Management System) copyrighted by Pascal & Associates of Chapel Hill, NC.

Setting Up a Computer Data File

The data collection instrument shown in Table 11.2 and reproduced here as Table 15.2 will illustrate our discussion of data management during project implementation. This and other examples involve the use of a computerized data management system, since this represents the most typical and useful method of handling research data. The examples are used to illustrate and add substance to important data management concepts.

The average user of a data management system may think of data sets simply as nice, neat files consisting of rows and columns of information. Rows usually represent cases or subjects, and the columns are fields containing measurements on each variable. Setting up a PDMS data file will clarify this further. Referring to the example in Table 15.2, each SNILP questionnaire (one per research subject) will represent a row, whereas each field (columns or set of columns) will represent an item of information (i.e., a variable).

In creating a PDMS data file, it is necessary first to provide the computer with some information about the architecture of the data file. This is done by means of a descriptor file that becomes incorporated into the data file and that serves as an "input statement." Each variable is given a name, the computer is told how many columns are to be allocated to each variable, whether it is an alphabetic or numeric variable, and how many decimal places it contains (if any).

In PDMS, a procedure called CREATE is invoked to initiate the descriptor file (see Table 15.3A). Information about the variables is then put into this file under the headings. For the first several items of the SNILP instrument, the descriptor information would be as shown in Table 15.3B. The first variable (field name) Protocol, is designated as a character variable (type C) and allotted two columns (width) and no decimal places (dec.). The second variable is equivalent to an observation number, is named OBS, and allotted three spaces. Notice that, although numbers will be used for filling in values of OBS, it is designated as a

TABLE 15.3A. Headings for PDMS descriptor file.

Field name	Type	Width	Dec.

TABLE 15-3B. PDMS descriptor file for first several variables of SNILP questionnaire.

Field name	Type	Width	Dec.
Protocol	C	2	0
OBS	C	3	0
Chart#	C	6	0
Code	C	3	0
In	C	2	0
Nurse	C	2	0
Date	C	6	0
T#	C	2	0
Q5	C	1	0
DOB	C	6	0
GA	N	2	0
(etc.)			

character variable. The reason for doing this is that a character variable is easier to edit *in this particular package.* Since it would make no sense to perform any arithmetic operations on this field, it can be dealt with as a character variable. This will not interfere with a sorting procedure. It is not necessary to designate a variable as a true numeric until the question about gestational age (no. 7 on the instrument, Table 15.2). For this variable, one might want, for example, to add and divide at a later time to derive an average.

After the descriptor file is created, it is converted by a simple command (CONVERT) into the actual data file, which contains the descriptor table in its structure. The descriptor file thus serves as a skeleton for the data, giving them form and shape, while the data serve as the flesh. In converting the descriptor file, the computer is told to allocate a space on the disk for, say, 300 rows in the data file and to name the data file "SNILP."

Before any data are entered, the empty SNILP data file resembles Table 15.4A. The variable names that were listed down the left side of the descriptor file now appear across the top as column headings in the data file. Note that the variable names (column headings) are truncated on the screen according to the field width assigned to them. The rest of the name is still there, but it is "hidden" in the computer's memory.

Data from the questionnaires are then entered into the file. Table 15.4B shows the SNILP file after four (fictitious) observations have been entered. One important principle is to minimize the number of times the data are transferred. This will reduce the number of errors that creep into the data and will help a great deal to hold down costs. Reducing data transfer saves time when files are first set up and, especially, in having to correct fewer errors after a transfer occurs. One data transfer step that is often done unnecessarily is to copy response information from a data form onto a standard computer coding sheet (Fig. 15.2). It is often possible to reduce these two transfer steps to one by entering the data into the computer directly from the original form. This can occur only if the data collection form is thoughtfully designed and has been laid out in a

TABLE 15.4A. SNILP data file before data entry, showing variable names as column headings

Pr	OBS	Chart#	Cod	In	Nu	Date	T#	Q5	DOB	GA

Note truncation of names according to column widths.

TABLE 15.4B. SNILP data file containing four observations.

Pr	OBS	Chart#	Cod	In	Nu	Date	T#	Q5	DOB	GA
A1	001	312741	001	A1	JS	101984	1		110663	37
A1	002	266751	001	A1	SA	102184	1		061059	39
A1	003	256882	001	A1	JS	102584	2	A	021165	38
A1	004	423371	001	A1	LH	102884	1		093061	39

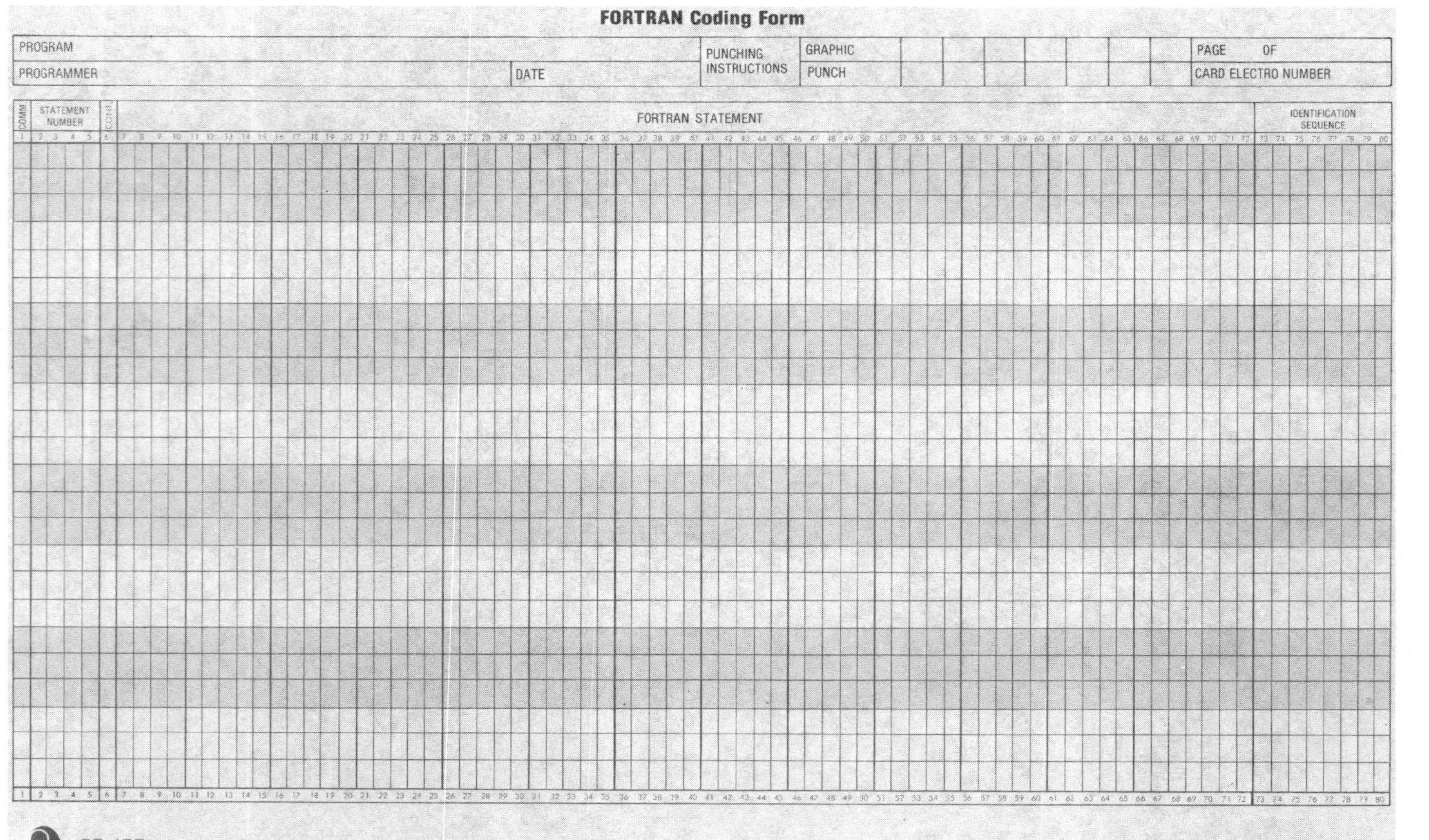

FORTRAN Coding Form

PROGRAM			PUNCHING INSTRUCTIONS	GRAPHIC							PAGE OF
PROGRAMMER		DATE		PUNCH							CARD ELECTRO NUMBER

COMM	STATEMENT NUMBER	CONT.	FORTRAN STATEMENT	IDENTIFICATION SEQUENCE
1	2 3 4 5	6	7 8 9 10 11 12 13 14 15 16 17 18 19 20 21 22 23 24 25 26 27 28 29 30 31 32 33 34 35 36 37 38 39 40 41 42 43 44 45 46 47 48 49 50 51 52 53 54 55 56 57 58 59 60 61 62 63 64 65 66 67 68 69 70 71 72	73 74 75 76 77 78 79 80

AMPAD 23-400

Figure 15.2. FORTRAN coding form. (Reproduced with the permission of Ampad Corporation.)

clear and uncomplicated manner so that a data entry person can follow it easily and accurately.

In the example shown in Table 15.2, the instrument is set up to have the responses transferred to the coding boxes along the right-hand side of the page. This is easier than transferring to a standard coding sheet and is reasonable if the data are to be entered onto computer cards or into a "raw" or unstructured computer file (i.e., the data are not entered directly into a data management system, but will become structured files after a descriptor or input statement has been made up and applied to them). In the example, the data are to be "uploaded" (transferred) over a telephone line to a mainframe computer where they will be analyzed by a statistical package. The numbered boxes in Table 15.2 allow the data entry people to know exactly what column they should be in at any given point and to compare it to a numbered scale on the computer screen that identifies the corresponding columns.

If the analysis will be done with a microcomputer data management system it would be more efficient if, instead of transferring the responses to the right-hand column, they were keyed directly into the structured file of a data management system. This would eliminate one data transfer step and reduce the likelihood of error. Some data management packages feature the capacity to design a facsimile of the data instrument right on the computer screen, and to fill in the blanks.

Error Detection

After the data are entered into a computer file, it is important to check for errors. The most dependable way is also the most obvious: a printout of the data is produced and the "eyeball technique" is employed to compare the printout to the original data forms.

Another type of error checking involves having the computer identify values that are "out of range" for each variable. One method for this is to prepare frequency distributions for the variables. Any values that are "illegal" (e.g., diastolic BP $>$ 500 mm Hg) can quickly be spotted and the file searched to correct them. This method has a drawback: incorrect values that are within the legal range cannot be noticed.

Summarizing Data

Data summaries are best explained by example. This one deals with diagnostic data from patient encounters with resident physicians. For simplicity, only one diagnosis will be given for each encounter. The example will show some of the most common and useful data summarization procedures including sorting, frequency counts by diagnosis, selecting, labelling, and reporting. Experience suggests that much, if not most, primary care research is conducted using procedures no more complicated than these.

SORTING, TABULATING, AND ANALYZING DATA

Table 15.5 shows the descriptor file for the diagnostic encounter example. None of the variables is numeric. The definitions of the variables are as follows:

RCTNO: Receipt number of the patient visit
YYMM: Year and month of the visit
DD: Day of the month
MD: Resident identification number
TM: Resident's clinic team or module
PT NUM: Patient identification number
DX: Primary diagnosis (ICD–9 code)

The reason for expressing the data in this form is to allow for computer sorts, if that is desired.

A small part of the data file derived from the above descriptor table is shown in Table 15.6. There are 23 observations in the file involving three residents whose clinical experiences are to be monitored by the faculty for purposes of educational management. For example, one objective might be to assure that each resident gets exposure to as wide a variety of cases as possible during training, or to a minimum number of certain diagnoses. Faculty may want to control tendencies of some residents to "specialize," as pointed out in Chapter 8. Also, for some diagnoses that do not present very often in the clinic, it may be desirable to make sure that each resident handles at least one case.

The data represent actual diagnoses from patient encounters by the three residents during April of 1983. For education management purposes, the data might be summarized every 3–6 months, and the files, of course, would be considerably larger.

Assume an investigator wants to know the diagnoses seen by each resident along with the frequency of each diagnosis. This may be accomplished in PDMS by invoking a procedure called SUM. The program module allows the user to select easily the variables for which counts are to be made (similar analyses can be done in other DMS packages). In this

TABLE 15.5. Descriptor file for diagnostic data.

Field name	Type	Width	Scale
RCTNO	CHAR	5	0
YYMM	CHAR	4	0
DD	CHAR	2	0
MD	CHAR	2	0
TM	CHAR	1	0
PT NUM	CHAR	7	0
DX	CHAR	4	0

TABLE 15.6. Diagnostic data file (DX).

RCTNO	YYMM	DD	MD	TM	PT NUM	DX
82153	8304	22	51	A	0320902	250–
82163	8304	22	62	B	0423001	685–
82164	8304	22	62	B	0501802	V220
82166	8304	22	62	B	0357701	250–
82167	8304	22	62	B	1009501	429–
82168	8304	22	73	C	1078301	V202
82171	8304	22	73	C	0999002	V220
82172	8304	22	73	C	195611	V202
82173	8304	22	73	C	0842911	595–
82175	8304	22	73	C	0932502	595–
82177	8304	22	73	C	0848801	V645
82179	8304	25	51	C	0277501	0799
82260	8304	25	51	A	1034702	3000
82295	8304	25	73	C	0999002	V220
82499	8304	29	62	B	0423002	136–
82501	8304	29	62	B	1008501	401–
82505	8304	29	73	C	0976302	491–
82509	8304	29	73	C	0837802	V255
82512	8304	29	73	C	0999002	V220
82514	8304	29	51	A	0203001	7890
82515	8304	29	51	A	0203002	455–
82516	8304	29	51	A	0091902	7242
82524	8304	29	62	B	1082901	429–

case MD will be the first-level variable and DX will be the second-level variable, so that we will end up with diagnostic counts "within" MD. The resulting summary table may be viewed on the computer's screen or printed out as shown in Table 15.7. The output variable "LL" stands for "lexicographic level," which is simply the level of summary. All values of count at LL 2 add up to the corresponding subtotal value of LL 1; all values of LL 1 sum of the grand total of LL 0. Hence, for MD #51, all counts at LL 2 add up to 13 for LL 1. The three values of count at LL 1 for the three MDs (i.e., 13, 14 and 16) add up to 43 at LL 0. The LL variable is very useful in controlling report formats, as will be seen later. For future reference, the file containing the summary table is named "MDDXSUM."

Notice that in Table 15.7 the frequency counts are ordered by diagnostic code for each resident, i.e., smaller diagnostic code numbers appear first. It might be preferred that the diagnostic codes be rearranged by order of frequency for each resident. To do this, one executes the SORT procedure and calls for MDDXSUM as the file to be sorted. The user is then allowed to select the variables in MDDXSUM on which to sort. In this case, MD is selected as the first-level variable and COUNT as the second-level variable. In this way, the MD order will be preserved and only the diagnoses will be rearranged according to frequency counts

TABLE 15.7. Frequency of diagnoses by resident (MDDXSUM).

LL	MD	Count	DX
2	51	1	0799
2	51	1	250–
2	51	1	3000
2	51	2	3820
2	51	1	455–
2	51	1	493–
2	51	1	614–
2	51	1	7242
2	51	1	7890
2	51	1	889–
2	51	1	959–
2	51	1	V202
1	51	13	
2	62	1	136–
2	62	1	250–
2	62	1	378–
2	62	1	401–
2	62	2	429–
2	62	1	6161
2	62	1	685–
2	62	1	7840
2	62	1	829–
2	62	1	VO3–
2	62	1	V202
2	62	1	V220
2	62	1	V70–
1	62	14	
2	73	1	3027
2	73	1	3820
2	73	1	491–
2	73	2	595–
2	73	1	739–
2	73	1	848–
2	73	2	V202
2	73	3	V220
2	73	1	V253
2	73	1	V255
2	73	1	V654
2	73	1	V70–
1	73	16	
0		43	

TABLE 15.8. Sorted frequency of diagnoses by resident (MDDXSUM).

LL	MD	Count	Dx
2	51	1	0799
2	51	1	250–
2	51	1	3000
2	51	1	455–
2	51	1	493–
2	51	1	614–
2	51	1	7242
2	51	1	7890
2	51	1	889–
2	51	1	959–
2	51	1	V202
2	51	2	3820
1	51	13	
2	62	1	136–
2	62	1	250–
2	62	1	378–
2	62	1	401–
2	62	1	6161
2	62	1	685–
2	62	1	7840
2	62	1	829–
2	62	1	VO3–
2	62	1	V202
2	62	1	V220
2	62	1	V70–
2	62	2	429–
1	62	14	
2	73	1	3027
2	73	1	3820
2	73	1	491–
2	73	1	739–
2	73	1	848–
2	73	1	V253
2	73	1	V255
2	73	1	V654
2	73	1	V70–
2	73	2	595–
2	73	2	V202
2	73	3	V220
1	73	16	
0		43	

within MD. The output is shown in Table 15.8. The output can be arranged into a simpler report that is easier to read, as we shall see later.

RELATIONAL PROCEDURES

Data management that uses information from more than one file, based on a common variable, is an essential feature of a "relational" system. A simple example follows.

Since few individuals commit diagnostic codes to memory, computer printouts of data such as those given in Tables 15.7 and 15.8 are more useful if labels are attached to the diagnoses. This is done by using a separate file, DXLBL, that contains diagnostic codes and their corresponding labels. Table 15.9 shows such a file. For the sake of brevity, this illustration uses a label file containing just those codes that are found in the original data file. There might be hundreds of codes in a complete label file. In PDMS, the diagnostic labels are attached to the frequency distribution in Table 15.8 by executing the JOIN procedure. The procedure requests the identity of the file containing the variable to which the labels are to be attached (MDDXSUM) and then the name of the label file (DXLBL). The common or joint variable found in both files is "DX," and this provides the basis of the join. The resulting output is put into a file (named MDDXSUML) and printed out as shown in Table 15.10.

Other label files containing physician names and patient names could also be attached to the data in the original file. The reason for having label information in separate files is not to obscure the identity of the coded variables, but to save data entry time and storage space by using abbreviated codes. A common use of label files is for keeping a complete array of demographic, financial, or other information on individuals, such as patients, which can be joined to the patients' codes when the need arises.

The label file can be used to monitor the match of a resident's patient care (diagnostic) experience with a desired distribution. For example, those diagnoses that are considered especially important for a resident to experience could be capitalized in the label file, or kept in a separate file which might even contain an additional variable indicating how many encounters of each diagnosis each resident should experience. By exercising the SELECT option to identify a resident, then joining that resident's diagnostic frequency file to the label file (i.e., executing the join in the opposite direction from that done before) the resident's patient contacts can be compared to a set of expected frequencies. Gaps will be particularly conspicuous. (See Chapter 8 for a discussion of the educational reasons for keeping files of patient encounter data.)

The overall rankings of each diagnosis for, say, an entire residency program's practice during a certain time period could be included in the label file. By joining the labels to the MDDXSUM file (which was sorted by diagnostic frequency within MD), a comparison can also be made between each resident's rankings and the overall rankings.

Reports

The earlier examples have included printouts of the files being illustrated. The printouts are produced by the REPORT procedure in the PDMS package. This procedure is quite versatile and enables the user to obtain reports in a variety of formats. Several are demonstrated below. Some report formats can be quite analytical.

TABLE 15.9. Diagnostic codes and labels (DXLBL).

DX	Diagnosis
0799	Viral syndrome
110–	Dermatophytisis
136–	Other infective
250–	Diabetes
3000	Anxiety
3004	Depression
3027	Sexual problems
3720	Conjunctivitis
378–	Other eyes/ears
3804	Wax in ear
3811	Otitis media, chronic
3820	Otitis media, acute
401–	Hypertension
429–	Other cardiac
455–	Hemorrhoids
460–	URI
491–	Bronchitis, chronic
493–	Asthma
595–	UTI
598–	Other genito/urinary
614–	PID/salpingitis
6161	Vaginitis
685–	Subcutaneous infection
708–	Urticaria
709–	Other skin
7242	Pain, lower back
7295	Pain, limb
739–	Other bones/joints
7807	Malaise/fatigue
7821	Rash
7840	Headache
7851	Palpitation
7890	Abdominal pain
829–	Fracture
848–	Sprain/strain
889–	Laceration
929–	Contusion
959–	Other acid/poison
VO3–	Immunization
V10–	High-risk patient
V201	Payment only
V202	Lab work
V220	Prenatal care
V253	Other prophylactic
V255	Birth control pill
V654	Patient education
V70–	Medical exam

When the REPORT procedure is executed, a "menu" appears that presents a number of options. Table 15.11 contains a menu showing some of the options, followed by a description of what each option controls.

TABLE 15.10. Labeled diagnostic frequencies by resident (MDXSUML).

LL	MD	Count	DX	Diagnosis
2	51	1	0799	Viral syndrome
2	51	1	250–	Diabetes
2	51	1	3000	Anxiety
2	51	1	4551–	Hemorrhoids
2	51	1	493–	Asthma
2	51	1	614–	PID/salpingitis
2	51	1	7242	Pain, lower back
2	51	1	7890	Abdominal pain
2	51	1	889–	Laceration
2	51	1	959–	Other acid/poison
2	51	1	V202	Lab work
2	51	2	3820	Otitis media, acute
1	51	13		
2	62	1	136–	Other infective
2	62	1	250–	Diabetes
2	62	1	378–	Other eyes/ears
2	62	1	401–	Hypertension
2	62	1	6161	Vaginitis
2	62	1	685–	Subcutaneous infection
2	62	1	7840	Headache
2	62	1	829–	Fracture
2	62	1	VO3–	Immunization
2	62	1	V202	Lab work
2	62	1	V220	Prenatal care
2	62	1	V70–	Medical exam
2	62	2	429–	Other cardiac
1	62	14		
2	73	1	3027	Sexual problems
2	73	1	3820	Otitis media, acute
2	73	1	491–	Bronchitis, chronic
2	73	1	739–	Other bones/joints
2	73	1	848–	Sprain/strain
2	73	1	V253	Other prophylactic
2	73	1	V255	Birth control pill
2	73	1	V654	Patient education
2	73	1	V70–	Medical exam
2	73	2	595–	UTI
2	73	2	V202	Lab work
2	73	3	V220	Prenatal care
1	73	16		
0		43		

The values shown are the defaults. The table resembles what the user actually sees on the computer screen after executing the REPORT command and calling for MDDXSUM as the file to be reported.

Selecting menu option 1 allows the user to type in a title (Heading) for the report. The titles of Tables 15.5 through 15.10 and Table 15.12 were produced this way.

Option 2 allows rows to be selected according to specified values or

TABLE 15.11. PDMS report menu screen.

Reporting MDDXSUM
Date: November 3, 1984
[hit (CR) to print, (−1) to escape]

Option no.	Description
1	HEADING:
2	PRINT ALL ROWS
3	PRINT ALL FIELDS
4	NO CONTROL BREAKS
5	NO TOTALS
6	OUTPUT TO: printer
7	FOOTER: page number
8	PRINTER AND PAGE PARAMETERS
9	PARAMETER FILE

ranges of values of designated variables. For instance, if an investigator wanted a line-by-line printout of each encounter of a single resident from the original data file, this option would be used. Selection is made on the variable "MD," typing in the resident's identification number. The report would then omit all other residents.

Option 3 allows the user to print out only certain fields (columns), or to rearrange the fields. Hence, combining options 2 and 3 can be used to select rows and columns.

Option 4 may be used to control breaks for subtotals. An illustration of this is seen in Table 15.12, where the LL variable is used for the break, after selecting only those rows for which LL=2 (refer also to Table 15.10).

Option 5 produces subtotals at each level of control break and a grand total at the bottom, as in Table 15.12.

Option 6 usually directs the report output to the printer, but can also direct it to the screen for viewing or to the disk for storage.

Option 7 can be used to control or suppress page numbering.

Option 8 brings to the screen another menu that allows control of margins, typing pitch, page length, etc.

Option 9 allows the user to store the report format from the above items in a file for future use whenever a similar report is to be produced. This saves much time and is a very convenient feature.

GRAPHICS

A data management procedure that coincides with data analysis and reporting is the production of charts and graphs to express research data visually. Since a discussion of graphics is included elsewhere in this volume (Chapter 19), just a simple example will be given here. The capability to create graphics may be contained in one of the modules of a data management package or may be contained in a separate program. Graphs

TABLE 15.12. Labeled diagnostic frequencies by resident (MDDXSUML).

MD = 51		
DX	Count	Diagnosis
0799	1	Viral syndrome
250–	1	Diabetes
3000	1	Anxiety
455–	1	Hemorrhoids
493–	1	Asthma
614–	1	PID/salpingitis
7242	1	Pain, lower back
7890	1	Abdominal pain
889–	1	Laceration
959–	1	Other acid/poison
V202	1	Lab work
3820	2	Otitis media, acute
	13	
MD = 62		
DX	Count	Diagnosis
136–	1	Other infective
250–	1	Diabetes
378–	1	Other eyes/ears
401–	1	Hypertension
6161	1	Vaginitis
685–	1	Subcutaneous infection
7840	1	Headache
829–	1	Fracture
VO3–	1	Immunization
V202	1	Lab work
V220	1	Prenatal care
V70–	1	Medical exam
429–	2	Other cardiac
	14	
MD = 73		
DX	Count	Diagnosis
3027	1	Sexual problems
3820	1	Otitis media, acute
491–	1	Bronchitis, chron.
739–	1	Other bones/joints
848–	1	Sprain/strain
V253	1	Other prophylactic
V255	1	Birth control pill
V654	1	Patient education
V70–	1	Medical exam
595–	2	UTI
V202	2	Lab work
V220	3	Prenatal care
	16	
	43	

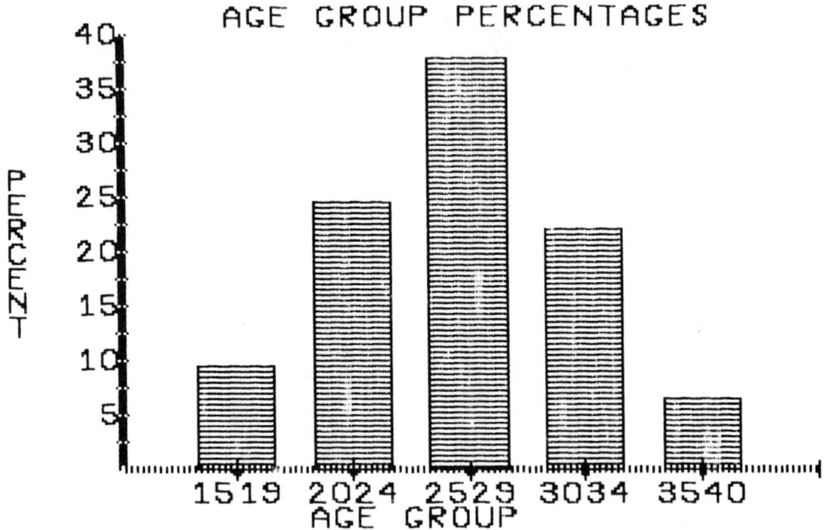

FIGURE 15.3. Example of computer graphics on ordinary dot-matrix printer.

are drawn after a computer reads through a data file and makes frequency counts or from frequencies that have been produced earlier and "plugged into" a graphics program module. The graphs are printed on a dot-matrix printer or drawn on a plotter. Computer graphics are drawn with astonishing speed. They allow the researcher to experiment quickly with different types of graphs to best visualize information contained in the data.

The vast majority of analytical graphs are either bar, line, pie, or scatter plots. As an illustration of a bar graph, a histogram constructed on an inexpensive dot-matrix printer from data representing gestational age for obstetrics cases is shown in Figure 15.3.

Conclusion

The purpose of this chapter has been to present several data management concepts and principles, and illustrate some basic procedures. The intent has been to give more than a short overview of research data management, yet not to serve as a full-fledged manual. Thus, the chapter should be a practical guide for academic physicians who wish to become acquainted with research data management. However, no written material is an adequate substitute for practice, and novice investigators are urged to "get on with it." By developing data management skills from experience, academic physicians will quickly recognize how much these competencies can contribute to research productivity.

Suggested Readings on Clinical Research

Fletcher R H, Fletcher S W, Wagner E H. *Clinical Epidemiology: The Essentials.* Baltimore: Williams & Wilkins, 1982.

Friedman C H. *Primer of Epidemiology,* 2nd ed. New York: McGraw-Hill, 1980.

Riegelman R K. *Studying a Study and Testing a Test.* Boston: Little, Brown, 1981.

Sackett D L, Hayes R B, Tugwell P. *Clinical Epidemiology: A Basic Science for Clinical Medicine.* Boston: Little, Brown, 1985.

At least one of these basic textbooks on clinical epidemiology belong on the academic physician's bookshelf. These books are informative and useful because each is written by clinicians for clinicians. As such, concepts, language, and examples are particularly relevant for physicians who wish to increase their skill at reading the medical literature or designing research studies.

Department of Clinical Epidemiology and Biostatistics, McMaster University. How to read clinical journals, Parts I to V. *Canadian Medical Association Journal 124*: March 1 to May 1, 1981 (successive issues).

In this series of articles clinicians are taught basic skills of critical journal reading about diagnostic tests, course and prognosis of disease, etiology and causation, and distinguishing useful from useless therapy. The series is highly recommended for its clear presentation, clinical utility, and its direct applicability to posing and answering research questions.

Kleinbaum D G, Kupper L L, Morganstern H. *Epidemiologic Research: Principles and Quantitative Methods.* Belmont, California: Lifetime Learning Publications, 1982.

This is an exhaustive epidemiology textbook that most physicians will find an excellent back-up for the more clinically oriented books cited earlier in this list. Academic clinicians will find the volume to be useful chiefly as a reference work.

Lilienfeld A M, Lilienfeld D E. *Foundations of Epidemiology,* 2nd ed. New York: Oxford University Press, 1980.

This is a comprehensive epidemiology textbook that is somewhat more advanced than the other selections given in this list, especially Sloane et al. Research designs are covered in detail along with such topics as morbidity and motality statistics, epidemiologic history, and selected statistical procedures.

Marks R G. *Designing a Research Project.* Belmont, California: Lifetime Learning Publications, 1982.

The steps involved in the design and execution of a clinical research project are identified and discussed in this short yet excellent book. The text is particularly useful because the author urges readers to "think through" all of the phases of a study—from formulating a research question to assigning treatments to subjects—before data are collected. Beginning researchers will find this book very informative.

Marks R G. *Analyzing Research Data.* Belmont, California: Lifetime Learning Publications, 1982.

This is a companion volume to *Designing a Research Project,* also authored by Marks. It covers basic information about descriptive statistics and proceeds to address hypothesis testing and inferential statistics. This book and its companion are highly recommended resources for academic physicians.

Sloane C, Brogman, D R, Eyres, S J, Lodnar, W. *Basic Epidemiological Methods and Biostatistics: A Workbook.* Belmont, California: Wadsworth, 1982.

This is a highly readable basic text on epidemiology that actively engages readers by calling for written responses to questions and exercises. A variety of research designs are covered, as are methodological pitfalls in cinical research. The book is strongly recommended to physicians who have no background in clinical research methods.

Manuals for computer programs, or packages that contain many separate programs, are a valuable research aid. Data management and statistical analysis of data are facilitated when clinical investigators have some knowledge of computer operations. Two of the most popular statistical packages are the *Statistical Analysis System* (SAS) and the *Statistical Package for the Social Sciences* (SPSSX). The User's Guide for either package is an useful resource for the physician researcher.

IV Professional Communications

Communication in its many forms and modes is integral to the professional life of the modern physician. From the subtleties of interpersonal communication between practitioner and patient—forming one of the bases of the art of medicine—to the complexities of computer-assisted record-keeping, diagnosis, and clinical management, the process of transferring information and ideas clearly and accurately is central to our effectiveness as professionals. This is true in some special ways for the academic physician, whose work in research and teaching, as well as patient care, makes specific demands for an ability to communicate well orally and in writing. Of course, one of the most common requirements for professional advancement is publication and presentation of new knowledge. But beyond that practical consideration, most academic physicians take very seriously their commitment to share ideas and information with their students, their colleagues, and the public at large.

Earlier chapters of this *Handbook* have focused upon the academic physician in his or her professional environment. To that discussion of the academic physician as an organism within the organization, we would now add that of the academic physician as speaker and writer. We have already noted the common career requirement that members of medical faculties publish the results of their scientific efforts, either through articles in professional journals or by oral presentation of prepared papers at professional meetings. The skills that produce the successful chairperson, clinical team leader, or organizational manager are immeasurably enhanced by the skills of effective spoken and written communication. The ability to organize ideas and information, to present them logically and concisely, and to understand the impact that the presentation makes on the audience has clear value in the academic organization.It's an ability well worth cultivating, whether your objective is to lengthen your bibliography or to lead your department.

The next four chapters will deal with the techniques of effective communication in oral and written forms for the academic physician. Chapter 16 outlines the practical principles of professional communication.

We consider the process of planning any written or oral presentation; the most common structures and formats of professional presentations; and the development of the basic structure into a full presentation draft. The chapter then discusses techniques for editing and refining the language, style and content of your draft in order to achieve clarity and conciseness.

Chapter 17 applies the principles of professional communication directly to the preparation of articles, letters, and other written forms. We discuss the identification of the audience for which you are writing and the analysis of the publication through which you hope to reach that audience. Manuscript preparation—from planning to problems of coauthorship—is reviewed, and so are several considerations in manuscript submission and dealing with editors. Then Chapter 18 applies professional communication principles to effective oral presentation in the context of professional meetings and conferences. There are several techniques of planning, writing, and rehearsing that apply particularly to oral presentation, and the chapter discusses some of the more important ones.

Chapter 19 focuses on the planning and use of visual materials to strengthen both written and oral professional communication. In this chapter, we analyze the need for and appropriateness of effective visual communication, and we suggest some guidelines for the physician author in the selection, planning, production, and use of slides, overhead transparencies, charts, graphs, video, computer-generated visuals, and other materials.

We'd like the chapters in this section of the *Handbook* to be as immediately and concretely applicable to your needs as possible. To achieve that, it will help a great deal if you will use these chapters as step-by-step guidelines for the development of a real presentation. We'll begin now with your idea for the subject of a professional presentation: your current research, an interesting patient case that should be published for its educational benefits, an analysis of a commuunity preventive care program, etc. As we discuss the principles of professional communication in Chapter 16, try to develop the conceptual framework of your presentation subject. Then in Chapters 17, 18, and 19, add to the framework with specific notes on the techniques of writing for publication, presenting to meetings, and using visual communication effectively. You may find at the conclusion of this section that you have the form of an article or paper that, with a bit more work, will be ready for public presentation.

We hope that you will find these chapters useful immediately as a working introduction to the techniques of written and oral professional communication. We also hope you will find it helpful to continue to refer to them from time to time as guidelines and refreshers.

16
Principles of Professional Communication

James W. Lea

In this chapter, we will examine principles of effective communication that apply both to written articles and to papers prepared for oral presentation. Considerations of type and format are essentially the same for both oral and written modes of presentation. In later chapters, we'll differentiate between the distinctive requirements of written communication and those of oral communication.

We'll begin with the first principle of any professional activity—planning.

Planning a Written or Oral Presentation

The first step in preparing a written or oral presentation is the preparation of a sound *plan* for the presentation. Such a plan includes a number of elements:

Determination of purpose
Characterization of audience
Formulation of objectives
Identification of presentation type
Selection of format

With the subject of your own presentation in mind, let's work through these elements.

Purpose

Even though it may sound too obvious to merit much attention, the first important step in planning a presentation is to determine to your own satisfaction your purpose in doing it. Each manuscript or oral presentation is developed for a specific purpose: to report on a piece of research, to review and interpret a body of literature, to present an interesting and instructive patient case. There are undoubtedly subsets of such purposes as well, but your thinking about any professional presentation should begin with a conscious decision regarding its purpose.

Audience

Closely linked to the purpose of your presentation is a description of the audience that you are trying to reach. A clear understanding of your audience—their specialized professional interests, their probable level of knowledge of your subject, their vocabulary—is important to the planning of your presentation. A paper describing a clinical study of kidney disorders in children should have one emphasis if the intended audience is made up of pediatric nephrologists, but a different one if the audience is to be primary care physicians. It is often easier to characterize your audience for an oral presentation at a professional meeting than to predict readership of an article published in a medical journal. But you should at least try to develop an idea of the audience for whom your presentation is intended.

Objectives

Although the objectives of a professional presentation may differ from those in classroom or clinical teaching, they should nevertheless be clear statements of what the presentation is intended to accomplish, both for the author and for the audience. Perhaps you are planning an article that will report the results of a series of clinical trials that you have conducted. Your objective might be to highlight the innovative methods that you employed in the handling of specimens or material. The article should be prepared to reflect this.

On the other hand, perhaps you feel that your findings offer significant new approaches to a widely recognized therapeutic problem. In this case, your objective might be to persuade readers of the potential efficacy of your new approach. Or, your investigation may refute the findings of earlier investigators. An important objective then becomes to develop the context in which one can question the continuing efficacy of previous approaches. Your presentation would have still different emphases.

The same raw materials for a professional presentation may be developed in several different ways, depending on the author's objective, and the intended effect of the presentation on the audience. Before beginning to outline your first paragraph, a clear statement of the objective of your work will give you a standard by which to gauge your progress in developing the article or paper. It may make quite a difference in your presentation's success and your recognition of its success.

Types of Presentations

There are several presentation types common to the health professions and to the written and oral communications of health professionals.The type of presentation and the choice of format for a paper or article may follow standards long established by the scientific community. Of course, there are times when academic physicians find themselves writing flowing essays or speaking eloquently outside the bounds of scientific tradition. For the purposes of this *Handbook,* however, we will concentrate on the mastery of the common types of medical professional presentations: the *clinical investigation report,* the *patient case report,* and the *review.* We will also make some comments on the techniques of writing the "letter to the editor," which is a recognized form of scholarly publication. In most cases, the nature of the professional activity—biomedical research, patient care, education—will influence the choice from among these common types of formats.

Structuring the Presentation

Consideration of format provides a good introduction to the basic principles of structure in a professional presentation. We are indebted to Margaret McCaffery's six-part series on "Writing That's Worth Reading," published in *Canadian Family Physician,* for the acronym IMRAD, which stands for *Introduction, Methods, Results, and Discussion* (see p. 334). In order to focus attention on these structural components, we'll discuss them first in terms of a report of a research study. Open one of your favorite journals now to an article reporting a clinical or laboratory investigation and analyze it for the IMRAD components as we discuss them.

The *Introduction* presents the background and the rationale for the study being reported. (It does not present the rationale for the publication of the report itself, although in some cases that would be very enlightening information.) In an article or a paper that reports a clinical or laboratory investigation, the Introduction might begin by citing previous research that has dealt with related issues and has either provided the basis for the author's research or produced results that are substantiated or refuted by the author's work. Often the author will cite a scientific or clinical observation that sparked the investigation. The Introduction may then state the purpose, the hypothesis, or the research question of the present study.

The *Methods* section of the investigative report describes the author's research method or protocol, as well as the conditions under which the study was carried out. Descriptions of the physical site, the subjects or materials and the criteria for their inclusion, a chronological narrative of the investigative process, and a description of data analysis are typical parts of the Methods section of the report.

The *Results* of the study should be exactly that: a factual statement of the findings of the investigation, usually reported in order of importance. The major research question presented in the Introduction is answered first. Then other outcomes of the study are reported. Some authors are tempted to editorialize at this point, but that should be saved for the concluding section. The Results section should present only results, stated objectively and concisely.

It is in the *Discussion* section that the author should comment on and provide interpretation of the reported results. It is here, too, that any weaknesses in the study design or its implementation should be acknowledged (if the author doesn't, someone else certainly will), as well as the presence in the study of any objectively inexplicable phenomena. The author might suggest implications of the findings for clinical practice or for further research.

Here's a quick-reference summary of the common components of the investigative report, structured according to McCaffery's IMRAD:

Introduction
  Background and rationale
  Brief references to the literature
  Observation that sparked the study
  Purpose, hypothesis, or problem

Methods
  Research method or protocol
  Conditions of the study
  Site, materials, criteria
  Narrative of investigative process
  Data analysis

Results
  Material or statistical outcomes

Discussion
  Commentary, interpretation
  Weaknesses, unexplainable phenomena
  Implications for clinical practice or further research.

In this discussion of IMRAD, we've been referring to the article or paper reporting on a research study. The components of the *Patient Case Report* are essentially the same, but with some variations. Look now at a patient case reported in a professional journal. The Introduction establishes the background and context for the report. The Methods section, and often the Results section as well, may follow the standard format for a medical workup. The Discussion section of the case report focuses on what makes this case distinctive and instructive and what implications it may have for patient care. If the planned presentation is a patient case report, think for a moment about how the patient data can be shaped by the IMRAD structure, as we review it in a quick-reference summary.

Introduction
- Previous reports of this problem
- Rationale for this case report

Methods
- Description of patient, history
- Clinical presentation
- Diagnostic procedures and results
- Therapeutic procedures and patient responses

Results
- Outcomes of therapeutic interventions
- Patient status: death, discharge
- Patient follow-up, if appropriate

Discussion
- Commentary on this case compared with other cases reported
- Implications for clinical management of similar cases
- Implications for research, professional training, etc.

Before we look at other common types of professional presentations, we should offer a caveat that applies especially to preparing the investigative report and the case report. These two types of presentations traditionally demand a style of writing or speaking that many people consider the worst type of medical communication—the terse, rigidly objective, "scientific" style that seems to be trying grimly to cram two tons of information into a one-ton paragraph. Our suggestion—that you structure your information and ideas efficiently by following proven formats—should not be taken as a suggestion that you also follow the tradition of stultified style that is too often associated with these formats. More on that later.

The *Review* is a type of presentation that collects information on a given subject from a variety of sources, categorizes it, analyzes it, and draws conclusions from it. By definition the review includes salient points from many articles published in the professional literature. Consequently, the review most often (but not always) takes the form of written rather than oral presentation. Take a moment to find a review article in your favorite professional journal.

The review does not follow a fixed or standardized format, as investigative and patient case reports generally do. However, the review can be characterized by certain common features. The review may open with an Introduction, which usually states the presumed need for the review, the author's purpose in preparing it, and the structure of the material to be reviewed. From that point onward, the typical review will adhere to a clear pattern of organization. The author may employ a *chronological* organization, especially if the stated purpose is to show an historical progression in research or patient care as evidenced by the literature. For example, the review might cover the past 10 years of reported research on endorphins for the purpose of showing how recognition of their role in pain tolerance has changed.

The author may alternatively choose a *topical* organization for the review, especially if the stated purpose is to review several different approaches or categories relative to a given subject. For example, previous publications on perioperative use of antibiotics might be grouped according to preoperative, intraoperative, and postoperative coverage. One might alternatively separate the organization into groups of antibiotics, discussing the indications and uses of each group.

Topical organization would also be a good tool if the review is addressing a subject that normally includes a number of categories, as would be the case in a review of clinical experiences with the major classes of antidepressant medications. Of course, there are other patterns of organization that can be applied to a review, but these two are the most common.

The review typically ends with a *Conclusion,* in which the author summarizes the trends that the review has brought to light and states their significance. Like the Conclusion section of other presentations, the review's final paragraphs suggest implications for future research or patient care.

In quick-reference summary form, the common format of a review looks like this:

Introduction
- Presumed need and purpose
- Brief overview of materials to be reviewed

Organization
- Chronological
- Topical

Conclusion
- Summary of content: trends, patterns
- Significance
- Implications for research, patient care

The *Letter to the Editor* is a common type of written professional presentation, a quite legitimate way to publish scientific findings as well as clinical observations, "mini-reviews," and opinions. Virtually every professional publication has a Letters department; pause now to read a few of the letters in a recent issue of your favorite journal.

Like a review, a letter to the editor does not follow a prescribed format.

Most journals provide guidelines only about letter length in their "Information for Contributors" section. An informative letter should, however, include a few basic features. In the opening sentence, the letter states exactly what the writer wishes to bring to the attention of the editor and the readers. In some cases, the writer makes reference (by title, author, and date of publication) to an article published previously in the journal to which the letter is addressed, and the aspect of that article with which he or she agrees or disagrees. In other cases, the writer cites a professional issue on which the letter will state an opinion or shed further light. In any case, the letter's *purpose* is stated immediately and clearly in the opening paragraph.

The next paragraph contains the *substance* of the letter. It may describe a clinical trial, patient case, or some other episode in the author's own professional experience. Or it may cite research reported elsewhere that amplifies or contradicts the article or issue cited in the opening paragraph of the letter. The writer may state an interpretation of the subject or point out the weaknesses of someone else's interpretation. Since letters to the editor must usually be confined to just a few lines of print, the statement of substance should be very much to the point.

The letter's concluding paragraph is a *summary* of the writer's point of view, perhaps recommending further study or action relevant to the subject of the letter. Like preceding paragraphs, the summary is concise and concrete. A letter to the editor may be less rigid in format than a presentation of clinical or biomedical activity, but it is seldom chatty.

For quick reference, the letter to the editor usually includes:

Purpose
- Reference to previous article or professional issue

Substance
- Writer's own scientific, clinical experience
- Findings published elsewhere
- Interpretation or comment

Summary
- Writer's point of view
- Recommendation of further study or action

Developing the Presentation

Having made the important decisions about presentation type, format, and structure, the next step in developing your presentation is the *distribution of data* on the structural framework that you have prepared. Some people have what might be called a "narrative intelligence" that enables them to start at the beginning of a presentation and work right through to the end. If you are among the majority who are not so blessed, you might find it easier to begin with the objectively reportable data of the Methods and Results sections before attempting the more subjective and judgmental Introduction and Discussion.

The Methods section of the presentation should open with a description, in concise but complete terms, of the physical site and of the subjects or materials of your study along with the criteria for their inclusion. Next comes a chronological narrative of the investigative or diagnostic/therapeutic process, including laboratory measures, drug dosages, time devoted to each step of the intervention, and the subject's interim responses at each step. The methods employed in data analysis should also be described in this section. The Methods section should deal only with the subject and its treatment.

The Results section closely follows the Methods section. This section is also objective and straightforward. In it, you state factually, and usually in order of importance, what biological, clinical, or statistical effects were produced by your methods. A stated research problem or hypothesis is usually addressed first in reporting results, followed by other findings. Don't be tempted to comment on the results as you report them. As Sergeant Friday of television's *Dragnet* used to say, "Just give us the facts, ma'am, just the facts."

Having poured your notebook into Methods and Results, you can now pour your mind into the discussion of the findings, in fact, of the entire process of investigation or case management. The discussion should interpret and comment on the results, putting them into the perspective established by previous work in the field. If the results were not up to expectation, say so and try to explain why without making excuses. If the study was weak in certain areas, confess it. State the conclusions to which the work has led you. Most important, identify ways in which your findings or the outcomes of your case should affect further research and future patient care. When you have said these things clearly and adequately, stop writing.

The Introduction should be developed last, for much the same reason that we feel more comfortable introducing someone we know well than someone who's a total stranger. The opening sentences of the Introduction may cite major previously published work on the subject. Or it may refer to a scientific observation or controversy that suggested the investigation. Next comes a very brief synopsis of the study case, with an indication of its agreement or disagreement with previously published findings. You may want to state the research problem or hypothesis that shaped or guided the study. The Introduction should be developed to provide the background and context of the presentation, to gain the attention of the audience, and to do all of this with a few well-chosen words.

A well-organized oral or written professional presentation is characterized by *continuity,* by a logical flow of ideas and information. We have seen that for several of the various types of professional presentations with which you may be working, the overall structure, or format, proceeds from introduction, to information (in the form of methods and results), to conclusion. Within each section and within each paragraph, the same logical sequence can be applied. The opening sentence of the paragraph makes a statement that introduces the principal thought of the paragraph. The next several sentences communicate the information of the paragraph. The closing sentence provides a conclusion, and perhaps some discussion, relevant to the information of the paragraph.

This sequence provides each paragraph with internal coherence and continuity, just as the elements of structure discussed earlier provide continuity to the entire presentation.

To test the idea of continuity in a paragraph—from introduction to information to conclusion—let's analyze the second paragraph above.

A well-organized written or oral professional presentation is characterized by:

Introduction

A well-organized written or oral professional presentation is characterized by *continuity,* by a logical flow of ideas and information.

Information

We have seen that for several of the various types of professional presentations with which you may be working, the overall structure, or format, proceeds from introduction, to information (in the form of methods and results), to conclusion. Within each section and within each paragraph, the same logical sequence can be applied. The opening sentence of the paragraph makes a statement that introduces the principal thought of the paragraph. The next several sentences communicate the information of the paragraph. The closing sentence provides a conclusion, and perhaps some discussion, relevant to the information of the paragraph.

Conclusion

This sequence provides each paragraph with an internal coherence and continuity, just as the elements of structure discussed earlier provide continuity to the entire presentation.

Continuity is an important element in any professional presentation.

Refining the Presentation

The last stage of development of your presentation is its refinement. This stage may be called polishing or editing, but it is actually more substantive than either of those. In refining a manuscript, for example, one is concerned with the quality of *directness,* with the choice of *vocabulary,* and with learning from informal but important *peer review* of the completed presentation. Thorough and thoughtful refining of the presentation may make the difference between an adequate piece of work and a report of which you can be truly proud.

Directness

A vital element of effective verbal communication is the proper use of the right tools of language—grammar, prose style, and sentence and paragraph structure—to achieve directness. A direct statement is the shortest distance between two minds, your own and that of your audience. And the most direct form of statement is the simple, declarative sentence.

The treatment relieves the symptoms.

This sentence demonstrates the *subject-predicate-object,* or noun-verb-noun, order that you probably learned in elementary school, and that remains the clearest way of communicating information verbally. The verb in the sentence is an *active voice* verb. A passive voice verb, which has become an unfortunate standard of much medical writing, would produce:

The symptoms are relieved by the treatment.

The result is a sentence that communicates the same information, but with more words and less authority. The point is that the right choice of grammatical elements, subject-predicate-object order, and use of an active verb creates a more directly communicative sentence.

A direct and effective statement employs as *few modifiers,* adjectives and adverbs, as necessary. Try to avoid the temptation to use clusters of modifiers to say in one sentence what should be said in two. And if you have begun to believe that a swarm of modifiers increases the forcefulness of a noun or a sentence, consider this example:

The previously described treatment directly relieves the patient's symptoms.

In this sentence, the statment is direct, the verb active, and the words few. Whereas a well-chosen modifier can enhance the meaning of a noun or verb, indiscriminate use of a number of them will quickly rob a word and a sentence of clarity. Review your presentation for choice and use of modifiers, and cut out the unnecessary ones.

Finally, the clarity and directness of your professional presentation is improved by concentrating on making declarative statements instead of *hedging.* Research situations, especially clinical investigations, are difficult to control. Readers who are clinicians and scientists themselves recognize this fact. Nevertheless, the profession has developed a tradition of hedging, of qualifying descriptions of interventions and results, to avoid making direct statements that can be challenged by listeners or readers. Here is an example of overqualifying, or hedging, a statement:

If appropriately selected, the treatment, when administered according to proven protocols, will in most cases probably relieve a majority of the symptoms.

In this example, the writer is self-protected against virtually all claims that the treatment doesn't relieve the symptoms. But the statement carries no sense of professional confidence; in fact, it's downright wishy-washy. You can't figure-skate in a sleeping bag, and you can't communicate with a professional audience if your ideas and information are bundled up in a series of *hedges*. For the sake of clarity in your presentation, refine out the hedges and leave the direct declarative statements.

Vocabulary

Another important consideration in refining your professional presentation is reviewing the choice of vocabulary. Words are tools, just as are scalpels and stethescopes, and they should be selected with care and used with precision. In reviewing the use of vocabulary in your presentation, pay particular attention to professional terminology and jargon, to concrete words and conceptual words, and to necessary words and unnecessary words.

The *terminology* of the academic medical world is both respectable and necessary. The terms that describe human anatomy, physiological processes, diagnostic and therapeutic procedures, and educational interventions are universally used and understood among professionals. Many of those terms have been thousands of years in the making. Professional terminology is direct and precise, and effective professional communication demands and profits from its use.

However, *jargon* is another matter. Jargon in professional communication can be defined as the unnecessary use of esoteric words or the use of stylistic devices that are traditional but unproductive. Jargon is usually faddish and only narrowly understood within the profession. It is often vague and therefore subject to misinterpretation. When reviewing vocabulary choices in your presentation, separate the jargon from the legitimate professional terminology by applying this question: is this term or phrase precise, universally understood, and helpful?

There is no established acceptable ratio of *conceptual* to *concrete* terms for professional presentations. But refining your presentation should include assuring that you are not using conceptual terms where concrete terms would be more productive. Conceptual terms are often broadly descriptive "umbrella" terms, and they can get out of hand. Reference to "the widespread use of the drug" communicates only a vague perception and very little tangible information. Concrete terms have more limited definitions, and are usually direct and precise. Concrete words are not intrinsically better than words that express concepts, but a presentation should use both types accurately.

The distinction between *necessary* and *unnecessary* words or phrases in a professional presentation is often highly subjective. But the effect of clutter created by unnecessary verbiage is clear when the reader is forced to reread sentences or paragraphs repeatedly to understand them. The listener, however, has no similar way of replaying oral statements in order to decipher their meaning, so he or she may just doze off instead of making the effort. Remember that necessary words are those required to communicate ideas and information concretely and completely. These are examples of some kinds of *necessary* words in a sentence:

Basic words:

Pelvic inflammatory disease has been a problem among women.

Essential modifiers:

Pelvic inflammatory disease has been a problem among women *using the intrauterine device.*

Elaborators:

Pelvic inflammatory disease has been *one of the most serious* problems among women using the intrauterine device.

Words are classified as *unnecessary* if they can be deleted without changing the meaning of the sentence. Here are some common examples:

Redundant words:

Principles that *underline and support* this hypothesis *may now be said to* include *those that are* the most *basic and fundamental* of *our civilization's* science.

Fillers:

The patient for whom such antidepressant and antipsychotic medications have been prescribed, either in combination or in sequence, will be found to have been provided with even more effective and significant relief.

Redundant words actually weaken rather than strengthen a single sentence and an entire presentation. Fillers, used by writers who have either time or pages they feel they must fill, have a remarkably soporific effect. If you encounter these unnecessary uses of vocabulary in refining your presentation, be ruthless about their removal.

Peer Review

When you have done as much as you can in planning, developing and refining your presentation, an essential last step is recruiting a professional colleague as an unofficial reviewer. In fact, it's a good idea to solicit interim reviews during the process of developing the presentation. But at the very least, you should have the advantage of feedback on the completed piece from someone who (a) knows enough about your study to understand what you are attempting to communicate, (b) knows as much about the field as the audience you are trying to communicate with, and (c) is both honest and sympathetic enough to provide constructive criticism. In later chapters, we'll discuss the specific kinds of feedback that are helpful to writing for publication and to practicing an oral presentation. For now, make a note always to solicit, accept, and make use of informal review of your fully developed presentation by a qualified peer.

17
Writing for Publication

JAMES W. LEA

In this chapter, we will discuss the most common written forms of professional scientific presentation—the published article and the letter to the editor—and we will review some techniques of administrative writing as well. We will be applying to the written mode the general principles of professional presentation discussed in the preceding chapter, and we will offer guidelines for succeeding in some specific professional writing situations.

Tailoring the Manuscript

In the preceding chapter, we discussed the importance of characterizing the audience as a step in planning the professional presentation. This concept has special application in the tailoring of a manuscript for publication. Before beginning the exacting and often tedious process of shaping information—patient records, research notes, literature references - into a publishable manuscript, you should have clearly established what your message is and to whom you want it delivered. Then match the message to the intended audience by tailoring the manuscript to fit the requirements of those outlets—journals, books, lay publications—that (a) publish material in your topic area and (b) are read by the readers you want to reach.

Tailoring your manuscript involves three factors:

- Message
- Intended audience
- Outlet

The tailoring process might be describe in three steps and at each step in this process, you should bear in mind some important questions. (Figure 17.1)

Step 1: "What type of researcher, specialist, or practitioner is most immediately interested in what I have to say?" (Possible answers: The

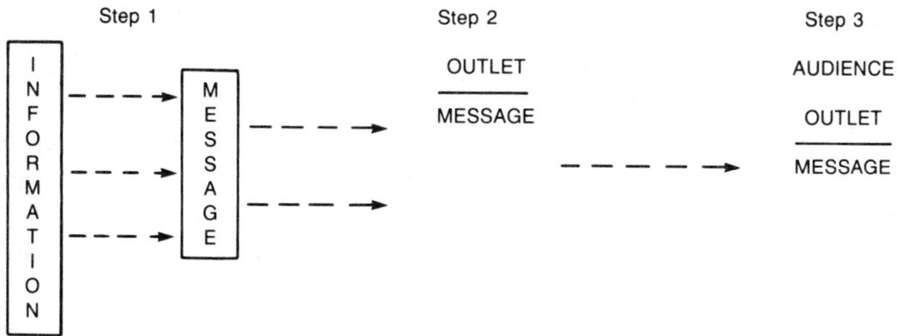

Figure 17.1. Steps in manuscript preparation.

person who's doing research on similar questions. The person who sees patients similar to mine. The person who needs to know what the literature says on this topic, but who doesn't have the time to do a complete review.)

Step 2: "What shape should I give to my message in order to emphasize the things that that reader most wants to know?" (Possible answers: The Methods are especially important to the biomedical or clinical investigator. Results and Discussion will interest the physician who may be scrambling for a diagnosis on a similar patient next week.)

Step 3: "What outlet is most likely to carry my message to my intended audience?" (Possible answers: The journal that you find most relevant to your work. The journals cited in the literature review in your Introduction.)

Outlets and Their Audiences

It is easy to match some categories of outlets with the readers for whom they are published. There are always exceptions to the general rules, of course, but the following three categories of outlets and audiences are a good start.

Scientific Journals for Scholar-Researchers

Scholar-researchers, men and women whose primary or even exclusive professional focus is on laboratory or clinical research and the publication of findings, probably communicate with one another most often through the scientific journals. Their reading interests are likely to lie in topics that extend the body of knowledge relevant to their particular areas of research. The journals published for them are consequently most interested in writing on hard, well-documented clinical or laboratory investigation. Some examples: *Proceedings of the National Academy of Sciences, Journal of Clinical Hematology and Oncology.*

Professional Journals for Scholar-Practitioners

Physicians whose professional responsibilities include a large component of patient care, teaching, or administrative medicine, as well as research, are categorized for our purposes as scholar-practitioners. The professional journals published for them include the general medical journals, such as *JAMA,* and specialty-specific journals, such as the *Journal of Family Practice* and the *Annals of Internal Medicine.* These journals publish articles reporting hard research, but are generally quite interested in the patient care implications of such research. Your investigative methodology must be sound and your discussion must be scientifically sensible if you want to reach your intended audience through the professional journals. But the utility of your findings for the clinical practitioner should also be clear.

There is one generic distinction among journals of which you should be aware, and that is the difference between *refereed* and *nonrefereed* journals. The refereed journal accepts or rejects a submitted manuscript largely on the basis of recommendations by reviewers, or "referees," who are representative of the author's professional peers. Because publication in a refereed journal is assumed to reflect judgment by a panel of knowledgeable reviewers, rather than by the journal's editor alone, it is generally viewed as more prestigious than publication in a nonrefereed journal.

Textbooks and Chapters for Various Audiences

Textbooks and book chapters are seldom published as a result of blind submissions by hopeful authors. Preliminary detailed negotiation with a publisher precedes the writing or editing of a book. A chapter in a book edited by someone else is usually written by invitation only. Although there are special considerations in using these outlets, if you find an opportunity to publish through them, always ask at the outset for whom the book is intended and to whom it will be marketed. The readership may eventually be quite broad, including students, professors, practitioners, and others, but the book will be edited and marketed with specific readers in mind. The characteristics of those readers will directly affect whether or not you have something to say to them and how you should say it.

Too many rejection slips accumulate on authors' desks simply because they didn't take the time to identify their most likely outlets and then to become thoroughly acquainted with them *before* starting to write. After making preliminary choices of journals that you feel would be likely outlets for your article, take the time to read several issues of each one. The "Information for Authors" section that most journals publish on their masthead pages includes very specific guidance on the topics each journal is likely to publish, as well as manuscript style and length, and specifications for charts and tables. Ask the opinions of colleagues who read those journals regularly. Time invested in becoming thoroughly familiar with prospective outlets will save much more time and frustration in the future.

Once you have chosen the most likely journal for your manuscript, take careful note of length, organization and suggested format for submissions. Much good work is returned because it doesn't follow the "house style" of a particular journal.

Writing and Revising a Manuscript

Planning

Preparing a manuscript for publication has one very important thing in common with good patient management: it must follow a plan. In Chapter 16, we discussed the value of knowing the objectives of any professional presentation. In writing for publication, you should have your objectives clearly in mind when you begin to plan your manuscript. In this case, stating your objectives takes into account the nature of your audience and the requirements of the selected outlet as considerations basic to your writing plan.

The choice of format for your article may follow—though not necessarily slavishly—the type of topic you're treating. We have already discussed the common types of professional presentations and their usual formats: the clinical or laboratory investigation, the case report, the literature review, and the letter to the editor. The components of the investigative and patient case report formats—Introduction, Methods, Results, and Discussion—are generally accepted by authors, editors, and readers because they distribute information in a way that tends to assure logical organization and appropriate detail in each component.

When you begin to plan your manuscript, the raw information should be in some form that allows it to be handled, literally moved around. Perhaps you have collected your information by topic or subtopic on note cards, separate sheets of paper, or stacks of computer printouts. Whatever the form, it should be flexible enough to be shifted as it's shaped and reshaped. Use of a word processor for manuscript planning, development, and editing greatly increases a writer's flexibility.

Drafting an *outline* of the manuscript is a tedious step, but for most authors it is essential. Your outline may vary in complexity and exactness from a straight list of contents to the classic form that uses Roman numerals, capital letters, Arabic numbers, lowercase letters, and so on. The outline should be as complete as possible. It should be divided into the major headings and then into subheadings required by your choice of format. For example, under the Results heading you should organize the data that belong in that component (including tables and figures) in the order in which they should appear, and the notes or comments that may be required to clarify them. In the Discussion section, the outline should order your principal points consistent with their importance or applicability. Finally, you may wish to put the finished outline aside for a while, then return to it when your mind is fresh to review its completeness and the logic of its organization. You should have confidence in your outline before you begin to write. It is the skeleton on which your manuscript will grow, so its strength and accuracy are essential.

Beginning to Write

You are now at one of the most challenging points in the entire process of preparing your manuscript. The objectives are set, the data are organized, and the outline is before you. Now there is nothing between you and the satisfaction of seeing your work in print—except that first dreadfully empty sheet of paper. It is at this point that the greatest scientific mind goes blank, the workaholic physician stares out the window, the star researcher calls for a lowly but literate graduate assistant. There are no secret paths around this obstacle, and no universal formulae for overcoming it. Writing experience seems to diminish writer's block, but not much. When you're well into the writing, you won't even remember this moment when the pen has to be forced onto the paper, the fingers forced onto the keys. The only comfort to be taken in this first confrontation with the empty page is in the knowledge that it happens to practically everybody.

As you proceed to put words on paper, keep at hand the notes that you made on the mechanisms of professional presentation in the preceding chapter. Be conscious of your use of appropriate professional vocabulary, respectable grammar, language that communicates directly, and a style of writing that is suitable to the topic and the audience. Attention to these factors will result in an article that communicates a lot of information in a relatively few words. Written presentations should not be turgid. They should be as concise as they are complete. Your finished manuscript may weigh a little less than its early versions, but it will be more readable.

References and Citations

Citations of sources of information outside your original work—references to previous journal articles or books, published letters, unsigned or even unpublished papers—should follow established forms. Individual journals may require slight variations according to their own styles, but the most commonly accepted citation forms are those established in the *Uniform Requirements for Manuscripts Submitted to Biomedical Jour-*

nals, often referred to as the Vancouver style. (Other formats include those used by the American Medical Association and the American Psychological Association.)

For the Vancouver style, citations are numbered within the text:

The work of Bobb and Reh (13) first identified the etiology of. . .

The number in parentheses refers to a completed citation of the source, in this case a journal article, which appears in the list of references at the end of the article in this form:

13. Bobb B, Reh S: Etiologic factors in chronic hypothermia. *J Med* 1978; 5:229–230.

This citation contains the name(s) of the author(s), the title of the cited article, the abbreviation of the name of the journal as used in the Index Medicus, the volume year, the volume number, and the inclusive page numbers. Note the use of capitalization and punctuation in the citation. Only the first four authors of a coauthored article are named; the others are indicated by *et al.* This connection between the reference number in the text and the complete citation at the end of the article enables the reader to go to the cited source for more information.

Here are some other standard reference forms:

A Book

Bobb B, Reh S: *Chronic Hypothermia.* Boston, MA: Brothers and Sons, 1975.

A Chapter in A Book:

Little JP: Signs and symptoms of hypothermia. In Bobb B, Reh S, eds. *Chronic Hypothermia.* Boston, MA: Brothers and Sons, 1975; 456–98.

An Unsigned Article:

Case 41223: Pedal hypothermia in a medical writer. *N Engl J Med* 1969; 283:91–93.

An Editorial:

An epidemic of hypothermia (Editorial). *JAMA* 1974; 337:504.

Preparing an Abstract

The last component of the article to be drafted is the abstract. The Information for Authors page of most journals defines the requirements for abstracts. Usually, the abstract is a 135–150 word synopsis of the article that states its main factual points and that is intended to give readers a quick but accurate overview of the article's contents. A very handy guide for the preparation of your abstract is the IMRAD scheme of organization, with each component presented in no more than a sentence or two:

Introduction—First sentence: States the research problem or otherwise characterizes the study.
Methods—Second sentence: Describes in one sentence (two at the most) the major features of the method of investigation or intervention.
Results—third (and/or fourth) sentence: A direct factual statement of the most important finding or outcome.
Discussion—Last sentence: The article's interpretive conclusion in one sentence.

Before drafting your abstract, read a few that appear in major journals. Although the numbers of total words and the numbers of sentences devoted to each component may vary, you should be able to see how the IMRAD structure applies in those abstracts that are most communicative.

Seeking Local Critique

With the attachment of the last page of references, or footnotes, the first draft of your article is finished. But every new product should be tested in-house before it's released to the public, and the same is true of your paper. Probably the best sources of presubmission criticism of your article are selected local colleagues, because they are (a) knowledgeable enough to understand whether you're saying what you think you're saying, (b) close at hand and available for give-and-take discussion, and (c) usually friendly and constructive in their comments. When you have read through the draft a few times, work up your courage and ask around the medical school or the department for some candid criticism. Tell the reviewers that you would appreciate their honest critique. Be as specific as possible in requesting comments. If you feel the flow may be a problem, the language unclear, or want a comment on the data organization, say so in your note to your colleagues. Then give them time and room to read. When you submit your work for consideration by a journal, the work will have to stand on its own merits. The objectives, organization, data distribution, and other aspects of the article must be self-evident. So try to resist the temptation to hang around and "narrate" the paper as it's being read; if the local reviewer can't comprehend your message without assistance, the message needs more work.

When your colleagues return the draft to you, ask them to devote a little more time to discussion of their findings. Ask them to identify very specifically the draft's deficits of clarity, organization, and language. Take seriously the suggestions for improvement, and take copious notes. But try to separate objective, documentable criticisms from vague points of divergent opinion. "If I were saying it, I would have said. . ." may represent the reviewer's very subjective sense of style, and that is simply a matter of authorial prerogative.

At this point, a few days "rest" for your manuscript may help you go about revisions in a more objective manner.

Revising the Draft

For some authors, making manuscript revisions is akin to taking out one's own gallbladder: psychologically wrenching and technically quite tricky. But revising your draft should not be that difficult a task, no more so than sharpening a good tool. Start by rereading the paper yourself with as objective an eye as possible. It's not unusual to spot obvious points of confusion and error in communication during such a rereading. Then turn to the informal reviews by your local colleagues. Reread your paper from the perspective of their comments; smooth out the wrinkles that they have identified. Next, go back to the principles of developing and refining a professional presentation given in Chapter 16. Test your manuscript for organization, coherence, and clarity according to the criteria suggested there, particularly in the section on distributing data. Although you may have lived with your investigation for months—or even longer—it's still possible to neglect some crucial information when you are writing about the work. Also be alert for those strange shapes that language can take when it's used for learned purposes. Remember that the simple declarative sentence produces the most effective communication, and be ruthless with writing that muddles communication—even if you wrote it yourself.

Finally, don't overrevise your manuscript. When you have corrected the major flaws and answered the questions raised by your colleagues' reviews and your own, you have probably done enough. Early in your professional publishing experience, you may often feel pressed to make "just one more change." Resist the temptation. The Great American Scientific Article will not be written with an overzealous blue pencil. Hand the manuscript over for final typing and turn your attention to the processes surrounding its publication.

Coauthorship

Because many manuscripts are coauthored, a few words about the sharing of authorial responsibility may be helpful here. Coauthorship is a legitimate scholarly way both of assuring recognition of all those who had major roles in the work being reported in the manuscript and of spreading the tasks of writing the article. One of the two or more authors, usually the one who led the research, is designated as senior author. The senior author, whose name appears first in the list of authors on the title page, has responsibility for coordinating the other authors' work, assuring the thoroughness and accuracy of the manuscript, and supervising preparation of it in final form. Editors to whom the manuscript is submitted will communicate with the senior author, and if the manuscript is published, requests for reprints and other correspondence about the article will usually be addressed to the senior author.

Who should be included as a coauthor and what constitutes sufficient contribution to a piece of research to justify inclusion as coauthor are hotly debated questions in academic life. Recent discoveries of flagrantly manufactured data published by young researchers to which highly regarded senior researchers had appended their names raises the question of the responsibility of *all* authors for the integrity of the research reported. The increasing trend to include any and all names of "contributors" as coauthors may work against each in the long run as research articles appear more and more to be "written" by the last four pages of the telephone directory.

A solution to the problem of hurt feelings or spurious data in research lies in authorship negotiation for manuscripts *before* the research project begins. If this is impossible, negotiation should occur before the presentation of results.

Coauthors have responsibilites, both to their profession and to themselves, for the quality and the accuracy of the manuscript. Never allow your name to be attached as coauthor to a manuscript that you have not read in final form. Regardless of which contributing coauthor made the blunder, a substantive inaccuracy in a published article reflects badly on all of them.

Submitting the Manuscript for Publication

The submission of a finished manuscript to the journal of your choice is a transition into another part of the process of professional publication. Once mailed, the manuscript begins a journey through editorial scrutiny, peer review, and, if it is accepted, proof printing and final printing. One important note: most journals state emphatically that a manuscript submitted to them cannot have been previously published or submitted simultaneously to another journal for consideration, either in its present form or a similar one. There are good reasons for those rules, and your credibility as a scientist or clinician can be damaged if you don't observe them.

The Review Process

The first review that a manuscript receives in the offices of the journal is given by the editor or the editorial staff to determine that basic submission requirements have been met. Generally, the manuscript will be checked to be sure that maximum length restrictions have been observed;

that the senior author and coauthors are clearly identified; that the original and the copies are typed double-spaced on the correct paper size; that tables, charts, and references are properly organized and labeled; and that any necessary transmittal letters and informed consent stipulations are attached. The basic criteria vary with individual journals. If the manuscript is seriously deficient in any of these areas, it may be returned to the author for correction. If the deficiency is minor, the editor will probably request that the author send the missing materials. Once the manuscript-submission is considered complete, the editor will distribute it for peer review.

The *peer review* of a manuscript is based on the academic tradition that the merits of scholarly endeavor are best judged by the community of scholars. The journal editor, therefore, selects a small number of reviewers (usually no more than three to five) from among a pool of people knowledgeable in the field or the subject about which the article is written. Reviewers may be members of the journal's editorial board, distinguished professionals, scholars who publish often in the journal or who are otherwise known to the editor, or those who have volunteered to review manuscripts dealing with certain topics. (This, incidentally, is an excellent way to broaden your own insight into the techniques of writing for publication; just write to your favorite journals and offer your services as a reviewer.) Manuscripts submitted by well-known authors or those dealing with controversial topics will generally be distributed to widely respected professionals for review.

Most journals conduct "blind" peer reviews. That is, the manuscript is sent to reviewers with all indentification of authorship removed in order to avoid any actual or perceived reviewer bias. Along with the manuscript, the editor sends each reviewer the specific criteria for review, which usually include scientific accuracy, technical quality of methods and analysis procedures, coherence and clarity, and appropriateness for publication in the journal. Many reviewers make a practice of confirming the accuracy and relevance of the author's bibliographic references. The reviewer is generally expected to write a page or more of commentary, focusing on the journal's criteria and suggesting ways in which the manuscript could be strengthened, and then to recommend that the manuscript be accepted for publication, accepted if specified revisions are made, or rejected.

If your manuscript is submitted to peer review, you may expect to receive the reviewers' comments and suggestions on your article, whether it is accepted or not. If the manuscript is accepted contingent on your making revisions that will satisfy the reviewers' objections, then the value of the reviews is obvious. But even if your manuscript is not accepted by that journal, you should still value the reviews as expert critiques of your research and your writing. Study them thoughtfully and incorporate their suggestions into your next effort.

If your manuscript is not accepted for publication, don't take it personally. Some people are "born writers," but for the rest of us the submission/acceptance ratio will improve only with time and experience. A rejection may not reflect the intrinsic merit of your article. There are many possible reasons for a paper's rejection by a given journal at a given time, ranging from a fundamental mismatch of message and outlet to an editor's decision that the journal has published enough articles on your particular topic recently. If your manuscript is rejected but you believe in its merits, then rework it, get another round of local critiques, and submit to another journal.

Publication

Despite rumors to the contrary, many manuscripts are accepted for publication. When this happens to your manuscript, it's a very good feeling, even if it has happened many times before. The editor's letter notifying you of acceptance usually also informs you when the article will be published and when you may expect to receive galley proofs for your inspection. The *galley proof* is your article set in type and printed in the journal's format. Galley proofs (which get their name from a printer's type tray) are produced to provide both editor and author with a last chance to check for errors in fact, grammar, or punctuation. When your galley proof arrives, you will be asked to review it and to indicate changes that should be made. Be careful about those changes. If there are typographical or spelling mistakes made by the typesetter, point them out clearly on the proof. But remember that the journal editor has probably made some revisions to your manuscript that will appear in the galley proof; unless you feel that the revisions damage the article, don't insist that they be changed. And don't try to use the galley proofing as an opportunity to add a few more data sets or a few good ideas that had not occurred to you earlier. Check the galley proof thoroughly, and return it to the editor by the specified deadline. Many journals have policies that interpret the author's failure to return proofs on time as acceptance by the author of the proofs as they stand.

This chapter has focused on basic techniques of writing for publication in professional journals and other outlets. Sound structure, careful grammar, proper use of language, and, of course, having something important to say in the first place are central elements of publication success. However, there is no substitute for experience in the use of the tools of written communication. Write and write often. And good luck.

18
Techniques of Oral Presentation

JAMES W. LEA

Several years ago, a major Midwestern newspaper, conducting man-on-the street interviews, surveyed passers-by about their greatest personal fears. The number one fear revealed by the most people was speaking before a large audience. Death ranked number four.

In previous chapters, we discussed the organizational and stylistic principles of professional communication, and how to apply them to writing for publication. In this chapter, we will apply the principles of professional communication to such tasks as presenting research papers at professional meetings and conferences. (We will not attempt to deal with effective lecturing per se, although many of the techniques are transferable). We will examine the adaptation of written material to spoken presentation. And we will review some of the physical and psychological elements of face-to-face communication with an audience of professional peers. When you complete this chapter, you should be able to use these proven techniques of oral presentation effectively and confidently.

Preparation and Practice

Effective oral presentation begins with something to say, that is, a well-developed paper that is consistent with the principles of format selection, structure, and style established in Chapter 16 of this *Handbook*. IMRAD is as useful for planning oral presentations as for planning a manuscript for publication. The same attention to details of outlining and data distribution should be paid to the preparation of oral and written presentations. Certainly, as a speaker at a professional meeting, you should try to get to know the background and expectations of the listeners as well as you get to know the readers of a journal to which you submit a manuscript. But there are also some very important differences between presentations made in print and those made orally. It is a deadly mistake to read *to* an audience a paper which is written to be read *by* an audience.

Writing for the Ear

There are subtle but important differences in language and sentence structure between material that is written to be read and that written to be heard. Good broadcast writers know that a clear and easily understood printed sentence can be utterly unintelligible when it is read. So they call writing for a listening audience "writing for the ear." For example, read the following sentence silently.

Sixty-five percent (114) of those receiving questionnaires responded, and 50% (57) of those respondents reported repeated epileptic episodes.

Although this sentence may not qualify as deathless prose, it is reasonably communicative in print. But now try reading the sentence aloud. If you wish further confirmation that the sentence was written for the eye and not the ear, read it aloud to someone else and ask your listener to repeat its substance to you.

Here is a short checklist to help assure that the paper you prepare for oral presentation is written for the ear:

- Short sentences are easier on the ear than are long sentences. A paragraph-long printed sentence is confusing enough, but it is virtually meaningless to the listener who can't reread it several times.
- Correct grammar is essential in oral presentations. Be particularly attentive to subject-verb-object agreement and to establishing clear antecedents for pronouns. One unclear use of "its," for example, can confuse your listeners entirely.
- Avoid parentheticals (the phrases of explanation or elaboration that look like this in print) because only a very skilled speaker can create parentheses with his voice.
- Alliteration, a series of words beginning with the same letter, can be interesting in small doses. But a long series of alliterative words may sound too "cute," and the listener can become lost in the word sequence.
- Pronunciation of words that you write for print is the reader's problem, if anyone's. Remember, however, when choosing words for an oral presentation, that you will have to pronounce them yourself.

Manuscripts and Notes

It is conventional to speak of making a presentation at a professional meeting as "reading a paper." That, unfortunately, is exactly what too many people do. The effect on the audience of a presenter standing behind a podium, head bent over a manuscript, reading page after monotonous page ranges from the mildly soporific to the totally anesthetic. Several suggestions are made in this chapter for avoiding such a problem. The first one concerns preparing your manuscript or notes for the presentation.

As a speaker, you are the medium for communication of your message to the audience, just as the printed page is the medium in a journal. The difference, however, is that the printed page is a passive, or even static, medium; you, one hopes, are not. Reinforce this distinction when preparing your presentation by translating your manuscript into an outline or note cards as the basis for your presentation. Some speakers write out full sentences for the opening of each paragraph, with key words or phrases to remind them of the important points that follow. If you will be using visuals to present data, there is no need to duplicate the visuals on the note cards. But important points of comparison or analysis should be noted. The form and amount of detail of podium notes will vary according to individual speakers' needs for prompting. But if you can comfortably rely on notes or an outline instead of a manuscript, your presentation should be a more active and interactive event.

Sometimes tradition or the need for exactness in the presentation will demand the use of a manuscript. In such a case, you should still make every effort to avoid simply reading it to the audience. Have the manuscript typed triple-spaced, and tell the typist not to break words with hyphens at the ends of the lines. Some speakers prefer to have their podium manuscript typed in all capital letters. Have the manuscript typed with wider than normal margins, especially on the left side, and make "director's notes" to yourself in the extra space. Indicate the points for slide changes or distribution of handouts. Note where you should pause for emphasis. If you are an inveterate manuscript-reader, jot down such occasional reminders as "look at audience." In the text itself, underline vital phrases. And don't staple the manuscript pages; leave them loose so that they can be easily and unobtrusively handled during the presentation. The manuscript that you use at the podium is your personal tool, so take every useful liberty in customizing it for your personal needs.

Rehearsal

Just as your local colleagues can provide valuable constructive critique of a manuscript you are preparing for publication, they can also help polish your oral presentation. Select one or two auditors from among your colleagues, and stage a rehearsal of your presentation. The auditors should be knowledgeable of your field, somewhat discriminating in their judgment of presentation styles, and willing to give honest and constructive criticism. As a general rule, spouses do not make the best auditors, although there are always exceptions.

Before beginning the rehearsal, take your notes or customized manuscript in hand and think through the presentation. Envision the setting and the make-up of the audience. Formulate an *objective* for your presentation. Try to simulate the immediate physical environment of the real presentation; set up a podium and a slide projector and pointer, for example, if you will be using them in the presentation. Then rehearse your presentation completely, timing it. Ask for the auditors' feedback on specific characteristics of your paper and your manner of presentation. Incorporate their suggestions into your note cards or the "director's notes" on your manuscript. Rehearse again to be sure that you have incorporated the correct suggestions and have actually strengthened the presentation. Don't be satisfied until the presentation sounds just like you (and the auditors, if they're patient enough) want it to sound.

Confirmation

The last important step in preparation comes just before the presentation. Before the meeting begins, review your materials carefully. Check your notes or manuscript pages to be certain they are in order. Check slides to be sure they're inserted in the tray properly. Check charts, overhead projector transparencies, and other support materials for order and accuracy. And check the equipment you will use in your presentation, too, to be sure it's in working order. When these details are confirmed, you're as ready as you can be.

Podium Presence

Like the first empty sheet of paper in the typewriter, the sight of a sea of expectant faces in an audience can produce some stirring effects. Pulse and respiratory rates change, the palms become moist as the mouth drys up, and diverse muscle groups begin to spasm. It's a form of stage fright experienced by most speakers at one time or another, to one degree or another. Some very experienced speakers say that it's a healthy, stimulating response and that they wouldn't be able to make a presentation without a touch of stage fright at the onset. Others find it embarrassing and uncomfortable, probably because they don't realize that the audience can rarely tell whether the speaker is nervous or not.

Like the young writer's block, beginning speaker's stage fright may be remedied by experience—or it may not. Since it happens to almost everyone, though, we'll discuss podium techniques to keep nervousness from interfering substantially in the effectiveness of your presentation.

Posture

Behind the podium, stand erect with your weight evenly balanced on both feet. Don't lean on the podium, and don't rest your weight on one leg with the hip protruding. You will feel and look more relaxed if you stand straight. Your chin should be slightly above the horizontal, except when glancing at your notes or manuscript, to help voice projection and breath control. Rest your hands naturally on the podium (no white-knuckle grip), or put them somewhere else that's both comfortable and reasonably casual. Avoid fiddling with papers, pens, necktie, or handkerchief. Such a self-contained, self-assured bearing will give you a sense of being in command of the situation, and it will reduce physical distractions that interfere with the audience's concentration on what you're saying.

Pause

Even if you are allotted only 20 minutes to describe 5 years of research, there's no need to rush into your opening sentence. When you first reach the podium, take a moment to arrange your papers. Look *at*—not over or through—members of your audience. Take a deep breath. Relax.

Speaking Style

Always *talk* to your audience, regardless of its size, even if you're relying very heavily on a manuscript. Use conversational "up-and-down" voice inflections. If you are comfortable using moderate hand gestures for emphasis of certain points, by all means do so. If you believe what you're saying and you have rehearsed it adequately, your presentation style will be authoritative and convincing, yet natural. One school of public speaking insists that an address of any kind must be delivered in a rhetorical, even bombastic, manner to be effective. It's not true. Abraham Lincoln carried the day at Gettysburg with a high, nasal voice and no podium-thumping at all. Your presentation will succeed with a straightforward, confident speaking style.

Even in the most serious scientific presentation, there is a role for controlled and tasteful humor. The audience didn't come to be entertained so don't feel that you must tell a joke to justify your place on the program. On the other hand, if a light remark feels natural to you and does not disrupt the flow of your presentation or distract the audience, include it. Just be careful to control an impluse to ad lib too freely or too often.

Handling Papers

Whether you are using note cards or a manuscript, keep your papers low on the podium and unobtrusive. Manuscript pages should not be stapled or bound, because turning pages is a physical distraction and it impedes the smooth, natural flow of your speaking style. Arrange the pages on the podium so that you are referring to the page on the left, with the next page in view on the right. Instead of turning to the next page, change pages by sliding them from right to left. Bending the upper-left corner of each page will make it easier to grasp and slide.

Communicating

Since communication is the essence of the oral presentation, remember that you are involved in an intellectual transaction with your audience. You share with them the goal of exchanging ideas and information. As you make your presentation, you are playing a special role in the group, but you are still a member of the group. The audience is on your side. Talk *to* them. Look them in the eyes, long and often. By doing so, you force their attention on you and what you're saying. But remember that your role is to communicate and interpret, not to entertain, so keep ideas and information—the intellectual content of your presentation—in the spotlight.

Handling Visuals

Visual materials—slides, overhead transparencies, charts, etc.—should be used to support the oral presentation, and not vice-versa. Don't let your attention and that of the audience become riveted to the visuals. If you, the speaker, must look at the screen or the chart, keep your glances brief. Don't turn your back on the audience or the podium microphone as you talk. And if the projector or other equipment breaks down in the middle of your presentation, don't be ruffled. What you have to say, not what you have to show, is the substantial thing. You can do without the slides.

Managing Questions

There are several reasons why members of an audience may ask questions following a presentation, and not all of them relate to desires for more information. Some questioners would like to satisfy their own egos by deflating the presenter. Others apparently didn't have papers accepted by the meeting's program committee, so they use the question period to make their presentations from the audience. In any case, if open questions are asked following your presentation, respond to them briefly, concisely, and in a professional manner. Don't be afraid to say that you don't know the answer to a question, or to admit to an obvious weakness in the study that you have presented. Remember that a question is not an invitation to present a lecture, so keep your responses short and direct. And if a questioner becomes argumentative, remember that you don't have to respond from the podium. "That's an interesting point, but it deserves a more complete answer than we have time for here. I'll discuss it with you following the session."

At this point, there's nothing more for you to do but graciously accept the applause. These techniques for effective oral presentation, in conjunction with the Chapter 16 guidelines for professional communication, should serve you well as you plan, develop, and deliver your presentation. Of course, there is no substitute for practice. As you become more experienced, you should become a more skilled presenter and you should feel more confident in your presentation style. You should even be able to relax and enjoy the intellectual challenge and the rewards of reporting and discussing your work before a group of colleagues. You should also continue to use the "tricks of the trade" that assure you of sound preparation and strong podium presence.

19 Visuals for Written and Oral Presentations

MADELINE P. BEERY

It has taken years to disseminate some of modern medicine's most original ideas because the originator did not communicate the significance of the information adequately. In a era when information is produced at exponential rates, it is vital that academic physicians formulate their ideas in ways that are communicated quickly and accurately. The previous chapters addressed important stylistic and professional considerations in developing the verbal component of professional presentations. This chapter will concentrate on the visual representation of such information, focusing on the textual, tabular or abstract forms possible.

Carl Jung said, "To the scientific mind it is most annoying to deal with phenomena that cannot be formulated in a way that is satisfactory to the intellect and logic." Some information is simply too complex, multifaceted, or abstract to be represented effectively by words alone. Visuals can be a tool to satisfy "intellect and logic."

In this chapter, we will explore a pragmatic approach to improving communication through the use of visuals. We will describe the different types of visuals, then the guidelines for the use of various types of graphic displays, whether for written or oral communication. We will offer specific suggestions for the use of visuals in written communications, and we will expand on the adaptation of the design principles for different media and visual display formats used in oral presentations.

We will suggest new options and alternatives for creating effective visuals. There is now an abundance of resources including software for the home computer; sophisticated display software for the mainframe; and an array of professional artists, photographers, and instructional designers skilled in the development and rendering of visual information. A working relationship with microcomputer technology or with someone skilled with that technology may also be needed to keep current.

Where Can Visuals Be Used?

Visuals have typically been used to amplify, clarify and even add comic relief to professional presentations. A less conventional use of visuals is to communicate a synthesis or "gestalt" of quantitative trends or abstract ideas. Visuals can make the abstract concrete by providing a visual referrent.

The Traditional Roles of the Visual

Journals, books, and presentations at professional meetings are filled with examples of visuals used to provide evidence by enumerating facts and statistics. Regular attendance at professional meetings might even suggest that presenters oversaturate medical audiences with unexplained data.

Realistic visuals illustrate or depict information. An iconic representation of a blood vessel or a molecule helps the audience comprehend accurately and quickly what a presenter means by such concepts.

Visuals can also be used to communicate background information that trace an idea from its origins. This can be useful in establishing the setting for a study or increasing its immediate relevance to the audience.

Another common use of visuals is to provide a logical overview of the subject matter. By presenting structures, directions, or procedures, visuals can depict a complex relationship that may have a spatial or temporal link.

Visuals are often used to gain attention, particularly at professional meetings. Such visuals present information in a provocatively topical or humorous manner.

The Less Traditional Roles of Visuals

Visuals can represent abstract theories and concepts in a way designed to provide a conceptual image. Dalton talks of the importance of providing others with a "cognitive structure" as a critical step in the integration and internalization of information. What better way to provide a cognitive structure than through visual representation that augments the language of the presentation?

Since most of us were trained in highly verbal settings, spending years refining our verbal skills, it is understandable why few of us communicate as effectively in this very different modality. Not all thought is rational, nor is all important information processing done in a linear, predictable manner. It is in this area that visuals can be most helpful.

Visuals are uniquely able to *represent nonlinear information* in a way that words cannot; they can simultaneously represent multivariate information such as data or abstract relationships that may be studied from different perspectives.

Recent research shows that the right cerebral hemisphere, which is more metaphoric and intuitive than the left cerebral hemisphere, processes most visual information. Since most of us have devoted years to refining our (analytical) "left brain," it is understandable that we are more comfortable with words. Yet, a good visual has a forthrightness that words often do not.

Visuals are able to make abstract concepts concrete by making them analogic in nature. Some applications, such as marketing and highway safety, capitalize on this by using visuals purposefully. Through the pictorial representation of a theory, research protocol, or model as shown in Figure 19.1 the physician can clarify his or her message and prepare the audience for the discussion of findings in a way that is more difficult with only the spoken word.

Visuals can also be useful in *comparing theory and practice.* The visual representation of the differences or similarities between hypothesis and results are highlighted and can make a powerful case for your conclusion. Although visuals can never change the outcome of research, the impact and clarity of the outcome can be amplified through well-conceived visual aids, as shown in Figure 19.2.

Visuals are a powerful means of *providing vicarious experience.* All the descriptive writing one can muster will not carry the effect of a photograph of the object described. The adage of a picture being worth a thousand words is axiomatic if you reflect on the effects of well chosen photographs in the course of your own education. There can be a strong emotional component to your visual aids that contrasts with the generally dry presentation of tables that characterize the majority of oral presentations. If selectively used, this emotional content can have an additive effect on other aspects of your presentation.

Providing critical information visually can help ensure that your audience has a *common grasp of important concepts,* allowing you to proceed more efficiently to results or discussion.

Oral communication has different visual needs from written communication, among which are *reinforcement and emphasis.* A group at a national meeting may require a restatement of the principal findings at the end of the results section to understand better the discussion of the findings. Visuals can be particularly effective in reinforcing and emphasizing the text through a different representation of the same information.

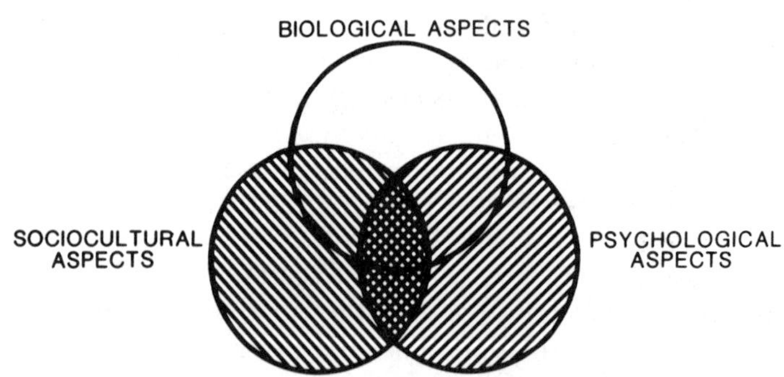

FIGURE 19.1. Biopsychosocial Model.

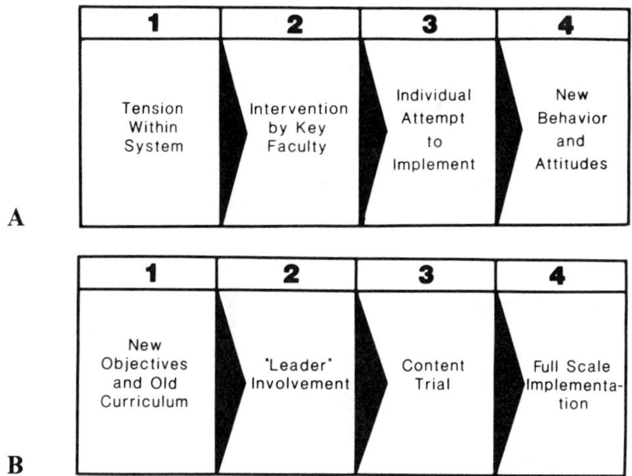

FIGURE 19.2. Dalton model: theory to practice. A: Dalton's model for induced change. B: Application of Dalton's model.

Many journal articles could also use the focus of visuals that emphasize critical findings or ideas.

A Definition of "Visual"

The subject of visuals often causes confusion. Are we talking of line drawings, representations of numerical data, abstract or representational graphics?

There are several classifications that can help clarify the subject. Merrill and Bunderson (1) break visuals into two categories, those that are word-and-number oriented, or alphanumeric, and those that are representational, or graphic, in nature; graphic visuals have varying degrees of fidelity to their subjects. In another classification, Wileman (2) puts information display on a continuum ranging from the most realistic to the most abstract. Table 19.1 attempts to combine those two ways of viewing visuals in a way that is useful for our purposes.

Alphanumeric Visuals

This category is comprised of words and numbers displayed in a textual or tabular form. The vast preponderance of visuals used in medical communications are of this category. Textual displays are very common in both written and oral communication, allowing easy study and review of content. In oral presentations, textual displays that take the form of reader slides are ubiquitous at professional meetings. In the written form it could be a chart or table.

TABLE 19.1. A definition of visuals

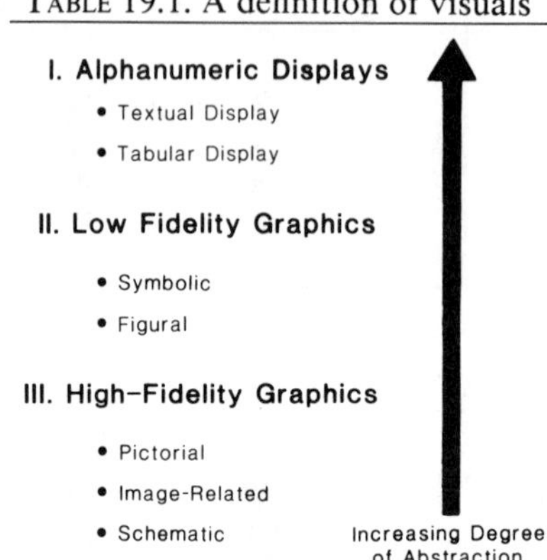

LOW-FIDELITY GRAPHICS

The graphics in this category are symbolic and figural in nature and bear no resemblance to the physical aspects of the object or event. The *symbolic* graphic presents an nonalphanumeric sign of the object or event (3) (for example, see Fig. 19.3). Such visuals can be powerful universal symbols for an entity that otherwise lacks an identifiable reference. Examples are the trade symbols used by AT&T, the AMA, and the Red Cross.

The *figural* category consists of graphic displays that create a visual analogy for information or data. Examples of these are line graphs, flowcharts, histograms, bar charts, and the bell-shaped curve. Figure 19.4 represents age and percent incidents visually.

HIGH-FIDELITY GRAPHICS

This graphics category includes two dimensional displays having some "degree of fidelity" or accuracy in representation. The first is the *pictorial* graphic; examples of these are photographs, films, and medical illustrations designed to resemble closely the appearance of the actual object. The second is the *image-related graphic* which builds on the basic shape of the entity but alters that shape, stylistically excluding details yet retaining the shape as readily identifiable. Examples of these include international signs for laboratories, restrooms, restaurants, and people as illustrated in the planned parenthood logo (Fig. 19.5).

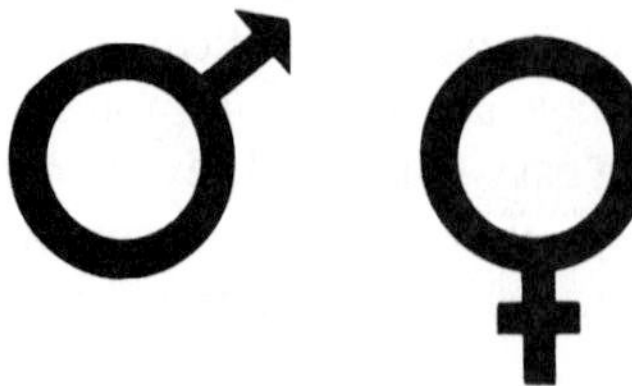

FIGURE 19.3. Symbolic graphic of man (left) and woman (right).

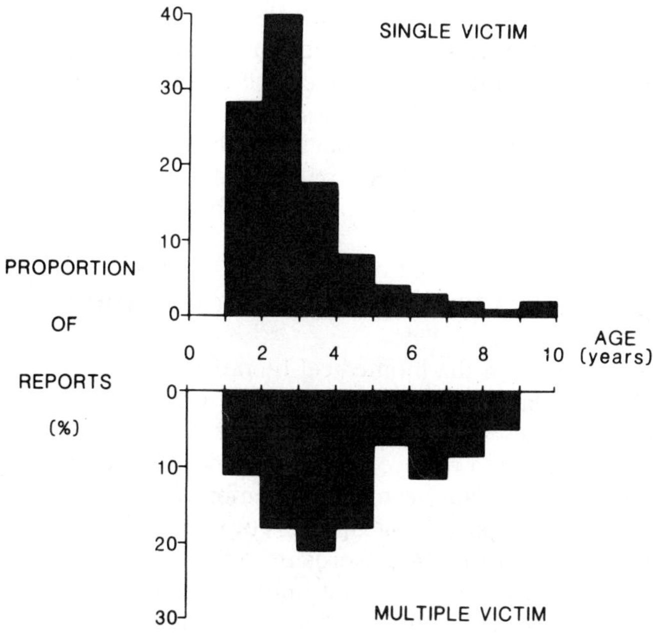

FIGURE 19.4. Histogram. (Reprinted from the *Journal of Toxicology: Clinical Toxicology* 19: 1075, 1982, with permission of Marcel Dekker, Inc.)

FIGURE 19.5. Image-related graphic. (Used by permission of Planned Parenthood of Orange County, North Carolina. Artist: Jacquelin Scharbrough.)

A third class of visuals having some degree of fidelity is the *schematic* which emphasizes the pattern and order of connections. Examples of schematics include a data flow model, a molecular model, an organizational chart, or a blueprint. Of these subtypes, the pictorial graphic has a greatest degree of resemblance or fidelity to the physical characteristics of the object; the schematic focuses on the pattern and interrelationships among component parts but has little resemblance of the entity itself.

Creating Visuals for Written Communication

A quick look through the biomedical literature finds visuals commonly used to present data, photographs, or a summary of findings. There are few visual examples of a process, an entity, a concept, or a theory.

Although unlikely to replace the verbal report, visual information can improve one's understanding of abstract information. Visuals are also an effective means of representing a process or relationship. Visuals, in fact, can work more efficiently than words in some cases.

Medicine relies heavily on visual information in the diagnostic phase but often seems to trust only numbers and words for professional communications. Lord Kelvin once said, "If it can't be counted, it isn't science." Science is not all statistics; it begins with hypotheses and theories that often deserve communicating, and can be represented by a schematic or symbolic visual.

Beginning to Think Visually

Communicating abstract ideas or relationships takes time. Much like writing an article, it requires experimentation with the ideas and conclusions to find the most effective way to convey your message.

Good Visuals Should Match the Focus

Choose an audience toward which you direct efforts; choose an idea that represents an important concept or relationship in your work; then begin over several days and weeks to think about the content; identify verbal analogies; and *design* various alternates for depicting your ideas. Finally, try designing a model or abstract symbol around your ideas.

Effective Visuals Require Trial and Error

When you are working with a visual idea, *develop* new visual alternatives each day but keep all the ideas together in a folder—tomorrow you may need the idea discarded yesterday. Look through journals, books, and other publications for ideas; look to see how others in your field communicate abstract ideas; scan the lay literature for graphic stimuli. Review the various solutions as you continue to develop or massage your information. *Try out* several preferred choices on a colleague, asking what is communicated or described. Is it what you intended? Finally, *revise* based on the response.

As you can infer from Fig. 19.6, this process is not unlike that used for the written text or manuscript. Refining may continue through the development cycle a second or third time with a revision of focus, a reworking of design, or a repositioning of elements.

Select the Format with Care

The selection of a visual format can be simple if you are experienced, and very confusing if not. Figure 19.7 is designed to aid you in asking the important questions and selecting the correct format for the information to be visualized.

Guidelines for Table Design

Tables are used frequently in professional journals, but often are either incomplete or too comprehensive. The following is a list of suggestions for the effective design of tables that builds on a list presented by Ehrenberg (4).

- Use columns for the display of numbers.
- Round numbers to two significant digits.
- Provide row and column averages.

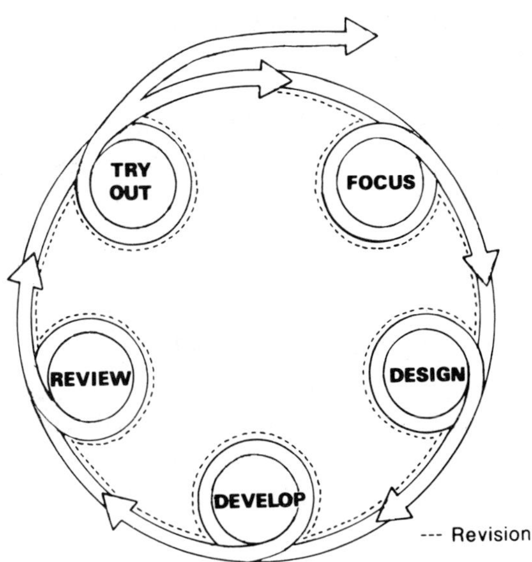

FIGURE 19.6. Cybernetic development model.

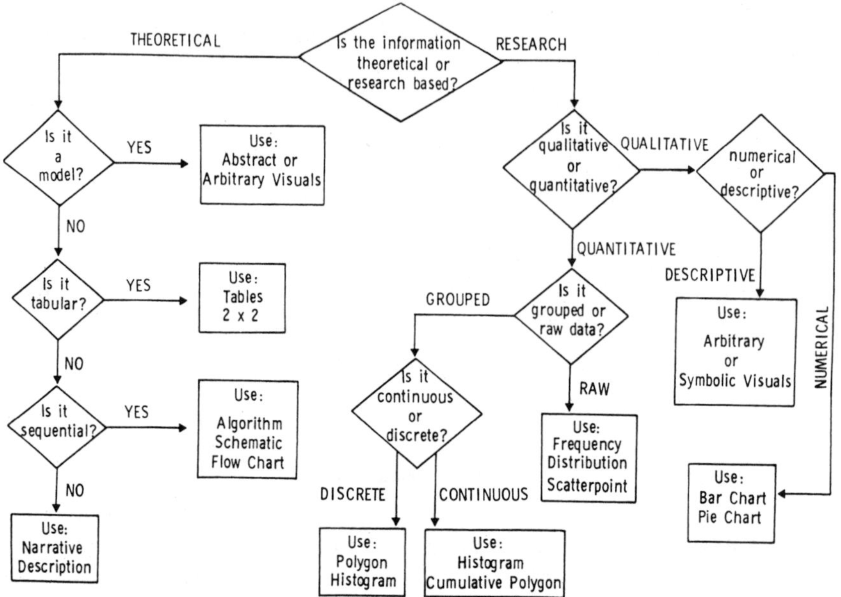

FIGURE 19.7. Decision tree; selection of visual format.

- Order the rows and columns by size of number.
- Use no more than five columns and six rows.
- Provide a succinct title for the entire table.
- Use a key to explain abbreviations.
- Provide references in footnote.
- Group the data wherever possible.
- Condense and summarize where possible.

The ultimate guideline should be to design a table to promote, not inhibit, understanding. As a compact means of communication, they require effort from the reader. The greater the clarity in layout, the greater the likelihood it will be studied.

Guidelines for Chart and Graph Design

Charts and graphs differ from tables in a number of ways. Although many of the rules are similar, there are also important differences. Basic guidelines for the formation of graphs and charts include:

- Provide a legend or title that is clear but brief.
- Label each axis.
- Use the x axis to record class intervals, and the y axis to record frequency.
- Place labels horizontally.

- Limit the information on each axis to one type.
- Limit the number of lines plotted on a polygon to five.
- Use a bar chart rather than pie chart if your data are precise.
- Use no more than seven slices on a pie chart.
- Support the chart in the text.
- Explain all abbreviations.
- Distinguish between continuous and noncontinuous data.

Guidelines for Designing an Abstract Visual

When developing arbitrary, "low-fidelity" graphics, certain principles of design can be used to enhance the message being sent.

- Be bold.
- Be conceptual.
- Be focused.
- Use shape purposefully.
- Be simple.

In the following example of Boolean Logic illustrating a search of the diabetes literature (Fig. 19.8), the visual was chosen because it is bold, it captures the concept of overlapping sets building on a known concept, and it is simple. In this case, we are searching the diabetes literature for that which focuses on the relationship between Type 1 diabetes and diet. Describing this process verbally can be confusing. A graphic can convey the message quickly and efficiently.

Creating Visuals for Oral Presentation

Visuals are commonly used in oral presentations to gain attention, reinforce verbal information, display data, and add variety. All are legitimate uses of the medium.

Visuals will benefit the presentation most, however, if they fit the overall purpose of the presentation. They should match the objectives and

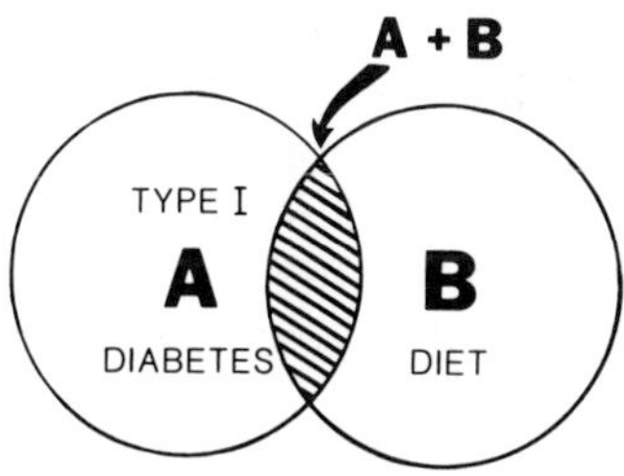

FIGURE 19.8. Boolean logic: searching the literature.

goals of the presentation in both the cognitive and affective areas. The medium should be selected that is most appropriate, based on stimuli, environment, audience, and speaker comfort.

Selection of the Medium

Visuals can be presented in a variety of formats including slides, videotape, overhead transparencies, posters and display. Slides are most commonly used at medical meetings because of advantages such as accuracy of color, magnification, and ease. Other formats offer other advantages that deserve consideration as well. Key criteria to consider include:

The Stimuli

Visuals can contain a number of important stimuli: color, magnification, movement, and shape. One way to determine the most effective medium is to identify an objective for the presentation. For example,

AT THE END OF THE PRESENTATION, THE PARTICIPANT WILL BE ABLE TO IDENTIFY THE DERMATITIS CLASS REQUIRING TREATMENT BY DRUG X.

Recognition of dermatitis requires that participants be provided with the accurate representation of the skin problem, showing color, shape, and size. The media that could accomplish this are slides, videotape, or film.

Consider the critical component of the visual, and then choose your medium based on that decision. Different media offer different stimuli. For example, slides offer color, magnification, and realism whereas video adds movement and sound. The color reproduction on video, however, is not as accurate as on the slide and, thus, might not be as appropriate, depending on the circumstance.

The Environment

The site of the presentation influences the effectiveness of the medium chosen.

At many national meetings, the speaker is forced to stand behind a podium. Use of a podium in this situation would limit the speaker's ability to interact with the overhead transparencies. Likewise, a poster for a formal presentation to an audience of 200 would not normally be effective, but would be fine for a panel or a poster session. Know the environment and audience size when making the final selection of the proper medium. Finalize arrangements with the program organizer prior to the beginning of a conference to alleviate problems and surprises.

The Audience

Audiences react differently to different types of visual media. Medical audiences seem to prefer the slide. Other subspecialty or special interest groups may prefer the use of overhead transparencies or videotapes. Know your audience, and design with that audience in mind; but don't be a slave to convention. Effective visual media that differ from the norm can work exceptionally well, especially when you are the fifth of 12 presenters in a 3-hour session. Using an alternative medium to provide variety and change of pace can help regain attention.

The medium should also provide for the depth of information judged to be appropriate for the audience. Specialists may require the finer detail afforded by the slide or film. In another circumstance the audience may need to see or hear with precision, as can be provided by a medium allowing for motion and sound.

Speaker Comfort

Nothing is worse than a speaker with material he/she does not know how to use. Likewise, nothing is worse than being the presenter whose slides are out of order or whose overhead projector is out of focus. If you are a squeamish large group presenter, select a medium that is familiar. Be wary of your ability to problem-solve a new technology "on your feet," in front of your esteemed colleagues. Do know your equipment; know how to use your material; know your limits.

Design Considerations

There are several important considerations when designing visual material. First, they should be planned, tested and revised before use. Second, they should meet the principles set out for good communication of all types: clarity, focus, appropriateness, accuracy, purpose. Third, use professionals for consultation and the final production of the materials.

Use of the Development Model

As with the preparation of visuals for written communication, the cybernetic development model (Fig. 19.6) should be applied here. Those critical steps include: focus, design, develop, review, tryout. This will require time—time for planning, time for revision.

Adaptation of Principles by Medium

In the first chapter of this section, a number of principles were outlined that apply to verbal skills required for effective speaking and writing; these same principles apply to the process of presenting information visually. The following is a brief overview of the different media and the application of the principles for each.

The Slide Medium

This medium seems to be preferred for many reasons including that it was one of the first to be truly portable. A second is that the large group-format at professional meetings often predicates its use. Third, it offers an accurate presentation of colored visual stimuli for much medical information. Designing for the slide medium requires the application of the following principles:

- *Clarity.* The important information must be clearly visible. Design the slide so that it addresses one point specifically; condense the description to ensure readability; limit the number of lines to five; alter the size and font for headings and critical points; present complex information over several slides; use color to highlight the essential information; use negative field (blue or black background) to help information stand out.
- *Focus.* Synthesize the data for the audience; use a slide series to present a critical sequence; be precise with the data—select only those that fit the objectives; match the verbal narrative to the slide content; present the slide when you are verbally addressing the point, and expand later.
- *Appropriateness.* The type of data presented should match the needs and sophistication of the audience. Match the degree of specificity to the level of current audience understanding; do not present highly detailed information to a general audience; be aware of oversimplifying the data for a sophisticated research audience.
- *Accuracy.* The slide should be correct, both in the presentation of data and the selection and spelling of words; proofread your slides as you do all your work.
- *Purpose.* Each slide should add something specific to the presentation. Select only that which matches the focus of a presentation; eliminate results or data not directly relevant or overly detailed; include detailed slides at the end to use if a question arises requiring their display.

The Overhead Transparency Medium

This is an overlooked medium that has many of the advantages of slides, and offers others. Three important advantages include: the lights can remain on while the material is displayed, making it less likely that your audience will go to sleep; the speaker can face the audience, which allows the pace to be adapted to the comprehension reflected in the sea of faces; "progressive disclosure" via gates or overlays allows introduction of information at the rate your audience can tolerate and when you can explain it. Figure 19.9 demonstrates revelation through the opening of cardboard gates.

Some design considerations for the overhead transparency that differ from those mentioned for slides include:

- *Clarity.* The content must be readable and understandable. To ensure this, limit the information on each layer of an overhead transparency; use gates (hinged cardboard) or overlays (additional transparencies that fold into place) to reveal information as it is needed; use a letter size that will be readable, such as above 20 point; space the lines apart so

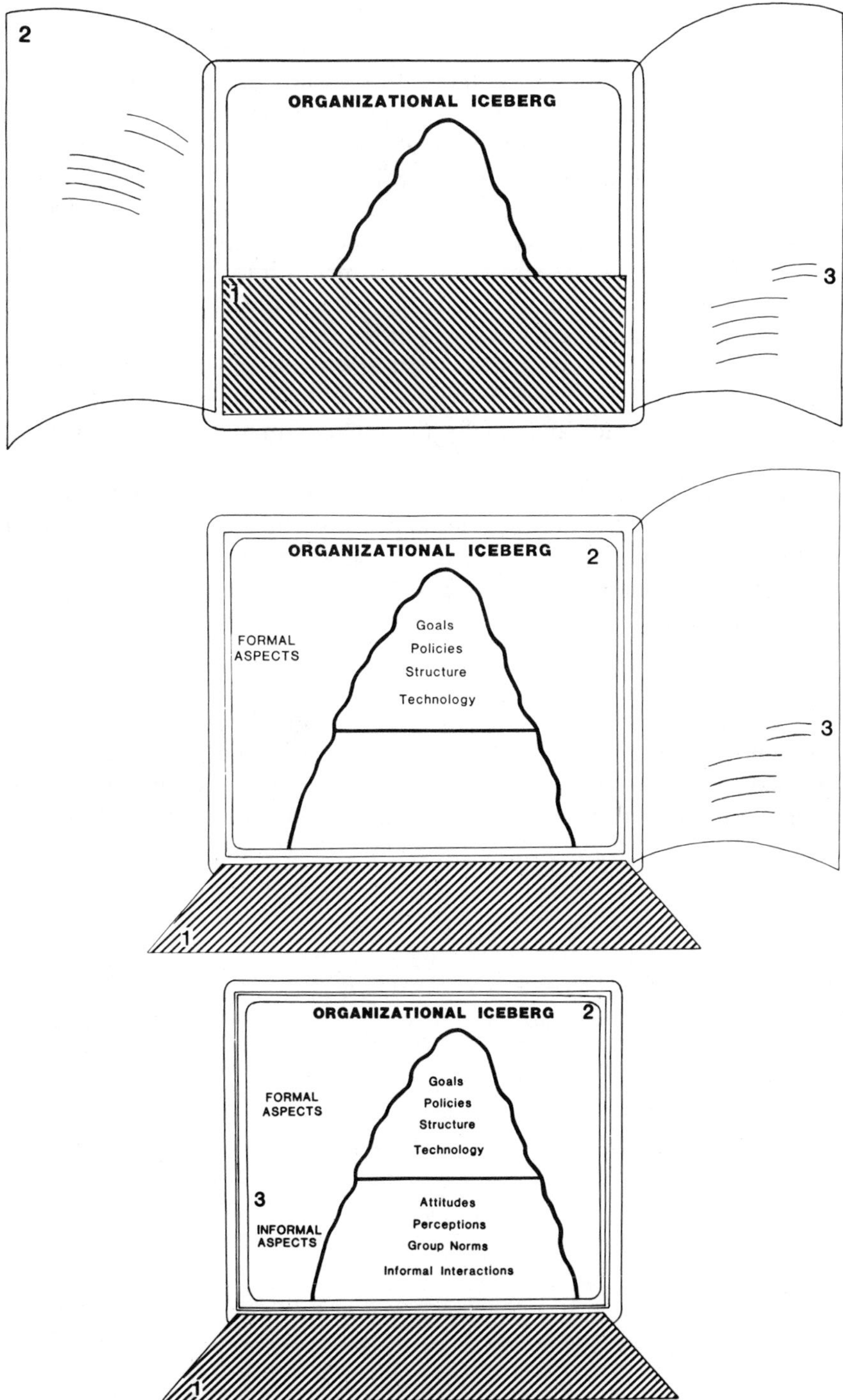

FIGURE 19.9. French and Bell: The organizational iceberg. (From Wendell L. French, Cecil H. Bell: *Organizational Development: Behavioral Science Interventions for Organization Improvement,* 3rd ed., © 1984, p. 19. Reprinted by permission of Prenctice-Hall, Inc., Englewood Cliffs, N.J.)

that the image remains distinct; use pressure letters, Kroytype letters, or a primary (large) typewriter; or type on an ordinary typewriter using double spacing and enlarge several times on copy machine; reproduce on a thermal or diazo overhead machine to produce a bolder image.

- *Focus.* A primary advantage of the overhead medium is that the sequence can be controlled by the presenter; use cardboard gates or the ubiquitous piece of paper to cover extraneous information to help focus attention on what you are discussing; place a pencil *on* the overhead to direct attention; use each layer to present one type of information. Figure 19.10 would be presented most effectively in oral presentation by sequencing with overlays: first present the mainframe information, then the mini computer and finally the microcomputer information using overlays.
- *Appropriateness.* Because overheads with overlays provide the speaker with the ability to alter the pace and depth of discussion, adaptation to the audience needs can occur on the spot. Select information that can be presented cleanly; select visual images that are bold, such as a bar graph, polygon, an image-related, or symbolic graphic.

The Poster Session Medium

This medium is increasingly becoming a professional forum and can be a roundtable discussion or an exhibition in which multiple topics are presented concurrently. The unique nature of this medium, particularly when it is an exhibit, is that the audience is a mass audience. Those pres-

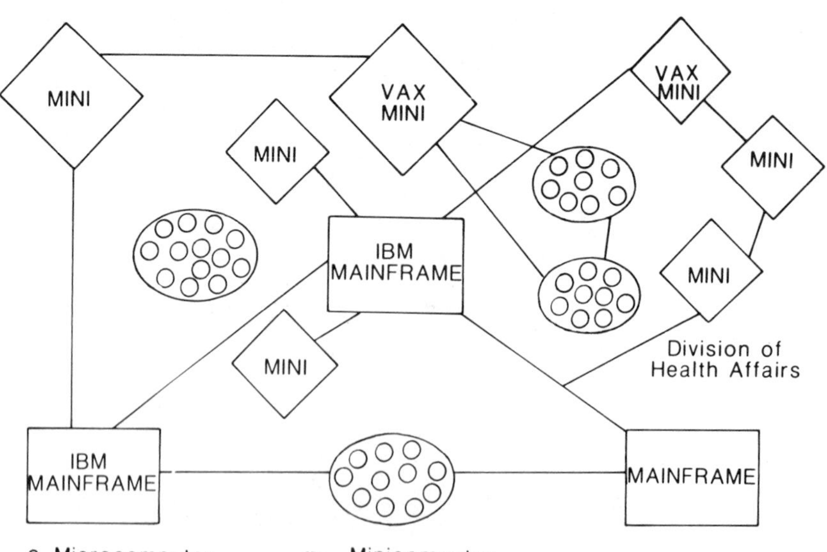

FIGURE 19.10. Computing environment.

ent will stop and listen only if they understand the intent and focus of the session. Some unique considerations when designing for this medium include:

- *Purpose.* Within one exhibit there may be several media used concurrently: handouts, a speaker, and posters. Different posters may have different purposes. In a session in which there may be 500 participants, you may have one poster serving as an *attention-getter,* whose sole goal is to stop the passers-by, prompt them to read further, and decide if this is of interest. These should be designed with large letters, condensed verbiage aimed at capturing the essence, and color or visuals to enhance the message appeal. A second poster on an easel may then serve to provide more information—aimed at the person who has decided to participate. The presenter or speaker would provide specific information on a one-on-one or small group basis. Materials may be available for study and use. There may be handouts to convey detailed information or to allow *follow-up.* Each medium should be used precisely and carefully; know the purpose of each medium; design carefully to address that need only. Figure 19.11 shows a 4′ x 3′ attention getting poster hung from curtains.
- *Focus.* Once the purpose of each medium is determined, the message should be designed to focus precisely. For instance, a poster that is an attention-getter should deal only with the broad issues. Design each medium to complement the other, not simply repeat the same information. The speaker can augment and individualize information; direct participants to the appropriate poster or handout so that you can provide more in-depth information.

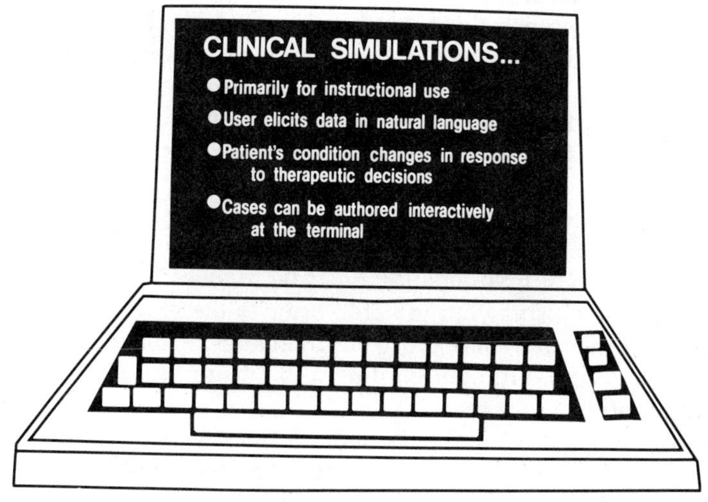

FIGURE 19.11. An attention getter.

- *Appropriateness.* This medium is very effective when used to present innovations and projects that are still under development. It also can be effective to display materials or products that require "hands-on" use. Once chosen, carefully select the medium that will allow easiest transfer of information; design so that the message flows easily from general to specific; allow ample room for the use of each material, i.e. computer software requires a space in which to work; hang displays appropriately so they can be seen.

Use of Design Professionals

Use the professionals of the field, consult with instructional designers, medical illustrators, graphic artists and video personnel. It is their field; they deal with these issues day in and day out and have undoubtedly seen many solutions to similar problems. Use of such personnel, however, requires time. It is rare that you will have a unique product if you put it through on the "rush list." Ask for drafts, respond to the drafts, let the design professional know why you prefer one version over the other, see the final copy, and be willing to send it back for revision. If you do not like it when you see it initially, you may not like it when you use it. If you must take a poor slide due to a deadline, bring it back for revision after the meeting. Designers are professionals, they appreciate a faculty member who cares about the material they create. Ultimately, the products you receive will be far superior if you participate and work through the cybernetic development process.

Effective Use

Effective use of visual material does not happen by accident, it comes about because of preparation, comfort, knowledge and rehearsal. First, select a medium that you know or will be familiar with in the allotted time. Second, cue the use of the visual within the outline or transcript (Table 19.2). Mark the beginning and end of each visual, so that nervousness or a sense of being rushed will not cause problems during the presentation.

Third, weave the content into the oral presentation. A slide, poster or overhead cannot stand alone. Verbally reinforce the important information: use a pointer or manipulation of a gate to focus attention.

Fourth, practice using your material prior to the meeting. Present the overhead or poster to a colleague to ensure that he or she understands the content. Revise the outline accordingly to ensure presentation of adequate information.

Fifth, arrive prepared. Have the slides arranged in the carousel: number the overheads with the overlays and gates open or closed as appropriate; know where, how, and with what the posters will be hung; have the videotape set properly.

TABLE 19.2. Sample outline

	SELECTION AND USE OF MEDIA
	I. ERA OF CHANGING STRATEGIES
Overhead #1	A. Technological changes
	B. Message REQUIRING THE MEDIUM not BEING the medium
	C. Evolution of the "shaping of the message"
Brainstorm using flipchart	II. CAUSE OF THE CHANGE
	A. Availability
	B. Diversity
	C. Population Explosion
	D. Time
	E. Accountability
	III. CRITERIA ON WHICH TO BASE DECISION
Overhead #2	A. Content
	B. Environment
	C. Objective

Sixth, do not panic. If it breaks, it will not be the first time. If the bulb blows, it can be fixed or replaced—there is often someone around willing to help. Let the participants take a 2-minute stretch and fix the hardware with less anxiety. If the image stops projecting abruptly, unplug the cord. If it is more than that, ask for help. We live in a technological world, and no one is expected to know all equipment and solutions. If your presentation depends on visuals and there are others to follow, ask the next speaker if he or she can speak without the support of hardware, and then take the intervening time to solve the problem. And if all this fails, summarize your findings and skip to your carefully prepared closing.

Using the Computer in Visual Communication

The computer is radically changing how we think, how we write, how we analyze quantitative and qualitative data, and how we formulate and display information. Even the simplest of the 8-bit microcomputers has the capability to support graphics programs that create bar charts, pie charts, line graphs, and histograms. The manipulation of data using these programs allows easy experimentation, adaptation, and change—a step that has been tedious and costly when preparing graphs manually.

Via the graphics packages, data are entered once and are then available for display in a multitude of forms. The ease of creating and changing the format of information allows the user to select the format appropriately, based on clarity, effectiveness, and communicability. These software "charting programs" deal with data of various types: that which changes over time, that which reflects discrete points in time, and that which is conceptual or graphic in nature.

Figure 19.12 is a sample of two variations of a bar chart that can be created using a common "charting" or graphics program.

The production process has also changed considerably with the advent of these graphics packages. Nearly every printer, from the simplest dot matrix printer to the most sophisticated graphics printers, can create a black-and-white print of a graph that can be enlarged, reduced, or used as created. Multi-color printers and plotters can be added to create multi-color prints or transparencies of a high quality.

One of the recent developments in this area includes relatively inexpensive photographic equipment that allows you to take a color slide or

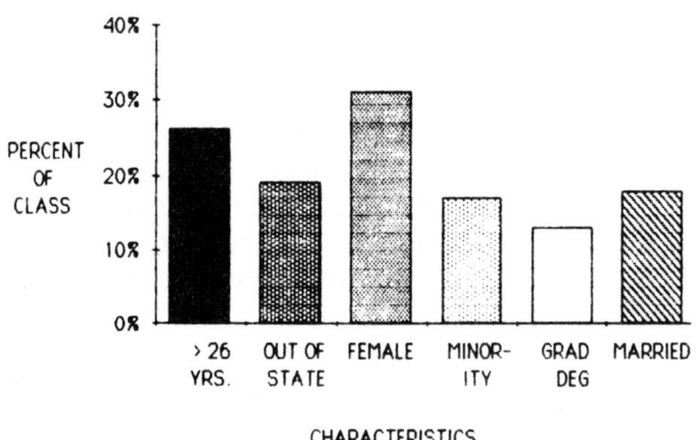

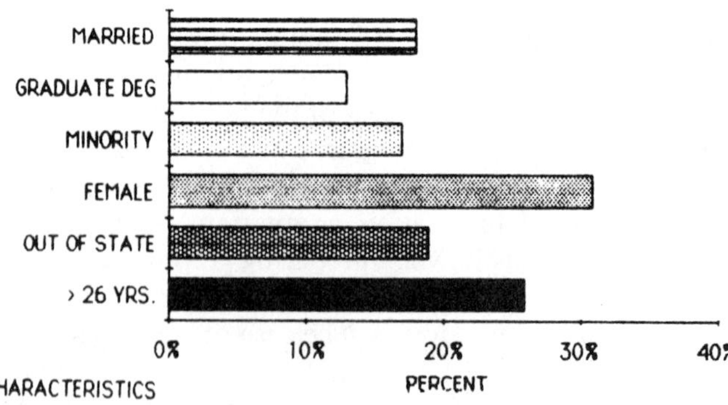

FIGURE 19.12. Computer graphics bar charts.

picture of the information displayed on the screen. Some systems will produce quality prints or color slides in 5 seconds, making it feasible to generate quickly a good quality print or slide of results that have just been pumped out of the computer.

Access to this technology is quickly becoming plentiful. Most computer units within health institutions have "charting programs" or graphics packages with good quality computer plotter and printers available. Other units such as the traditional medical photography and illustration units may already have such capabilities and may have sophisticated hardware that can do far more than what has just been described. Units such as libraries, learning resource centers and computer laboratories often have the technology and software for public use. Colleagues may have this software at home or next door in their office. Becoming aware of local "user's groups" will help put you in touch with people interested and competent in this area. Many of those involved with computers will have the enthusiasm of the newly converted and will relish the opportunity to help.

These computer capabilities can be controlled and manipulated by the faculty member who best understands the data. It is feasible to experiment with the format while at home and then have it rendered professionally by medical illustrators once you are certain of the content and format you prefer.

Other "graphics" packages allow the simple creation of conceptual graphics of the "low-fidelity" category. Some programs and systems have a reservoir of visual images ready to be used such as a circle, a map of the United States, people, and scientific symbols. By using a variety of type fonts, sizes, and color, words can often be manipulated. Incorporating these into visual images you create can easily give birth to an artistic freedom many of us have seldom experienced with the use of hands and eyes alone.

Figure 19.13 uses both computer-generated and artist-generated copy. With appropriate software both parts could be created at the terminal. Other management software creates schematics such as Program Evaluation Review Techniques (PERTs), flow charts, and timetables with ease and incredible speed.

Summary

Graphic representations will never replace the verbal report but can be viewed as a means of enhancing the communication of ideas and research. In some instances visuals work better than words: they can be more efficient, more effective, and more conceptual. Visuals can, in fact, be the vehicle that allows the "intellect and logic" to be satisfied by effecting a more comprehensive transmittal of information.

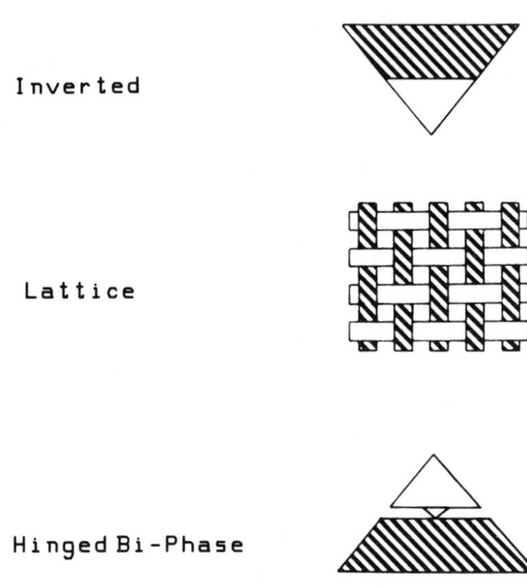

FIGURE 19.13. Computer graphics: image.

Through careful layout, visuals allow easy study of information from a variety of perspectives.

Although the academic physician is familiar with the use of figural graphics such as bar charts and polygons, there is a great deal of room for expansion via the use of schematics, conceptual, or "high-fidelity" graphics. Basic recommendations for improving visuals include: condense the information, eliminate the nonessential, attempt to represent abstract ideas and relationships visually, design the important information to stand out, be accurate.

Just as our first written drafts are usually problematic, so are our first visual conceptualizations. Our ability to think visually builds over time, as we experiment ourselves or work with others who specialize in this area. By applying the developmental model, allowing the development time a quality product requires, availing ourselves of emerging technology and using colleagues as a sounding board we can improve our ability to communicate visually. Working collaboratively with communications professionals will improve the ultimate product and make the long term result more satisfactory and effective.

References

1. Merrill PF, Bunderson CV. Preliminary guidelines for employing graphics in instruction. *J Instr Devel 4 (2):* 2–8, 1981.
2. Wileman R. *Exercises in Visual Thinking.* New York: Hastings House Publishers, 1980.
3. Duchastel PC. Research on illustrations in text: issues and perspectives. *Educ Comm Tech J 4:* 283–287, 1980.
4. Ehrenberg ASC. Rudiments of numeracy. *J Royal Stat Soc, Series A 39 (2):* 277–297, 1977.

Suggested Readings on Professional Communication

Evans M. The use of slides in teaching: a practical guide. *Med Educ 15* :186–191, 1981.

Evans offers a practical guide for the design, production, and use of slides for professional communication. The article discusses photographs, diagrams, tables, and line drawings.

King L. *Why Not Say It Clearly?* Boston: Little, Brown, 1978.

In clear and direct terms, King provides guidelines for clear and direct writing in a number of common contexts. The book discusses language choice, structure, style, and grammar for the writing professional.

Macdonald-Ross M. Graphics in text. In *Review of Educational Research, Vol. 5.* Shulman LS (Ed) Itasca, Illinois: FE Peacock, 1977, pp 49–85.

A scholarly summary of what is known about the use of graphics in text, based both on research and on the experience of communications professionals. The article covers presentation of quantitative data, algorithms, cartography and typography, and ends with an extensive reference list.

McCaffery M. Writing that's worth reading: a practical guide for writers of medical articles. *Can Fam Phys 26*:429–432, 585–587, 749–751, 872–878, 967–972, 1064–1066, 1980.

Probably one of the best-written and most useful simple guide for effective medical writing and publishing. With humor and insight, McCaffery covers the written presentation process from library research to distribution of reprints.

Reeder R. *Sourcebook of Medical Communication.* St. Louis: CV Mosby Company, 1981.

Reeder has compiled some of the best and most relevant thinking on communication in the medical profession. The book is an amply illustrated and stimulating reference for the development of data, the preparation of papers, and the presentation of ideas in the medical sciences.

White EB. *Elements of Style.* New York: Macmillan, 1979.

E.B. White's editorial update of the Strunk style guide is *the* basic reference in grammar, structure, and usage. White does not address medical writing in particular, but no one who writes English for educated readers should be without *Elements of Style.*

Wileman R. *Exercises in Visual Thinking.* New York: Hastings House Publishers, 1980.

The book explains the process of communicating visually. Wileman provides a theoretical basis for understanding visuals and illustrates theory with more than 50 visual examples. There are a number of activities that help the reader to visualize numerical, process, and conceptual information.

Williams J. Writing for the ear. In *Profiles in College Teaching.* Mathis BC, McGaghie WC (Eds), Evanston, Illinois. Center for the Teaching Professions, Northwestern University, 1972, pp 165–172.

The author discusses the differences between writing for printed publication and writing for broadcast presentation. Although his examples are from journalism, Williams makes several points of value to the academic physican preparing for oral presentations to students or professional groups.

V Ethics: Teaching and Patient Care

The purpose of this section is to engage academic physicians in aspects of ethics which they are likely to encounter in their daily work as both clinicians and teachers.

Chapter 20, "Ethical Decisions," focuses upon the physician as a moral agent who must make, and assist patients to make, hard ethical choices. Our approach here eschews "schools" of ethical theory, such as utilitarianism or Kantianism, in favor of emphasis on ways of reasoning and processes of deciding. Our goal is not to argue for one way of moral thinking over others, but to put on display many of the basic elements of ethical reasoning and allow readers to discover what resources might be best suited to their own situation. There are no final, standard or universally "right" answers in ethics in our current cultural context; similarly, there is no single formula for decision-making which all physicians must use. Our hope is that physicians will recognize their own moral propensities expressed here, and also find new tools with which to work.

Chapter 21, "Teaching Ethics," focuses on the physician as teacher of students, residents and colleagues. First, we describe the moral values inherent in all teaching. Next, we consider three values which are indispensable to good teaching. Finally, we explore four essentials of teaching ethics which we believe any teacher should be able to do when assisting a learner to resolve a moral problem. We argue throughout that the teacher of ethics has a role which is distinctive from that of the moral problem-solver, a role which calls for special sensibilities and mastery of ethical and interpersonal skill.

Chapter 22, "Case Studies," is an effort to work through four fairly typical academic situations in some detail. Each case is presented and discussed with reference to its ethical significance and moral dimensions, using some of the tools and perspectives introduced in the two previous chapters.

The sequence of these chapters is designed to guide the reader from introductory material and concepts in ethics through consideration of its application in cases. Some readers may wish to begin with the cases in

Chapter 22 and then consult Chapters 20 and 21 for further refinement of their thinking and consideration of more theoretical aspects of ethics. We hold no brief in these chapters for an exclusive starting-point. What is essential to recognize is that both theoretical tools and practical applications are necessary to any adequate approach to medical ethics. That is how ethics employs both the science and the art of medicine.

20 Ethical Decisions

Larry R. Churchill, Harmon L. Smith, and John J. Frey

This chapter should assist academic physicians to deal with moral problems inherent in their practice. Our aim is not to supply answers, but rather to make grappling with the problems a more knowledgeable, thorough, and satisfying process. The chapter is not a "cookbook" of ethical recipes, providing simple answers for moral problems, but a stimulus to the physician's own moral sensibility and creativity. The thoughts, methods, and analyses presented here are more systematic expressions of the moral deliberations that characterize clinical medicine on a routine basis.

Ethics is a blend of the practical and the theoretical. It engages us as moral agents who must choose, but whose choices are grounded in conceptual ideals and principles of what is right or good. Hence we have attempted to punctuate each of the three chapters in this section with clinical material, through which the more theoretical discussions can be brought to bear on the reader's own practice. Moral problem-solving without the benefit of theoretical, ethical reflection is like exercising a moral reflex—a knee-jerk morality. Theoretical reflection untested by its application to real life "on the firing line" is equally useless. Our conviction is that an appreciation of the formal, theoretical tools of ethics will make a difference to the practice of medicine, and reciprocally, a deeper appreciation of the concrete concerns of the clinic will enhance the theoretical apparatus. The reader is asked to form a dialogue between practice and ideals, between actual choices and notions of good and right.

Ethics is not so much a static body of knowledge as a set of skills, a practical competence that one acquires over time and with practice. Like cooking, one must practice it with regularity in order to achieve a moderate amount of skill and craftsmanship. We do not intend to mandate how much oregano or basil is needed for the stew, but simply to point out the way other cooks have found it appropriate to combine their spices. By drawing an analogy with cooking we do not imply that ethics is merely a matter of taste. Yet like cooking, ethical acumen depends upon skills that are tacit—that one learns and improves by exercise. Like fine cooking, skill in ethical reasoning must be cultivated. One cannot cultivate culinary skills by making only omelettes, just as one cannot exercise sound moral judgment by invoking only a few pet moral dogmas.

Ethics in its broadest sense concerns how we live and what choices we make. *Medical ethics concerns how physicians choose to practice and the values embodied in their professional relationships with patients, families, colleagues, students, and the general public.* For the purpose of focus and clarity we will deal initially with the aspect of ethics that concerns how moral problems are solved. The problem-solving aspect of ethics, sometimes termed "quandary ethics," is only one side of the coin, but it is the most accessible side, so that is where we begin.

The complement to quandary ethics is less easily definable and concerns not so much our choices but our character, our sense of integrity, our ideals and virtues as these inform routine professional life—that is, when no critical problem is being contemplated. It is this aspect of ethics that concerns us as moral agents before and after the situation of choice. These two components of ethics, quandary ethics and character ethics, are not competing notions but complementary aspects of moral life. Physicians routinely draw on their character to resolve quandaries, just as the resolution of quandaries shapes and reinforces their ideals.

Identifying Ethical Problems

A recent study of medical education (1) concluded that "the moral perspective enters medical decision-making only as a last resort, after the clinical and legal aspects have been exhausted." Perhaps this is a result of the customs of tertiary-care training centers, where educational activities must share time with patient care and research priorities. Whatever the case, the fact is that many ethical problems never reach the stage of explicit formulation but remain invisible, or are sometimes mistakenly interpreted as technical or legal concerns.

The first task of medical ethics is to make ethical problems more visible. How do we know an ethical problem when we see one? What are its defining characteristics? Ethical problems can be defined both negatively and positively, both in terms of what they are and are not.

Moral difficulties are not scientific or empirical problems. They cannot be solved by recourse to more sophisticated scientific methods, or by gathering more evidence; with the best scientific knowledge applied and all the available evidence in, moral problems are still problematic because they call for a judgment based on values. That there are no effective remedies for some forms of cancer is a scientific problem, though the choices that ensue from this recognition in terms of patient care are frequently moral in nature.

In a similar way, ethics is not a matter of esthetics, manners, or taste. It may be objectionable to eat with one's fingers rather than to use a fork, but it is not a moral fault to do so. Moral questions carry a deeper sense of oughtness and obligation—they evoke a sense that something larger is at stake.

In short, moral questions are often mistaken for other species of problems and consequently misplaced. When this occurs, the "sense of something larger at stake" is lost as value issues are reduced to legal, technical, or scientific judgments. Frequently, this eliminates the moral edge of a problem by disguising it and thus prohibiting its consideration.

Let's look at another example. Radical feminists sometimes talk of abortion as if it were solely a political question, just as some gynecologists treat abortion as a purely medical issue. In both cases the moral weight of the act is reduced to a question of equal rights for women, equal access to service, or proper medical technique. There are, of course, political and medical aspects to abortion, but these modes of understanding do not encompass its moral significance.

Positively defined, a moral problem is one in which values conflict. Resolutions to moral problems occur as persons make choices based on their assessment of values; that is, when they choose a course of action because it is thought to be right or good, rather than technically correct or scientifically valid. Moral choices call on us to rank our values and espouse one value or ideal over others. For example, physicians are traditionally thought to value both the relief of suffering and the preservation of life. Yet there are situations in which the relief of suffering jeopardizes life, and the preservation of life may extend suffering. Moral difficulties are rarely this simple, but the heart of a moral problem will cluster around some set of mutually exclusive choices that embody different values in conflict.

Because they reflect the values of the moral agent, ethical quandaries carry a sense of significance or importance. The existential philosophers and novelists of this century have insisted that we *are* our choices, that we become what we choose. Although we resist this simple equation, it is accurate to say that part of the gravity we associate with moral choices stems from our recognition that who we are is reflected in our moral decisions. Moral choices affirm or violate our sense of self in a way that most other types of choices do not. Moral choices are accepted or rejected because they appeal to the sorts of persons we currently are, or aspire to become.

It is, of course, true that some persons perceive moral problems where others do not. Sometimes this is due to failure of moral imagination or to lack of experience, or may be the result of sheer callousness. More frequently this difference results from differing values and, ultimately, from differing visions of the world.

Obstacles to Ethics

Two attitudes about ethics sometimes keep us from perceiving moral problems. These attitudes embody assumptions about what ethics is and what it means to be a moral agent.

Moral Absolutism

This attitude usually takes the form of rigid adherence to a set of formulas or action. For example, a moral absolutist might hold that abortion is morally illicit, even to save the life of a mother, and that it is wrong for everyone no matter what their circumstances. Here moral argument is reduced to militancy for a cause, and reason is reduced to rhetoric in the service of dogma. This attitude is a barrier to the perception of moral problems because the problematic character of moral choices is not acknowledged. Instead of working *toward* solutions, the moral absolutist works *from* solutions. The problem, then, is transformed from a problem about what choice is right to an issue of how to bring others to the solution already held. The strategy is conversion rather than persuasion. Moral absolutists do not practice a particular type of ethics. Rather, their stance is anti-ethics in that they allow ethical reflection and deliberation no place.

Moral Relativism

Perhaps a more prevalent attitude today than moral absolutism is moral relativism. The moral relativist believes moral choices are no more than opinions, and, like opinions about the taste of foods or preferences in hair styles, they "can't be proved or disapproved." To the relativist, morals are like esthetic preferences, natural biases, or other idiosyncratic choices. You like chocolate whereas I prefer vanilla, and who can say that one of us is more correct than the other? The contentions that there are no answers to ethical questions, and that ethical discussions are just bull sessions, are statements of moral relativism.

Like absolutism, relativism is inimical to ethics. By reducing moral convictions to subjective opinions, relativism eliminates the need for reasoning, or more precisely, it renders moral reasoning *meaningless.* Whereas absolutism takes moral judgments to be perceptions of fact, relativism makes these seem to be sheer, subjective opinion. Both attitudes, if taken seriously, eliminate ethics, because both refuse to acknowledge any tension in making choices. Absolutists foreclose on that tension from the beginning. Relativists mortgage it off to the realm of private opinion.

It is, of course, rare to find these specimens in a pure form. Most of us tend to be relativists on some issues and absolutists on others. Ethics entails an attitude of suspicion about areas of morality that are entirely closed to us, especially if we cannot argue for our convictions or seem reticent to discuss or reexamine them. One central task of ethics is to help us reflect more deeply or fully, even in areas in which our convictions are firm and seem well-founded.

Why Ethics? Some Rationales

Why study ethics and learn the skills of ethical reasoning? What do we hope to achieve? What good will it do?

Three rationales can be given. First, ethical reflection provides the space to reflect critically and examine our moral hunches and intuitions. It protects us from a knee-jerk morality. Ethics requires us to consider our moral reflexes in the light of other values and other possible actions.

Second, and more practically, ethical deliberation is a way of dealing with disputes between persons without resorting to recriminations or, at worst, violence. Moral discussion can preserve civility between disputing parties, and at its best, nourish a sense of community. Ethical discussion may not reduce conflict, but it puts conflict in the larger context of reason and reflection.

Third, ethical reflection can be an aid in avoiding moral tragedy. Moral tragedy occurs when, having acted, we realize we would have acted differently if more care had been given to the choice. It is the irreversible character of many of the choices in medicine that makes ethical deliberation so important.

Ethics has a forward and backward movement—to illumine both the choices we face and the value system from which those choices are made. In the forward movement, the consequences of alternatives are played out in our imagination. In the backward movement our moral choices affirm (or repudiate) our past and seek consonance with our most deeply held values, while at the same time testing the application of those values in the light of the present moral dilemma.

Making Decisions: Some Movements in the Process

The making of decisions does not exhaust ethics, but it is an important part. Many texts on ethics begin either with a list of principles that are subsequently explored or with a list of difficult cases or situations that the reader is invited to resolve. We begin with neither principles nor problems, but with an emphasis on process. Although focus on principles and problems is essential, both must find their appropriate place in a larger process of moral reasoning. Beginning with principles is too abstract and may leave the impression that solving a moral problem is no more than the calculus of applying principles. Beginning with cases or problems appeals to our need for relevance but frequently clouds the larger process of problem-solving by inviting fixation on the fine points of the situation. It tempts us to think the facts of the case—if considered with enough care—will finally yield the moral solution. To begin with either principles or problems is to deemphasize the essential feature of moral problem-solving: human judgment about values. It is in the final analysis value judgments that determine how we understand and apply principles and how we frame and analyze a problematic situation.

The process outlined below is not offered as a paradigm for moral reasoning. It is neither exhaustive in its description nor to be followed in a serial fashion. Rather, our effort is to outline *some* of the elements we believe are essential to any moral decision-making. Our aim is twofold: first, to explicate these elements themselves; and second, to stimulate readers to consider what elements and what larger process are most fitting to their practice.

Focusing on the Central Problem

As indicated earlier, not all problems are moral in nature. Unlike scientific or technical problems, moral difficulties reflect conflicting values in which human agents must espouse one value or set of values over others. Any given situation may contain a host of moral conflicts. Sorting them out to recognize their interaction without fusing them into a single issue takes care and patience. Moreover, what makes for a moral problem is seldom self-evident in the alternative courses of action that are considered. To cite a typical example, consider the situation of a physician who must decide whether to tell his cancer patient a discouraging prognosis. Should the physician be truthful in discussion with the patient, or should he soften the bad news, delay discussion of this knowledge, or even blatantly lie to the patient?

These options for action may constitute the practical edge of the problem, but they do not encompass the problem as a moral difficulty. The moral conflict is about the significance of these choices in terms of the values they embody. Carrying our situation further, let us suppose that for the physician the choices between disclosing and withholding reflect two moral norms, both of which he wishes to honor. Truthfulness could be seen to embody respect for the self-determination of the patient. Without an accurate sense of his predicament, no patient can make choices for himself about the ordering of his life. It could be argued that the physician who holds such knowledge of a patient's condition has a special obligation of full disclosure. Alternatively, withholding the truth can be construed as honoring the value of beneficence, in short, the effort to do what is best for the patient. In this situation, withholding a discouraging prognosis might be said to avoid unnecessary depression, or suffering. The moral conflict is, to be sure, a conflict between alternatives for action. Yet morally this is a meaningless choice, unless these differing choices signify differing values. The moral choice here is between the good of honoring self-determination versus the good of beneficent protection. In clarifying the moral problem it is essential to know not only the options but also to know what these options designate, morally.

Levels of Engagement

It is an ongoing debate in moral philosophy as to whether ethical judgments are grounded in our reasoning abilities or in our affective sensibilities. We have no wish to further this dispute. Reason and the affective dimensions are both involved in moral judgments. At a minimum, it should be clear that moral problems are not simply intellectual puzzles. They are frequently disturbing and challenging precisely because they threaten to bring into question our most cherished values. Our values are not merely ideas that we believe, but ways of life to which we are committed. To ignore our affective responses to moral problems is to become victimized by them.

The Western tradition of rationalism encourages a divorce between the cognitive and affective dimensions, and this may take the form of a disenfranchizing judgment on the emotions. Medical customs teach that "professional distance" is a virtue, and that affective judgments are a distortion of the necessary objectivity in patient care. Yet ethical judgments, unlike differential diagnosis or other clinical judgments, reflect our human presence. Moral judgments are not simply reducible to our affective senses, but neither are they dispassionate, logical analyses. Acknowledging one's gut reaction to a dilemma is not the last word but it is an important initial consideration.

The philosopher Henry Aiken has designated four levels of moral activity, which he calls the expressive level, the level of moral rules, the level of ethical principles, and the postethical level (2). These can be illustrated using the problem of truth-telling discussed above.

Expressive level	"I deplore the telling of lies."
Level of moral rules	"Do not tell lies."
Level of ethical principles	"The patient's right to the truth is more fundamental than the doctor's obligation to beneficently protect the patient from harmful truths."
Postethical level	"It is important to tell the truth otherwise the basic trust needed for, physicians to help patients is damaged."

Acknowledging the repugnance one feels about deception does not solve the moral problem: one sometimes must choose between unsavory alternatives. Yet registering the distaste can be useful in the large process of reaching a decision that truly reflects one's values.

Equally important is moving beyond the emotive level of moral sensibilities to a careful delineation of principles and consequences. All too frequently *argument* is never engaged because the disputing parties choose to remain at a purely emotional level of engagement. Such an exchange can be cathartic, but it is rarely, if ever, persuasive.

The cathartic function can be an important one. It not only makes us feel better, but gets our feelings in a position where rational discourse can deal with them, and bring them into the conversation at a proper stage. If left out, our feelings may carry the day for all the wrong reasons, or rather for no reason at all. Moral argument is a matter of giving reasons, making inferences, and arguing for one position over others. The aim is to convince, or at least to make one's own convictions and choices coherent, even if only to oneself. This can take place fully only if what Aiken calls "the level of moral principles" is taken as the arena of exchange.

In short, each level of moral engagement has an important place in ethical reasoning. No one level can exhaust our moral concerns or contain our moral judgments. The wise moral agent knows the difference between the levels and acknowledges the role of all levels. The whole person is the locus of ethical inquiry; one can speak of the place or role of different aspects, but fundamentally we are moral creatures because we are human, not merely because we can think or feel.

Roles and Responsibilities

Professional self-definitions are central to the distinction between right/wrong, good/evil, and appropriate/inappropriate. No doubt differing roles and degrees of responsibility are fitting for different patients and different circumstances.These need careful identification. Primary care pediatricians and family doctors, for example, may have special complications in terms of responsibilities to families as a whole, or to children too young to make their own voice heard in a decision.

Additional factors that influence this dimension of moral judgments are the customs of the community, one's professional peer group, legal considerations, and the seriousness of the consequences for the patient or the family.

To return to our case, the physician might decide that responsibility requires complete truthfulness at all times. Recognizing that harmful consequences may ensue is no deterrent, if the physician's responsibilities do not extend beyond what happens in the clinic. "Doing what is best" is here thought to be carefully circumscribed and confined to a specific, limited role that physicians must assume, that is, responsibility for the "medical" aspects of the patient. Clearly, other more expansive notions of role and responsibility will call for different responses, and perhaps even novel formulations of the moral problem. Indeed the entire process of moral decision-making is one of going forward and then doubling back to check the nature of the problem. Clarifying roles and responsibilities can even settle the issue if it means that prerogatives for action should rest with the patient, parents, or others.

Consulting Personal Principles and Professional Codes

Personal principles and professional codes may differ. Contemporary medical codes of ethics tend to focus narrowly on obligations and prohibitions and are couched in a legalistic idiom. This is not true of some of the older medical codes. For example, the 1847 AMA Code is concerned not only with a list of prohibitions and obligations but also with the larger character, ideals, and purposes of professionals. Unfortunately the current AMA Code is but a skeleton of the 1847 statement.

Generally, medical codes have included norms such as nonmaleficence (do not harm), beneficence, and confidentiality. Other norms of equal importance have been neglected in codal formulations. Among these, patient autonomy, equality of treatment, and concern for social justice in health care distribution are the most glaring omissions.

In general it should be noted that ancient codes of ethics are preferable to modern ones. Hippocratic ethics requires the swearing of an oath and the evocation of a binding commitment to professional integrity and a chaste personal life. It is also realistic about human frailties, reminding physicians that it is both wrong and harmful to seduce patients. The current AMA "Statement of Principles"is anemic by comparison, speaking in cautious, legalistic tones about the most common virtues.

A full discussion of codes and principles is not possible here. It is enough to say that personal principles of ethics are usually more expansive than professional codes and that both need to be consulted to find the norms applicable to a given situation. In order to become useful, professional codes must be studied in terms of their explicit principles and the moral ethos of the professional they evoke.

Weighing Consequences

This aspect of decision-making is straightforward and needs little explication. Central questions may be how to weigh long- versus short-term consequences, and the degree of probability to be assigned to each event that might ensue. For example, it is useful to differentiate between those consequences that are almost certain (If I pull the plug, the patient dies) and those that are mere possibilities (If I pull the plug, the family will misunderstand my actions, become angry, or bring suit).

The short-term consequences of disclosing a discouraging prognosis may be to upset or depress the patient. Yet, the long-term consequences may be to give the patient sufficient time to put affairs in order prior to death. Anticipating consequences correctly without reacting to a single undesirable possibility requires knowledge of the individuals involved and some experience in relating to people. Also, physicians can occasionally alter consequences by the quality of the relationship they maintain with patients. A physician who withholds a discouraging prognosis for fear that the patient may "jump from a window" or become depressed may have underestimated his or her own ability to influence consequences by not abandoning the patient *in extremis*. It is also easy to underestimate the patient's ability to cope. In brief, assessing the consequences may involve self-assessments of roles, responsibilities, and obligations that shape those consequences. At best, the attempt to weigh consequences is educated guesswork. Ethical decisions based on anticipated consequences alone are suspect for just this reason. The moral significance of any given consequence is tempered by the other factors in the decision-making process.

Clarifying Intentions

Decisions are judged not only on the basis of their consequences but also by reference to the intentions of the moral agent. To return to our case, if consequences alone are important, one might reason that truthfulness is best because it can be expected to result in a greater proportion of good over evil than any other course of action. Truthfulness in sharing the prognosis will allow for preparation for death and decrease the need for pretense between doctor and patient, and between patient and family. From the point of view of judging consequences, truth is merely an instrumental value, that is, something of value because it leads to desirable results. Deception would be a better choice if it could be shown to lead to even better results. Truth—in itself and for its own sake—has no value. And from this point of view, one does not intend the truth *because it is* the truth, one intends the truth (intends not to deceive) because of the good that particular act of honesty produces.

Reasoning from intentions, the same course of action may be selected, but for very different reasons. For example, truthfulness with the patient could be deemed best, quite apart from its consequences, because it is good for its own sake.

One can, of course, tell the truth because of its presumed good consequences, *and* because the truth itself must be held as a high value. That is, truthfulness can have both an *instrumental* and an *intrinsic* value. The point here is that the value truth assumes in any given situation of choice will depend on a clear understanding of what one intends.

Frequently intentions are confused with motivations, desires, hopes, or wishes. To speak of intentions is not to speak of our psychological states of mind, or our conscience, or our hunches about outcomes. Intentions are quite simply our aims and purposes in undertaking an action. Properly speaking, intentions are not what motivates to act, but what we seek to accomplish through performing an action, or refraining from it.

Intentions can also be confused with causes. It is possible to cause harm, or even death, without intending to do so. We are, of course, responsible both for what we cause intentionally, and for what we both cause and intend, but in different ways. Intentions give prominence to our personal involvement in our actions; they place us squarely within them. Through intentions we explicitly claim our agency in our acts.

If the physician discloses a discouraging prognosis to the patient it may cause the patient harm, though unless the physician's manner is harsh, or brutal in a calculating way, we would not suppose the harm was intended. Doing harm was not the reason for telling the prognosis, though it may have been an unavoidable result. The law customarily recognizes differences among actions, even those with the same outcome. For example, actions resulting in death may be expressed as voluntary manslaughter, homicide, self-defense, or any of the varying degrees of murder, depending on the circumstances and the intentions of the agent. Ethical reflection about intentions should have no less sophistication and discernment.

Considering Options

In moral reasoning it is easy to fall prey to the fallacy of black-and-white thinking. This occurs when the problematic situation is framed so that only two choices are presented as alternatives for action. The dramatic "life or death" options that preoccupy the press are good examples of this tendency. This is usually fallacious reasoning because most situations admit to a wider range of possibilities than two either/or alternatives.

The options open to a physicain who has to relate a discouraging prognosis to a patient are more rich and varied than is indicated in the black-and-white option of truth-telling or lying. The truth can be told in a variety of ways and by a variety of means. One can tell artfully or clumsily, with deference or force, with tact and skill, or the reverse. One can relate the news in stages, in code, in symbols, stories, parables or metaphors, through a spouse or clergyman, or in innumerable other ways. Deception is susceptible to an equal variation in character, tone, and content. To focus on the literal saying of the words, as if this singular act could carry the entire moral weight of the action, is behind our tendency to set up truth-telling/lying as the only option.

The ability to consider multiple options requires patience and imagination—two neglected virtues of moral reasoning. We cannot act morally without some lively sense of the other human beings on whom our actions bear. This takes skill and insight at discerning their sense of themselves and patience in the ambiguity of the situation of choice in order to let a full range of options present themselves to us.

The presence of two competing options has a masking effect on our ability to think through a situation. When our thinking is confined to two options, our perception of the situation of choice can become blunted.

Certain aspects of a situation can be seen and thought about only if a third option and the things it signifies are no longer hidden from view. For example, many discussions of *dilemma* of truth-telling/lying overlook the fact that patients may already know, or have guessed, their predicament. If true, this would make truth-telling more like acknowledging and lying, at best ineffective and at worst conspiratorial. Alternative options reshape the situation and reformulate the nature of the choice before us. To confine the moral situation to only two choices is frequently to misperceive the situation. The word "dilemma," which means literally "two words," encourages us to think in this either/or fashion. The moral absolutist and the moral relativist are wedded to the notion of the dilemma and to black-and-white thinking. Most quandaries that occupy our moral thinking are usually not dilemmas at all, in this sense, and we prefer the more general option of "problem" to describe the experience of the moral agent.

Eliciting Assumptions

Fundamental to any process of moral reasoning (and resident within the other elements) is the activity of eliciting assumptions. Assumptions are the unquestioned grounds of a conviction or belief, that which goes unnoticed because it is the foundation for something else. Assumptions are usually tacit, rather than explicit, in our moral reasoning; proximal rather than distal in relation to our values. They are what we reach with, or where we reach from, rather than what we reach toward and seek to grasp.

Frequently what we assume is more important in guiding our actions than the reasons we give when asked. And because assumptions are not questioned in the routine conduct of our lives, they may become unduly powerful and occupy an unwarranted position in our moral thinking. Moral philosophy, or ethics, is therefore especially concerned with the effort to elicit assumptions, state them clearly, and critically evaluate their range of application.

Returning to our ongoing scenario, a physician might assume that his or her role is one in which paternalistic protection of patients from harm is part of "what it means to be a physician." This is, indeed, not an uncommon assumption, grounded in medical traditions and customs, and even occasionally specifically listed as a duty in medical codes. The close association between protective paternalism and the very meaning of professional conduct in medicine can so shape the choices, and the moral deliberations about these choices, that nonpaternalistic possibilities are neglected. It is not that such other choices are considered and rejected; rather they are preempted by assumptions about what is possible. Here the conflation of paternalism and professionalism could make a full disclosure of the diagnosis literally "unthinkable."

Assumptions present some possibilities while foreclosing on others. Different assumptions lead to different choices because they are different ways of representing the situation of choice. Not all assumptions are easily ferreted out, nor is it useful to keep probing the background indefinitely. But failure to do any probing into what we take for granted does reduce us to acting on accepted ideas, customs, and traditions. We are not suggesting that these customs and traditions are wrong, only that it is a necessary part of responsible decision-making to seek clarity about them and to assess them critically.

Giving Reasons

To engage in ethical reflection is, in some way, to offer reasons for one's choices. To give reasons for one's choices is to affirm the essentially public and social nature of morality. Private covenants and agreements can be useful, but they lack the crucial ingredients of public criticism and community affirmation. Moral choices must in some way be capable of public statement or expression. In giving reasons for one's choices one acknowledges the *social* character of actions, and the presence of others as moral agents in the world. Giving reasons is an act of seeking community with other moral selves, by appealing to their moral sensibilities. Ethical reflection is antithetical to radical individualism. In giving reasons we impute moral agency to others, and the reasons we give are calculated to elicit a certain moral sense between us. No reasons can be persuasive unless they appeal to the sorts of persons we are or aspire to become.

Moreover, it is only moral reasons that make sense of moral choices. One can, of course, appeal to the facts of the situation, but in so doing, there is always the supposition that these facts will have a particular moral weight—namely, that of supporting the choice. For example, if we argue that it is a right and good thing to disclose fully the prognosis to the cancer patient, one reason we might give is that it will allow him to put his financial affairs in order. In so arguing, the unstated assumption is that putting financial affairs in order prior to death is a good thing, and telling the patient is the means to achieving that good.

There are a wide variety of reasons that are typically given to make sense of a moral choice. Many of these have been discussed earlier in the sections on principles, consequences, roles and responsibilities, and intentions. Each of these may serve as a way of giving reasons. For example, the Golden Rule is frequently invoked as a reason for moral choice in the following way, "That's the action I would want to have taken if I were the patient," or "That's the degree of freedom (choice, self-determination) I would want if I were terminally ill." It is noteworthy that in making such an appeal, the moral agent not only invests the listener with moral sensibilities, but affirms a deep need to justify the choice or at least make it coherent. Giving reasons is a way of rendering sense where it may not be evident. Appealing to principles is one of the more powerful ways of doing this. This is because principles are like paint brushes in the hands of the moral artist. Principles paint the moral landscape and set the contours of our choices. They alter the world we see and in which our choices are set. The Golden Rule portrays a world in which our own self-regard and self-interest become the yardstick for measuring the imagined desire of others. Individual others occupy the foreground of the painting. Use of a different principle, say, social utility, portrays individual persons in less clear focus and relegates their presence in the finished painting to the background.

The introductory remarks of this chapter stated that ethics is concerned with both problem-solving and with moral character. In giving reasons for our moral choices we can appeal either to the process of thinking and set of principles, or to the capabilities and responsibilities of character. Ultimately these tactics converge, since principles are always held by persons of some given character, and persons must finally express their characteristic values through principles and ways of reasoning with them.

Finally, giving reasons presumes the *reasonableness* of the moral process. It places ethical reflection at a level above manners, or taste, and dissociates it entirely from whim or capriciousness.

Why Ethics Fails: Some Common Snares

The process of moral decision-making that has been described is not exhaustive but rather depicts some of the central elements. Likewise this section simply suggests some of the pitfalls experienced in the consideration of moral problems.

The Moral Quagmire

When multiple issues are raised, one after the other, with little or no discussion or differentiation, a quagmire develops. A miry bog of moral problems engulfs the discussion and threatens to submerge rational deliberation altogether. When five complex ethical problems all surface at once, the ground seems to shift under each and no firm footing is available.

Perhaps the moral quagmire is more likely to develop when those charged with the decision believe themselves (rightly or wrongly) to be under tremendous constraints of time, and when multiple perspectives are thought to be important. Decisions about the treatment of children can be especially difficult. It is possible in these settings to conflate questions of autonomy for the child, parental wishes, legal precedents, notions of professional duty, and therapeutic goals to such a degree that it becomes very difficult to perceive which issue is being addressed. The way out of the bog is only through a careful sorting out of problems and a resolve to take them one at a time. The moral quagmire results when no process of moral reasoning is undertaken, and more precisely, when the central moral problem is not identified.

Dominoes

This is sometimes called the slippery slope phenomenon. It runs something like this: "If we allow Karen Quinlan to die, we will be encouraging Nazi medical practices. We might as well kill everyone who is on a respirator." The fallacy is one of hasty generalization. It treats dissimilar cases in a similar way, or better, overlooks relevant differences among similar situations. This type of reasoning is fallacious because it jumps from a single instance to a general conclusion without careful attention to *just how,* for example, Karen Quinlan's situation is similar or different from that of European Jews who became experimental subjects at Auschwitz. Like dominoes lined up in rows, the argument goes, if the first death is allowed, all the others will eventually follow. Or to change the metaphor, one step on the slippery slope and one will soon be at the bottom. Again, only careful articulation of similarities and differences between cases will allow us to clarify which actions really do amount to the same thing, morally.

The Misuse of Rules

Ethical dominoes is a position most likely to be held by a rule-oriented moral agent, and by one who misuses rules. Rules are best seen as instruments for interpreting the moral facets of situations, and as guides to action. The misuse of rules we refer to occurs when circumstances are shaped (and frequently distorted) to fit a rule. This thinking is *from* the rule, instead of using the rule as one instrument among others to think *with.* For example, the rule "Thou shall not kill" can be made to gather and lump situations from first-trimester abortion to euthanasia for the "hopelessly" ill. These are all, in a sense, instances of killing. But in what sense? We believe a better use of rules holds them less as categories of understanding and more as tools for guiding actions, like probes in a dimly lit room. Rules are like cookbooks; they present recipes and moral formulae. Some cookbooks favor the use of garlic in every dish, but this does not mean that one cannot prepare a meal without it. There are dishes for which garlic is inappropriate, just as there are situations in which invoking a rule against killing distorts our perception more than the rule enhances it.

The proper use of rules is as probes or guides to orient us in situations of moral perplexity. But even in their appropriate use, rules are just shorthand reminders for more deep-seated convictions and commitments.

They should be used as instruments of exploration for moral perception and judgment, not as the measure of decisions.

The Fallacy of Exceptions

There is a saying in legal circles that "hard cases make bad laws." Likewise, the use of unusual situations as paradigms for moral action is suspect. Unique problems are not ordinarily precedent-setting, and should not be presumed to carry the weight of a moral tradition. Since Karen Quinlan's situation is atypical, it would be a mistake to take the decision made there as a paradigm for moral choice. Moral reasoning must serve to guide *routine* conduct and shape the mundane practice of medicine. In analyzing the process of moral decision-making it is more illuminating to consider everyday cases, such as whether to break a confidence, deceive a patient, or confront a colleague's mistake. Exotic life-or-death choices are fascinating and clearly important, but they are not usually the stuff out of which moral character is formed or nourished. We are not claiming that these exotic decisions are unimportant, only that they probably play too large a part in the moral imagination and should not establish moral precedents.

Hazards of Situation Ethics

One final pitfall is too common to go unnoticed, even in this short list, and that is the customary juxtaposing of "rule ethics" with "situation ethics." The former, we are led to believe, is the belief that moral choice is no more than the application of rules, with the addendum that there are no moral predicaments to which rules cannot be applied. To remedy this rigid notion, situation ethics is offered as a method that respects the unique characteristics of each situation. The attractiveness of situation ethics, so conceived, is obvious to the clinician who is caught in routines of diagnosis, prognosis, and therapy where individual judgments are numerous. No two patients are exactly the same; each disease process may present itself and behave differently. No two patients have identical reactions to drugs, or experience illness in the same way. It should not seem surprising that situation ethics, thus conceived, seems to fit clinical practice nicely.

Yet there are hazards, primarily derived from the oversimplified notions of ethics resident in the rule model and the situation model. In reality, ethics is meaningless without both rules and situations, and these elements are not competing but mutually interpretive elements. Rules, like moral principles, help us see what a situation is and what is at stake, morally, in our choices. Situations, likewise, when carefully understood, elicit or bespeak rules as frames of reference, or action guides.

One of the hazards of situation ethics, especially when interpreted superficially, is that it tempts us to think that one can experience and describe a situation in the raw—uninterpreted, unfiltered, and free of embellishment by our past moral experiences, our values and commitments.That is, situation ethics so conceived assumes a spatiotemporal isolation, a discontinuous slice-of-life, as the context of deliberation and choice. A situation is, in this sense, thought to be without precedent, unique, wholly novel, and presumably, one to which no rules apply.

In clinical medicine we are sometimes encouraged to believe that our technological expertise has created novel dilemmas of moral sovereignty over life's meaning and definition. A serious study of ethics should tell us that whatever new wrinkles upon old problems may present themselves, those problems that are entirely new to human experience are few. Usually other doctors, and other persons, have been in our place before us and have struggled as we struggle with the meanings and definitions of human life and death or more commonly with the meaning of confidentiality, or establishing a helping relationship.

Thus the problem with situation ethics is that it cannot do justice to our moral experience as human beings. Our practice as moral agents is richer, more diverse, and complex than the thin portrait of "the situation" leads us to believe. We are not isolated, or discontinuous in our moral struggles, and our experience of the predicament of moral choice is depleted if our only interpretive tool is "the situation" itself, and not the other elements of the process depicted above.

The process and pitfalls presented here do not exhaust the list. There are as many ways of doing ethics as there are moral agents. Ethics is a definitively *human* enterprise. We enjoy, and suffer under, reciprocal relationships with our values. We define our values and, in turn, they define us. There is no single "right" way to do ethics, and to suggest that there is would itself be morally illicit. What we have attempted to do in describing a process is to indicate that the use of *some* systematic procedure is important. Also, we want to display the variety of elements that need to be in the process. The central issue is the willingness to examine our reasoning process and improve it in light of experience.

Ethics and Tragedy

The approach taken in this chapter suggests that the decision-making process is largely rational. We have expressed considerable faith in the power of reason to ferret out assumptions, weigh competing values, clarify principles, and even assess deep emotional responses to morally problematic situations. This approach assumes that at least some moral tragedy can be eliminated, and that much of it can be reduced in impact through reasoned reflection. But the tragic dimension tends, as anyone ever confronted by a morally serious choice knows, to be more or less omnipresent. This is especially true for those decisions reached in haste, by means of flawed understanding, or in which professional habits or personal biases have preempted a fuller range of deliberations.

Some problems seem, nevertheless, to be beyond the powers of even our best efforts to resolve. Indeed, some medical problems appear (at least in the present) to be genuinely intractable moral paradoxes. They exhaust the limits of our analytic abilities; they defy rational resolution; perhaps most importantly, they test the very intelligibility of our moral assumptions. They threaten our usual guides, such as principles or procedures, and by doing so threaten to destroy our moral vision. This sort of moral tragedy, we believe, is not remediable by more careful thought, more factual information, or more skillful application of rules. This sort of moral tragedy does not occur because of the failure of reason but because of its limits; it occurs because human moral concerns stretch where reason cannot go. Moral tragedy of this sort is not irrational, but nonrational.

Medicine, or better the practice of medicine, is tragic in this deep sense. Because it ventures to care for persons in distress and *in extremis*—and because of its power—medicine reflects, in startlingly clear ways for the larger society it serves, the boundaries of the human situation.

Perhaps only a single example will suffice. It has long been accepted that the primary moral maxim of Western medicine is *primum non nocere;* but all of us know that, even in the name of altruism and service and with beneficent motivation, physicians sometimes do harm. Patients sometimes have adverse reactions to drugs, and procedures of diagnosis sometimes cause distress or disability. Professional advice and counseling sometimes lead patients away from help or toward self-destructive behavior. To iterate our previous scenario, disclosure of a cancer diagnosis to a patient can result in avoidance and estrangement from family and friends. Or in a more technical arena, a surgical procedure designed to enhance a patient's chances for survival may put the patient in the surgical ICU, removed from family and friends, in pain or obtunded. On occasion physicians are obligated to inflict harm as the only known avenue to achieving a therapeutic good. Sometimes physicians are involved in tragedy insofar as "the good" is, in irreducible ways, sometimes mysterious and beyond calculation.

Modern American culture, including American medicine, takes rational self-consciousness to be a cherished virtue. "Goodness" is in large measure equated with knowledge, and evil is accordingly correlated with ignorance. The answer to moral evil is accordingly correlated with ignorance. The answer to moral evil must therefore be more knowledge! By emphasizing the tragic dimension, we argue that this image of the moral agent is deeply flawed, and that it is neither sufficient nor faithful to human experience. The sense of the tragic is bound up with our sense of limits, of boundaries, and especially with our acknowledgment of the borders of rational self-consciousness to know "the good."

By addressing the tragic dimension we do not intend to romanticize it or to make it a convenient category for dealing with difficult problems. Acknowledging tragedy and mystery is not an excuse for indifferent or slothful moral analysis. The point, instead, is that moral problems are not reducible to intellectual puzzles; they are not finally solved, like algebra problems, or answered like empirical questions. They are rather lived through, and lived out, in conscious awareness of the suffering and hurt that, in our common experience, so often accompany our best efforts to achieve the good. Stanley Hauerwas has put the point well:

> . . .The moral crisis in contemporary medicine is not the result of the development of scientific medicine, but of our failure to have a sufficient sense of the physical and moral limits involved in any attempt to help and care for one another (3).

There are, of course, alternatives to acknowledging the tragic dimension of human moral limitation—cynicism,or skepticism, or ostrich-like optimism, or blinding faith in limitless reason. But in the end all of these are ways of denying our common human condition, and of spurning the necessity of making some moral judgments that will inevitably be wrong. It is only with a profound and tender sense of the tragic dimension that we can have the courage to be decisive without the pretense of being definitive.

References

1. Carlton W. *In Our Professional Opinion. . .* South Bend, Indiana: University of Notre Dame Press, 1978.
2. Aiken H. *Reason and Conduct: New Beginnings in Moral Philosophy.* Westport, Connecticut: Greenwood Press, 1962.
3. Hauerwas S. *Truthfulness and Tragedy.* South Bend, Indiana, University of Notre Dame Press, 1977.

21
Teaching Ethics

HARMON L. SMITH, LARRY R. CHURCHILL,
AND JOHN J. FREY

The previous chapter dealt with physicians as moral agents and decision-makers. It presupposed physician–patient interaction as the context of moral deliberation and choice, and it addressed the physician as a clinician and provider of care.

This chapter treats an additionally important role for academic physicians, that of teacher. Like other people, academics make moral decisions; but, unlike other people, academics are continuously engaged in understanding and teaching how ethical reflection and moral agency are most appropriately carried out. As with other activities, there is surely a sense of personal satisfaction when this job is done well; beyond that, however, the real beneficiaries of this effort by academic physicians are medical students, residents, colleagues, and other health professionals.

Teaching ethics in the medical setting is, even when it is not acknowledged as "teaching ethics," a powerful intervention. Indeed, teaching ethics in this place tends not to be explicit or dogmatic; it is more frequently tacit and implicit than formal and didactic. But it is there, all the same, despite some resistance to recognizing that teaching moral values goes on in medical education. We will argue that noticing and appreciating the transmission of moral values is an essential and unavoidable dimension of medical education. We will also discuss teaching ethics as an academic discipline—the goals to be sought, the methods to be used, and certain snares to be avoided—and draw on some of the concepts and principles outlined in the preceding chapter. In contrast to that chapter, however, the focus here is on how the physician-teacher can convey what he or she knows to the apprentice who is seeking guidance with an ethical problem.

Moral Values in Teaching

It is sometimes said that ethics cannot be taught. On examination this assertion usually means that moral values cannot be forced down a student's throat, that coercive and domineering tactics are neither proper nor do they work well in human relationships. But this assertion, benign as it appears, also masks an effort to teach a specific moral value—in this case, that one must be free to formulate and act on his or her own values, without compelling regard for what others might have thought or said. But that, we will argue, is an effort that is bound invariably to fail.

In actual experience, however, and because value dimensions are present in every significant human relationship, ethics is taught and learned constantly. What we need, therefore, is a more accurate sense of how these processes occur.

Sometimes we tend to identify ethics and value formation exclusively with either the cognitive or the affective dimension of human personality; in fact, values characterize persons as a whole. To isolate values as cognitive or affective is to commit the fallacy of simple location (1). None of us is ruled entirely by reason or entirely by sentiment, and we are usually most uncomfortable with ourselves when one of these conflicts with the other; as, conversely, we are most at peace when our reason and our feelings concur.

The tacit and attitudinal elements of value transmission are more likely to be demonstrated than formally taught. In medical education, where clerkship experiences and patient care responsibilities provide a workshop in values for the apprentice physician, this sort of demonstration is ubiquitous. Medical educators speak frequently about the importance of good role models for students, and the impact of these models in the intense setting of clinical training is unarguable. Moreover, medical education is not unique in this regard. Hannah Arendt put the case succinctly and well:

> The fact that we usually treat matters of good and evil in courses in 'morals' or 'ethics' may indicate how little we know of them, for morals comes from *mores* and ethics from *ethos,* the Latin and Greek words for customs and habit. (2)

Aristotle made the same point. In the *Nicomachean Ethics* he distinguished betwen intellectual and moral virtue, and further insisted that whereas the former might be attained by formal study, the latter was characteristically acquired through custom, habit, and experience.

In addition—as Kant argued, and as Piaget and Kohlberg have subsequently shown—there is an important cognitive dimension to the learning of values (3,4). Though there are powerful affective forces in value formation, the more precise aspects of value judgments depend on careful discussion and reflection in formal teaching situations.

The point is that moral values are present in every teaching situation. Even when we are not teaching ethics (perhaps especially then!), moral values are taught. This is so only because values are integral to and embedded in teaching styles and strategies as well as the general environment of the classroom, laboratory, or hospital ward.

Consider this scenario: to the question of how to tell cancer patients their diagnosis, a senior physician remarked to a young medical student: "You'll have to work that one out for yourself." Our experience is that when a distressing or catastrophic diagnosis must be shared, the encounter between medical students and patients is one of the most anxiety-provoking in all of medical education. In this instance, the response of the senior physician is simultaneously unhelpful and objectionable, both for what is said and for what is not said. First, this medical student is left with the impression that how to relate a diagnosis is pretty much a private matter, about which collaboration with or learning from others is of no use. Secondly, the implied norm is that "if it works, it's okay," but there is no suggestion that questions of right and wrong, good and bad, appropriate and inappropriate, are at all involved. In sum, what is said by the senior physician implies two criteria for judging moral values—namely, individualism and pragmatism—and thereafter the student is presumably left to his own devices.

The senior physician's silence on other aspects of this student's problem is equally troublesome. Isn't there a body of wisdom, one might reasonably ask, that this professor could share with this student? Doesn't concern for the student (not to mention the patient) suggest a more responsive and supportive role for the professor? Aren't there, after all, ways in which the student can go seriously wrong in this task? He might, for example, be so overcome by anxiety that he would try to withhold the diagnosis; or espousing a rigid dogma about truth-telling, he might force a demoralizing list of grisly details of physical degeneration upon the patient. As a teacher, the senior physician would be expected routinely to say that actions of this sort are morally wrong and professionally inappropriate; furthermore, he would be expected to indicate why. Otherwise no teaching or learning takes place here.

The refusal of a teacher, however benevolently motivated, to challenge and critically evaluate the values of a student undermines their mutual appreciation of the moral life and any common sense of vocation that they might otherwise be presupposed to share. The student is specifically disadvantaged, handicapped, and crippled in his development toward becoming a self-affirming and self-critical moral agent.

Several years ago we were doing research on the moral values that characterize medical education. After the purpose of the study had been explained, a third-year medical student responded: "There's no time to learn values here; what we learn is how to be competent physicians. The old family doc was great with his patients, but he simply didn't know his medicine." Apart from being less than historically accurate, such a response is a vivid commentary on one set of moral values that was learned in medical education.

It would be naive to suppose that assertions of the absence of moral values in teaching would be taken as indication that teaching is value-free. In fact, such assertions are better understood as indices for judging precisely what values are being taught. A theory of education that refuses to acknowledge value transmissions in teaching is simply revelatory of the value system at work in that educational theory. Yet, even when a place for ethics in an educational process is affirmed or permitted, those values are frequently treated as a finishing touch—icing on the cake, or a desirable frill.

Actually, our thinking should be inverted if it is to reflect our practice accurately. Moral values permeate the entire structure of education, even those educational endeavors that claim for themselves a value-free agenda of "training for competence." They are there from the outset in the purpose for training of this sort; "competence" itself is not value-free. And that, preeminently, is why value dimensions are not merely an additional component of our teaching, but definitive of the intellectual and social ethos in which education takes place.

It is in this sense that values are not only present, but central in teaching. Moral values are taught—whether or not we acknowledge it—and only some contrived and inaccurate notion of teaching as the "transmission of formalized units of knowledge" would make it otherwise. There are values in teaching, and they are inescapable. The critical questions are, "Which values?" and "Is their presence acknowledged or do they go unacknowledged, and hence unexamined?"

Some Values Definitive of Teaching

In what follows we will explore three characteristics that we believe indispensable to good teaching. Others could be listed, but we think these three are essential. In their absence, the activity in classrooms, laboratories, and wards ceases to be teaching and becomes some other mode of relating to students. In the absence of these three characteristics, there is an ethical problem for the profession.

Respect for the Otherness of Students

We mean here, quite simply, respect for the way in which students differ among themselves and from their teachers. This means more than appreciation of the pluralism in American universities or acknowledgment of the wide variety of life-styles, ethnic groups, and cultural and religious traditions among students.

Students are "other than" teachers in recognition of the asymmetry of power between teacher and student, an asymmetry that is grounded in differences in knowledge, experience, and skills. This distinction and dissimilarity is appropriate, and we do not suggest that it should be diminished or abandoned. Yet, it may pose a barrier to teaching. The problem arises not so much because teachers abuse their authority, although this sometimes occurs; the more frequent occasion for professorial abuse of power occurs when teachers—like others in authority—forget what the dependency side of the learning relationship is like. We may forget that things that we now do routinely and with considerable ease were initially learned by laborious effort and sometimes in great anxiety.

So options that may appear obvious to the teacher can be unimagined to the student who is struggling to make sense of a problem. The physician who made the cryptic remark about how to relate a distressing diagnosis failed as a teacher because this advice was disrespectful, contemptuous of the student's situation. Instead of engaging the student within the student's perception of the problem, the physician-teacher left the student to struggle alone within the boundaries of his own very limited resources. This is not always a bad teaching strategy, and we can imagine situations in which the "sink or swim" challenge seems appropriate; yet in this case that strategy was a disservice rooted in a lack of appreciation of and respect for the student's predicament, perhaps together with an amnesia of the tension that a problem of this sort poses for students.

Frederick Perls has composed one of the most popular versions of another kind of "otherness":

> I do my thing, and you do your thing. I am not in this world to live up to your expectations. And you are not in this world to live up to mine. You are you, and I am I. And if by chance, we find each other, it's beautiful. If not, it can't be helped. (5)

Even in its general social context, Perls's view is a fantasy—in much the same way that the Tarzan myths of Edgar Rice Borroughs are entertaining but impossible. But especially in the teaching situation, that type of radical autonomy is nonsense. Mutuality is a key to good teaching; but it cannot be achieved apart from respect for otherness. In the nature of the case, the burden for that relationship between teacher and student lies chiefly with the teacher.

A Commitment to Objectivity

Here we mean the presentation of knowledge as knowledge, and of opinion as opinion. One of the teacher's temptations within the asymmetrical teacher–student relationship is to present one's own opinions, hunches, or intuitions as though they were established facts. That temptation, conversely, is to present established knowledge as merely opinion when it runs contrary to one's own views. The name for this prejudicial form of teaching, in which one's own ideas are given acritical status, is "indoctrination." Indoctrination not only converts hunches into facts; it also manipulates the methods of inquiry so that conclusions reached will nicely confirm the teacher's opinions and thereafter foreclose further inquiry.

One of the ways in which a teacher can fail to be objective is exemplified in the earlier scenario between the physician-teacher and the medical student. The teacher fails in objectivity because he omits to mention the experience or skills—his own or those of others—that might have assisted the student. In other words, the physician espoused a value relativism that supposes that it really doesn't matter how one breaks bad news to a patient. That it does matter a great deal, that there is greater or lesser skill to be employed in communicating this word to the patient, and that one shouldn't expect to do it perfectly or maybe even appropriately—the first or every time—these would be invaluable lessons for this student to learn. The neutrality of the physician, however, and his failure to indicate these things, constitutes a bias, a lack of objectivity about the student's problem.

It is sometimes thought that bias arises only in issues of adherence to proper methods of research; but bias can also occur in teaching. Only the extreme moral relativist, or the disappointed moral absolutist, would likely believe that in teaching we are left to guesswork, whereas in research we can be objective. To be objective in teaching, in terms of accuracy of empathy and in the choice of appropriate methods and strategies employed in instruction, is an essential characteristic of good teaching because it manifests the public and professional trust placed in those who teach.

Enabling Students to Learn

In examining the knowledge of those around him, Socrates found not only that people were making false claims to knowledge, but that those claims prevented them from thinking critically. He thus conceived his task to purge people of those unexamined prejudgments that prevented them from conducting a critical inquiry into the matters about which they had opinions. We still honor the Socratic maxim that addresses this point: "The unexamined life is not worth living." Socrates' task was to bring others to an acknowledgment of their ignorance; yet he claimed only ignorance for himself in this endeavor. Maurice Merleau-Ponty has stated succinctly what we want to emphasize:

> The story of Socrates is not to say less in order to win an advantage in showing great mental power, or in suggesting some esoteric knowledge. 'Whenever I convince anyone of his ignorance . . . my listener imagines that I know everything that he does not know.' Socrates does not know anymore than they know. He knows only that there is no absolute knowledge, and that it is by this absence that we are open to the truth. (6)

In contradistinction to indoctrination, or facilitation, or being a "neutral" resource person, or whatever other surrogate is currently fashionable—discernibly different from all of these, teaching invariably involves a type of Socratic dialogue in which the teacher, being knowledgeable of the range of his or her own knowledge and ignorance, enables the student to think, to examine critically, and to resist premature closure or tenured ideas.

Socrates suggested that teachers are essentially midwives who seek to help in the difficult process by which each student gives birth to the knowledge that gestates within. Educational philosophies since the Enlightenment have espoused another paradigm, in which knowledge is on the outside and somehow to be placed onto the blank receptacle of the mind. The efforts to place this external knowledge within have frequently done violence to both the student and the knowledge. It may be helpful to remember that the Socratic view is enabling, whereas the Lockean paradigm is imprinting. The Latin root of education is *educo,* which means to support growth of, to tend, to nurture; that is also what we mean by enabling.

Teaching Ethics

If students could learn ethics entirely from example and precept, there would be no need at all to give formal or explicit attention to ethics. An assembly of bright and observant students, together with skilled and exemplary faculty, would be all that we would require. But that more than this is wanted is eloquent testimony not only to the need to recognize cognitive and theoretical aspects of ethical reasoning; it attests also to the variety of values that exist in American culture. The plain fact is that a common cultural and moral heritage can no longer be assumed in this society; and this is a condition that applies as much to physician–student relationships as it does to physician–patient encounters.

Alasdair MacIntyre has argued convincingly that we do not have a rational method for resolving moral dispute in American society because we lack a shared tradition with shared notions about our nature and true end (7). We cannot agree on conclusions, he says, because we cannot agree on starting points. The result is that we possess fragments of several conceptual schemes, but those fragments yield only a simulacra of morality.

Much has been written recently about teaching ethics as part of medical education (8–10). Despite an enormous variety in methodologies and subject matter, one thing emerges as patently clear: teaching ethics in medical education, if it is to be useful, must be consensually purposive in its aim and intention, and disciplined in its approach. It cannot be done *ad hoc,* or "off the cuff," by instructors who are armed only with tender concerns and good intentions.

One popular approach to ethics in medical education is the problem-solving case-study. This approach has the merit of relevance, but it succeeds as a teaching vehicle only when it goes beyond *what to do* to considerations of *why*? and *by what authority*? The philosopher Edmund Pincoffs uses the disparaging term "quandary ethics" to describe the fixation on exotic dilemmas of life-or-death choices that ignores the reasoning that should inform such choices (11). When ethics teaching is confined to decision-making, the personal values of character are neglected. Focusing exclusively on choices in a decision misses the mark of teaching, which isn't the decision but the *person* who decides. There is certainly much in medical ethics that is problem-oriented; but the teaching of ethics—medical or otherwise—is always person-centered and concerned with what students know about their own values, how they are enculturated into a profession, how they reason with principles, and the manner in which they weigh consequences. This means that there is no formula or set of rules to convey to students that will guarantee appropriate moral agency. What we chiefly have to offer students is a disciplined mode of engaging them about their values.

Another popular approach to ethics goes under the name of "values clarification" (12). The aim here is simply to identify the value(s) that one employs in solving the problem posed by the case study. Although such an exercise has its uses, it is by no means a substitute for moral reasoning. When left by itself, it simply affirms in a *laissez faire* manner whatever values the student espouses at the time. Values clarification typically emphasizes empathy for diverse values in a group, without regard for traditional norms or for the fact that some values are thought to be normative in relation to others. Ethical reasoning prosecutes the search for the best choices and ways of life, together with how to achieve them. At its worst, values clarification endorses the belief that ethical decisions are just private preferences or matters of taste, and thereby leads to ethical relativism by uncritically affirming moral pluralism.

Daniel Callahan has said that the place of ethics is wherever there are problems in medicine that raise ethical questions and need ethical answers. In a frankly pluralistic society, we search for ways to bring at least some coherence to our understandings and attitudes and actions. So the tasks of the ethicist, in the face of that kind of ambiguity and variety, are principally three, according to Callahan: (1) to assist in identifying the defining morally problematic issues; (2) to offer a systematic way of thinking about and through these issues; and (3) to help medical professionals to make "right" decisions, i.e., decisions in which affirmation coincides with action, belief gets expressed in behavior, and character is consolidated in conduct (13). If this is so, we are back at our starting point: reflecting on a consensually universal context of value in order to say why we have decided to do one thing rather than another.

Four Essentials

The process of making a moral decision for oneself, of being a moral agent, is as complex as the people involved. The previous chapter described a wide variety of movements that are helpful in this process. It deserves emphasis here that the teacher's role is of a rather different sort. Helping a student, resident, or colleague with his or her moral problems is necessarily more limited in its aims. Time is usually short, contact is limited, and anxiety frequently runs high. We find it useful, therefore, to concentrate on four essentials that are the backbone of the more sophisticated process described earlier, and should be seen as the first step in helping learners to initiate their own processes of moral reflection.

Describing Moral Experience

Ethical problems are complex and involve, as we have said, the whole person. They usually affect us profoundly; and the choices we make alter or reaffirm our image of others and our concept of ourselves. Ethical problems are not just intellectual puzzles or emotional trips. Therefore, one important task of the teacher is to help students recognize how they feel about the issues—to acknowledge their gut responses or affective reactions (see Chapter 20, pp. 345–346.).

Skill in describing one's moral experience takes practice, and an essential role for a teacher here is affirmation of the student's predicament and sympathy with the student's dilemma. Responses such as "That's no problem" disenfranchise the experience of the student and discourage the sort of work that is necessary for moral self-understanding. The goal is not the "right" answer, but the student's achievement of satisfactory and appropriate moral resolution.

But descriptions that remain at the emotive level are as unproductive as descriptions of moral experience that neglect the emotive level altogether. Helping the student to move beyond an emotive response to the problem itself is another role that the teacher should assume. Frequently this can be done by focusing on the precise nature of the problem and identifying the conflicts and their locus. Consider, for example, the following situation:

> You are a preceptor for a group of students who are learning to do cardiac exams. One of the male patients on your service has a "good" heart murmur and you want your students to hear it. When you ask the patient if your students can listen to his heart, he agrees. By the time the fifth student is ready to take his turn, the patient says to the student, "Please don't practice on me anymore."

The fifth student may feel like a mere observer, a voyeur, with nothing to offer the patient, and at the same time be angry at the thought of being deprived of the learning experience.

Any student in this situation would feel some conflict between his obligation and zeal to learn all he can and his duty to respect the patient's wishes. Although this conflict ordinarily cannot be sorted out on the spot, the teaching situation that it presents for later conversation is typical of one of the important levels of medical education. Identifying precisely *why* the student feels conflict and what the competing values are (zeal for knowledge versus respect for patient wishes) ought to be a major help, even if the difficulty itself cannot be immediately resolved. Taking the time to describe the conflict affirms its validity as a genuine problem, and discourages the temptation to press ahead and ignore the patient (or possibly to succumb entirely to every patient request).

Finally, helping students to describe their moral experience can entail describing one's own similar experiences. Drawing on one's own experience without presenting it as *the* solution, however, is difficult. Frequently, "This is what I do" carries a tone that implies, "This is what you should do also." Such an implication is what students sometimes want and even need, so long as it does not arrest but rather enhances their own process of moral deliberation. By helping students describe their moral experience, the teacher assists them in appreciating the richness and fecundity of their moral life. This, in turn, will help them to appreciate the moral quandries they will encounter with their colleagues and patients.

Eliciting Assumptions

Eliciting assumptions and holding them up to critical scrutiny is the heart of sound reasoning generally, and moral reasoning is no exception (see Chapter 20, pp. 350–351.). Assumptions, by definition, are not explicit but typically underlie the matter currently being considered; they are out of focus and largely unacknowledged. Moral dilemmas, like icebergs, have an impact because of their submerged weight. Thinking and judging and acting appropriately require taking these hidden aspects into account.

Eliciting assumptions means asking questions such as, "Why is this particular issue a moral problem?" or "Why do others not feel this difficult or disagree with you?" Consider the following situation:

A first-year resident comes to you seeking help about a patient he is seeing for the first time. The patient is a 16-year-old unmarried woman, previously under the care of a graduate senior resident. She wants her birth control pills refilled. The patient's exam, including a pelvic, is normal. This resident does not think the physician should prescribe contraceptive measures to unmarried women.

Here the teacher's role is to ferret out precisely why the resident disapproves of this action. Does he, for example, believe that to grant this patient's request would amount to his condoning—professionally or personally—premarital sexual relationships? Does he have religious convictions about sexuality and procreation that are ingredient to his professional opinion? The goal here is not to change the resident's mind, or to alter his behavior, but to help him understand why he thinks as he does. What assumptions are at work here? Are they assumptions of fact (e.g., prescriptions for contraceptives relieve anxiety about and therefore encourage sexual activity) or assumptions of value (e.g., premarital sexual relationships are wrong)? Assumptions of fact are sometimes false and easily corrected; assumptions of value usually pose knottier problems. A helpful way to initiate a conversation about these matters might be for the teacher to remind the learner that there are no uninterpreted facts, that all facts are the product of human perception and imagination.

Assumptions about the status, office, or responsibilities of physicians are another place to focus. Do physicians really have much power or influence over adolescent behavior? And if they do, should it be exercised?

Is this resident conflating, or lumping together, personal convictions with professional behavior? Is he imposing his values on others—perhaps physicians generally, and this patient in particular? Or does he have some specific reason to believe that, regardless of his other convictions, contraceptives would not be good for this patient because of her medical circumstances?

The sensitive teacher will also be aware that the resident may feel threatened by several other factors: (a) the less traditional, more liberated, attitude of some women; (b) his ability effectively to relate to such women in the future; and (c) his success in the residency program if he does not treat patients assigned to him and assume his share of the clinical work. The teacher cannot honestly be totally reassuring on any of these points; but again, the teacher's goal for this encounter is not, *qua* teacher, to get the clinical work done, but to help the resident reach a satisfactory resolution that he can live with. Given this goal, the drawing out of assumptions is essential.

Considering Multiple Alternatives

The most prevalent moral fallacy is "black-and-white" thinking, the almost irresistible temptation to deal with moral problems by dividing the options for action into two mutually exclusive camps at opposite extremes of the spectrum (see Chapter 20, pp. 349–350.). The exotic choices of tertiary care and high technology medicine encourage this tendency, and typical examples are familiar and easy to cite: to save or let die, to abort or continue to term, to tell or not to tell (e.g., the truth about cancer diagnosis). Assisting students in problem-solving means helping them beyond the dichotomies of either/or ethics.

Consider the following situation:

Your resident seems deeply troubled, and when you inquire, you discover that her concern is about Mr. J. in Room 305. Yesterday Mrs. J. pleaded with the resident not to tell Mr. J. of his condition: cancer of the pancreas, widely metastasized. "I want our last days to be happy," she said. Although she generally disapproved, the resident reluctantly agreed not to tell Mr. J. just how severely sick he actually is. Today the resident regrets her decision and is confronted by Mr. J., who asks if he is going to die.

The clinician-teacher who devises a solution that is centered on the options of truth-telling/lying will be of little help to this resident. Everyone will probably agree that truth-telling is generally preferable to lying, and almost everyone will likely agree that lying is ordinarily to be avoided. Besides, the first problem here is not whether to tell the truth or lie; it is rather a problem of divided loyalties. Who is the patient? Mr. J. or Mrs. J., or both? Can this resident care for both? Of what possible use might the old maxim be that teaches that the physician's duty is to the well-being of the primary patient without regard to spin-off benefits to dependent populations? Who is the best judge of what Mr. J. needs or wants? Of what Mrs. J. needs or wants? Who decides whether the needs and wants of Mr. J. or Mrs. J. take priority? Who decides what will provide for the "happiest" days for Mr. and Mrs. J.?

These are questions that allow the experience and sophistication of the teacher to be of enormous help to the resident. Truth-telling can take a variety of forms, and be done by a variety of means. One does not always have to choose between the sledge hammer of truth and the beneficent deception of a "little white lie." There are subtle but important differences between and among several alternatives: forcing the truth on patients; directly responding to a question; responding in story, parable, or metaphor; redirecting the question; delaying a response; or giving the news in concert with or through a chaplain or other appropriate person. Each of these alternatives can span a wide range of truthfulness or deception. In this case, we suggest that it may be possible to honor the requests of both Mr. J. and Mrs. J.; but we recognize that it is not always possible to achieve a resolution of that sort.

The principal reason for posing multiple alternatives is that doing so frequently leads to a reformulation of the original problem. In this case, the initial choice of truth-telling/lying is better formulated as one of physician loyalty. This puts the truth-telling issue in a different light and accentuates the *truthfulness* of the resident rather than the factual truth of dispensing information. Not acting on the dictates of Mrs. J. will help the resident rely on her own capacity to distinguish legitimate family requests from those that are merely self-serving or harmful. Alternatives directed toward short-range expediency can have undesirable long-range consequences. Weighing the consequences of each alternative accurately is learned through experience, and here the teacher's failure to advise (such as the case with the surgeon discussed earlier) would be a breach of the ethics of teaching, as well as of the teaching of ethics.

Justifying Choices

Giving account for one's choices is the way in which appeal is made to the larger community of peers, patients, and society. Justifications are the backing given to actions, the reasons that actions are warranted. To give justification is to go public, to appeal to a broader realm than individual conscience and to seek affirmation and consensus among others. Because ethics is the attempt to give account of moral agency, it is inherently public and communal, not private and individual.

There are a wide variety of ways by which people seek to justify or account for their actions. The ways in which justifications are given tell us as much (and frequently more) about the person as the actual choices they make. There are three basic ways of offering justifications: (1) by appeals to principle or rules, (2) by appeals to consequences, or (3) by appeals to character. Frequently more than one, and sometimes all, of these three backings are given. The teacher's role is to make learners aware of the justifications they offer, and to test these accounts in terms of their cogency and support for a favored choice.

The following example illustrates the possibilities:

A senior medical student serving an Acting Internship at your hospital is caring for a 3-week-old infant who is brought to the emergency room. The mother stated that the baby began screaming this morning and she noted swelling in the right thigh. An x-ray film revealed a greenstick fracture of the right femur. The student tells you that he suspects child and spouse abuse, since last week he saw the mother for a cold and noted that she had a black eye. The mother has been very complimentary toward this student, and he says that he feels she won't return at all if she is confronted with his suspicion.

The student would like to probe the potential family abuse matter further but he is fearful of doing so. The options and their justifications might include:

1. Confront the mother, because the physician has a duty to prevent harm (principle):
2. Forget what he has seen, because the chances for successful intervention are slim (consequences):
3. Confide in the mother about these concerns but promise to do nothing without her approval (character).

Other options are also possible; indeed, the same option might be taken for different reasons. For example, the student might confront the mother and report the situation to legal authorities because he feels morally obligated to obey the law in all cases. Or, he might elect to forget what he has seen because bringing attention to it could bring even more harm to the mother and child from a battering spouse and father.

Ethics is not just what we do, but what we do and why. Agreement between people in action does not always mean agreement in moral rationale; the meaning of any action is not self-evident from a simple observation of the action itself. Some interpretation is required if we are to understand what is going on. In the process of building moral reasoning, it is awareness of the justification and accounting for choices that nourishes the moral life.

Another way in which teachers may ask for the rationale is through the question, "Who are your heroes?" Here is a direct appeal to character morals. Another, more sophisticated, question is, "To what community do you belong?" Do you belong to a community that recognizes (either by condoning or by working to alleviate) family violence? Communities of affiliation are indicators of those from whom we seek moral approval. "Whom do we want to validate in our choices?" "Whose opinion matters?" Responses to these questions do not tell us what to do, yet they give us insight into how a person's reasoning process finds its sources of authority. Such questions are essential to any serious attempt to teach ethics.

We ought to be aware, finally, that the reasons given for choices tend invariably to be recycled into our descriptions of moral experience. If it is true that behavior is a function of perception, awareness of the reasons we employ becomes a part of us and we tend to recapitulate those reasons in future decisions. It is true that we do who we are; it is equally true that we are (and become) what we do.

Epilogue

These four essentials are not isolated modes of engaging students; they overlap and are interrelated. The four essentials are not staged questions to be asked in a developmental sequence so much as they are windows into the moral sensibility. They will be of little or no use whatever if they are used by teachers who believe that ethical problems are an afterthought to science or matters of etiquette. They will be most helpful to those who, as teachers, are both fascinated with and concerned about the complexities of the moral lives of students, residents, and others affected by the teacher's moral vision.

The reader may be a little frustrated at this point because of a lack of resolution in the cases discussed, or because we have portrayed the teacher as more of a midwife than a mentor. The purpose here, however, is not to solve moral problems, but to show how ethics might be taught effectively in the setting of medical education. The role of the academic physician-teacher, we believe, is similar. The task of the teacher in clinical medical education is not to instruct in morals but to engage in dialectic; to probe, to encourage, to criticize; to teach what the student knows but has forgotten, knows but does not recognize, knows but does not like to acknowledge, or knows but does not know how to talk about, has no names for, or is afraid to confront. There are, of course, other ways to do this besides those we have discussed in this brief space. We think all the same, that what we have outlined will be helpfully suggestive for the physician-teacher who wants to engage students in serious ethical reflection.

References

1. Whitehead AN. *Science and the Modern World* (1925). New York: Free Press, 1967.
2. Arendt H. *The Life of the Mind,* Vol. 1. New York: Harcourt, Brace, Jovanovich, 1977.
3. Piaget J. *The Moral Judgment of the Child* (1932). New York: Free Press, 1965.
4. Kohlberg L. Stage and sequence: the cognitive-development approach to socialization. In *Handbook of Socialization Theory and Research.* Goslin D (Ed). Chicago: Rand McNally, 1969.
5. Perls FS. *Gestalt Therapy Verbatim.* New York: Bantam Books, 1981.
6. Merleau-Ponty M. In praise of philosophy. In *The Essential Writings of Merleau-Ponty.* Fisher AL (Ed). New York: Harcourt, Brace and World, 1969.
7. McIntyre A. *After Virtue.* South Bend, Indiana: University of Notre Dame Press, 1981.
8. Callahan D, Bok S. *Ethics Teaching in Higher Education.* Hastings-on-Hudson, New York: The Hastings Center, 1980.
9. Clouser KD. *Teaching Bioethics: Strategies, Problems and Resources.* Hastings-on-Hudson, New York: Institute of Society, Ethics and the Life Sciences, 1981.
10. McElhinney TK (Ed). *Human Values Teaching Programs for Health Professionals.* Philadelphia: Society for Health and Human Values, 1981.
11. Pincoffs E. Quandary ethics. *Mind 80*:552–571, 1971.
12. Simon S, Howe LW, Kirschenbaum H. *Values Clarification.* New York: Hart Publishing, 1972.
13. Callahan D. Bioethics as a discipline. *Hastings Center Studies, 1*:66–73, 1973.

22
Case Studies in Ethics

JOHN J. FREY, HARMON L. SMITH,
AND LARRY R. CHURCHILL

Just as any practicing physician, throughout a career the teacher of clinical medicine is faced with a variety of ethical dilemmas. There are, however, additional ethical issues that arise because of the structure and goals of the educational system. This chapter will address several cases that contain an ethically significant dimension. Although they are not unique, the cases illustrate the complex nature of the teaching setting and provide opportunities for discussion in an academic department, a medical school, or a teaching hospital.

CASE 1: CONFLICTING RESPONSIBILITIES

A married resident in the final year of her residency training tells her colleagues that she is pregnant. She is cared for by the senior member of the faculty who is also the director of the residency program. According to the dates of her pregnancy, she should deliver in the spring of the final year of the program. With three weeks for maternity leave she should be able to complete her program in sufficient time to qualify for boards which will be taken shortly after the completion of the residency program. Her family physician outlines this plan for the other faculty and residents and gets general approval.

During her 33rd week of pregnancy, she begins to develop increased tibial edema and her blood pressure has elevated 10 points over her baseline. There is no proteinuria and reflexes are normal. She is advised that her pressure is elevated and that she must spend more time with her feet up, and is asked to report to the practice if there are any headaches or visual changes. She is told to return for an appointment in 1 week.

At her next appointment the blood pressure has remained at the same level as during the previous week. At this point, her family physician is faced with the potential conflict between his responsibility as a physician and his responsibility as the director of the program. The patient's condition is such that he might make a different recommendation if the patient were other than a busy resident in her final year.

Discussion

This case illustrates a conflict between professional roles that commonly arises in academic settings. On the one hand, faculty members routinely accept responsibility for the educational development of learners. Many faculty also feel responsible for residents' personal welfare, including assurance that the health care needs of residents and their families are covered. On the other hand, residents may not be truly free to choose the source of their health care. Tacit factors including professional respect, convenience, and cost may compel residents to choose a faculty member as a primary care provider, rather than a community practitioner. Mistrust may even arise when a resident chooses to obtain personal and family care from a nonfaculty physician. Consequently, tension can emerge when a resident has to "trust" faculty members for medical care to avoid potential conflicts of interest.

The principle of confidentiality, which is central to the ethical basis of the doctor–patient relationship, is challenged in situations where a learner is a patient in his/her own educational institution. Confidentiality is always an issue. However, confidentiality takes on an additional dimension because educational programs are by their nature evaluative systems. Faculty are required to carry out summative and formative evaluations of learners, and the materials included in those evaluations range from self-assessments by learners to nationally standardized tests. Although physical and emotional health are usually not included in formal decisions about resident promotion, they may play an important role if there are questions about whether a resident will be able to cope with the rigorous nature of the training program. In this particular case, the issue of physical ability of the resident is compromised by the fact that her physician is the same person who is required to decide about her suitability for promotion or completion of the residency program. Whereas one role requires patient confidentiality and patient advocacy, the other requires the candidness of an educational evaluator.

Another concern in this case is justice. In the past few years, residents have been demanding, with varied success, that their lives be more reasonably organized in recognition of the need for personal time with families and other activities outside of medicine. This movement has been fueled in great part by the activism of women medical students who feel that there is a place for a different blend of training and family than what has been traditionally offered. Thus, residents in most programs have obtained the right to take time for maternity leave and other personal needs. There are now many experiences with interrupted and part time residency programs that show that they are compatible with the life of families and childbearing. Many women residents choose not to defer childbearing until after the completion of training and have thus created an increased frequency, if not a tradition, of starting families while still residents.

On the other hand residents who take maternity leave or who are pregnant during their training place a burden on the system and on other residents. Some programs have asked women medical students who are applying to them whether they have plans to start a family during residency. Questioning students about such matters creates not only an ethical problem but a legal one if discrimination on the basis of sex can be shown by applicants.

Similarly, there is great potential for women who have large patient care responsibilities to have their own health care compromised or to take risks that they would not want their own patients to take. The problem is increasing with the growing number of women who attempt to combine residencies and pregnancy. The burden of increased workload for her colleagues, owing to constraints on her performance related to her pregnancy, is an uneasy prospect for the woman in this case; in addition, and for similar reasons, her judgment regarding compliance for her medical care may be hindered. What is just for her and perhaps best for her fetus may not be just in light of her obligation to her fellow residents.

Case 1 [Two weeks later]

The Director of the family practice residency program who is also the physician for the patient/physician in question has tried to couple his administrative and patient care responsibilities by moving the resident to an outpatient service with minimal night call demands. The Director has also arranged to have the resident take the elective half time for 2 weeks. On the next visit to the practice the resident's blood pressure remains 12–15 points over baseline and there has been an additional 4 pound weight gain. There is no proteinuria. No other signs of preeclampsia are noted. The resident/patient is at present at 36 weeks of pregnancy. The Director feels that he is getting into an increasingly difficult position since he is fully aware of the extra administrative work that a resident, dropping out for 2 months, would put on the program and residents. In speaking with other directors of programs, he finds no consistent method of dealing with such issues but only individual solutions each time the problem has arisen. The resident/patient is also worried about being able to qualify for her boards and has to complete a prescribed number of rotations before qualifying. She asks her physician/residency director to help her arrange a mix of part-time and longitudinal electives, including a reading elective, to help her finish the program in enough time to take boards. The residency director feels more and more uncomfortable with this position and asks his colleagues to relieve him of his administrative responsibility. He remains only as a faculty member and health provider to the pregnant resident.

Discussion

An issue in this case is whether any action taken by the residency program is within established institutional or national norms. There are, of course, norms that set limits for individuals who have predictable problems as to the amount of "unscheduled time" that is available to them, such as sick time or personal time. In this situation, institutional maternity leave is 3 weeks; but this will not be sufficient to cover the resident in question, nor will a combination of sick time and personal time cover her needs. One way to address the issue of fairness is to appeal to standards or norms. Yet to do so may be unfair, as the complications of this pregnancy are not standard and in any case were not anticipated.

The goals of the management of this case are to assure proper medical care for the resident and to seek a measure of fairness in the administrative and educational decisions. But these goals conflict; and the conflict is ethical because the parties involved have different needs and wants and there are no accepted standards for decisions. Rational resolution of the conflict is not possible on these terms. The additional conflict for the residency director, between his professional and administrative responsibilities, can be partly resolved through taking action that absolves him of one of his responsibilities. In this way he can avoid the conflict by stepping out of one role.

An important question in this and other similar situations is for whom is the program director an advocate. He must set priorities when his role as physician/patient advocate and administrator/program advocate are in conflict. He cannot be all things to all people.

Although the situation we have described may be somewhat unusual, the problems it presents are not unique. Physicians are frequently put into situations that might create conflict with their administrative or evaluative role. For example, a great deal of "curbside" consultation occurs in hospital settings. Colleagues and health team workers—nurses, technicians, and others—often ask information about their own or their family's medical problems without any formal contract either to do what the physician suggests or to follow up care in any systematic way. A difficulty for the physician is to understand the nature of the "curbside consultation" and his/her responsibility. The patient/colleague can decide to change management plans without a formal record being made and with only a tacit sense of the meaning of the interaction. Medical records exist primarily as a means of communication of facts and thinking. If no record is available, then questions of responsibility and liability for bad outcomes become acute. Thus, the possible ethical dilemma created by caring for individuals with whom we also have a supervisory relationship is compounded by the informality of the hallway or coffeeshop consultation. Both sorts of situations rely on implicit factors in a previous nontherapeutic relationship and tend to stretch that relationship to encompass therapeutic goals. There is no one solution that can be applied to all situations; so the potential for problems should make us always careful to negotiate with our patient/colleague the parameters of, and the rules by which, each party will honor the relationship.

In summary, this case focuses primarily on conflicting roles in the relationship between a residency director/physician and a resident/patient. However, it also involves conflict between a variety of interests, including resident/resident conflict, similar to that in any group in which one group member requires special consideration. Beyond these, resident morale, standards of conduct, even-handed treatment of the resident in question, and larger issues of women in medicine are all involved.

Identification of the diverse cultural and ethical problem requires us to deal at the level of principles, probing the meaning of confidentiality and fairness, rather than stopping our deliberations at the more rigid level of rules. Rules can be helpful in this situation as guidelines for conduct. Yet when guidelines conflict, or in novel situations not anticipated by our usual guidelines, a judicious use of principles is called for. Here is where the discussion can yield both a solution to this particular situation and serve as the basis for understanding similar role conflicts in the future.

Case 2: Mixed messages about CPR

An elderly man with chronic renal insufficiency and arteriosclerotic heart disease is admitted to the hospital in congestive heart failure. He is widowed and at present is living with his son, who for some time has been discussing the possibility of putting his father in a nursing home. He is cared for by a community physician who admits and follows patients on the inpatient service of the teaching hospital. He is admitted for recurrent congestive heart failure and adjustment of his medication due to the increasing renal failure.

The management of his medical care is uneventful until the house officers are on rounds one day and the nurse calls their attention to the fact that the patient has had some mild chest pain for the past 30 minutes. On entering the room, the inpatient team is confronted with a man who is in obvious distress and clinically in pulmonary edema. They call the nursing staff, call the medical resident in the ICU, and arrange to take the man to the CCU once initial treatment has been started. There is some confusion with the nursing staff and the resident angrily accuses the nurses of being slow at getting the emergency cart to the man's room. After some delay, they begin to push him on the stretcher to the other part of the hospital where the intensive care units are located.

On the way, the man begins to experience increased chest pain and has a cardiopulmonary arrest. One of the residents calls a code for cardiac arrest and directs the code. During the code, the junior resident on the team expresses a strong feeling that they should not do an aggressive code. The reasons he gives are that the patient is in very bad health, has progressive disease, and is a placement problem for the family. The resident in charge of the code expresses her disagreement with this idea and the code continues. There is an obvious conflict arising between residents, which the junior resident tries to resolve by phoning the attending physician. After reaching him and discussing the situation, the resident gets the attending's opinion that the family and the patient would not want aggressive "heroic" medical treatment for this patient. However, since there is not a "No Code" or "Do Not Resuscitate" order on the chart, the code progresses as directed by the resident in charge despite the information gathered by the other resident. The patient is finally placed in the CCU on a ventilator with a stable cardiac rhythm.

Discussion

Much has been written about patients who experience cardiac arrest and resuscitation. The need for such a literature arose from the confused, highly subjective, and often arbitrary nature of decisions about the beginning or continuation of cardiopulmonary resuscitation. This case illustrates the confusion that may exist if the wishes of the patient, and that of the patient's family, are not sufficiently explored prior to any sudden development, or prior to the development of complications or poor outcomes of medical or surgical procedures. It is now evident from both management and legal perspectives—as well as an ethical one—that the wishes of the patient must not only be sought but also noted in the chart and the order sheet. Some questions that must be asked in this situation include:

1. What does the patient understand about his disease and its prognosis?
2. Who is responsible for the discussion of this with the patient and family, and how will the discussion be noted in the medical record?
3. Who makes decisions in emergency situations regarding the type and duration of treatment for an acute problem, and on what grounds?

The patient's understanding depends on his competence at the time of the explanation. To say he is competent implies that the patient understands the implications of the discussion about resuscitation and is able to indicate a preference clearly. We believe it is also important that the patient makes the decision. Often, the first discussion with the patient results in, "I'd like to think it over" or some similar lack of clear decision. In many cases, depending on the severity and acuteness of the situation, there is time to sit down again with the patient and discuss the issue. If that is the case, patients should be allowed to reflect on the matter and make a deliberate choice.

The tendency of many busy physicians to seek out family members and ask their opinion when a clear decision is not forthcoming from the patient ought to be avoided except in extreme situations. Such actions infantilize patients, and rob them of whatever degree of control remains for them to exercise. Patients need, and we think appropriately, to have the opportunity to decide for themselves. When possible, it is best to discuss the concrete issues of resuscitation with the patient when family members are present. Then the discussion will be clear and open, and every chance given to avoid misunderstanding in the future when information is passed through a third party.

Most physicians are neither trained in, nor are they very comfortable with, discussing resuscitation with patients or family. Thus a major part of the orientation of residents to the activities of a teaching hospital should be practicing, perhaps with role plays or video scenarios, the process of discussing the resuscitation decision with a patient and family members. Although the discussion of such an important part of one's life with relative strangers (such as a house officer) is less than optimal, it is common in academic medicine. It is distressing, but many attending physicians are not knowledgeable about the patient's life before hospitalization and often leave such activities to the house staff. Ideally, the house officer ought to be a participant, but not the leader, in this discussion which should be led by the patient's personal physician or the attending physician.

In the case under discussion, one difficulty central to the case is the lack of clarity regarding what should be done if the need for resuscitation arises. From the lack of "No Code" orders on the chart, one can assume that there was no formal decision by the patient, or his physician, or members of the inpatient team, to forego resuscitation. However, because CPR is of doubtful benefit and perhaps even detrimental to this patient, there is conflict among the house officers about whether to proceed. This is further complicated by the decision to call the attending as an arbiter when the question of continuing the code arises. The attending is the final legal authority, and has responsibility for all the decisions that take place on the teaching service, yet his physical absence from the situation together with the absence of written information about the patient's or family's wishes minimizes his rule. The house officers are thus in a difficult ethical position. They are clearly in conflict over the patient; they might also be in conflict over the attending physician.

Medical responsibility is to the patient, and his choice as a partner in the covenant between a patient and a physician is to be respected. Even when the patient's wishes are known (which is not the case here) conflict between physicians and patients is frequent, often representing differences between perceived costs and benefits to the patient and the larger society. Changes in the reimbursement process for hospital care are beginning to force institutions to examine resuscitation policies carefully and to create ways of approaching such decisions with more involvement of patients and less blind faith in the value of technology. Ethics implies making decisions and judgments. This is best done in a systematic way, as pointed out in previous chapters, and requires as much information as possible. Even in the best of circumstances it is a difficult task. It is the awkward situations such as this case which, on reflection, make the work of ethics, carefully considered and reasoned, so important to the patient and the physicians involved.

The President's Commission for the Study of Ethical Problems in Medicine and Biomedical and Behavioral Research has published extensive information on the state of decision-making regarding resuscitation in its report *Deciding to Forego Life-Sustaining Treatment*. In an important collation of thinking about the present legal status of "No Code," the Commission reviewed the elements of good decision-making, special issues involved in the care of the seriously ill newborn, and ethical problems associated with resuscitation such as competence, self-determination, and equity (1). In its report, the Commission reviewed decisions to proceed with CPR and outlined their recommendations (Table 22.1).

Given what we know about the patient's wishes in this situation, there is no expressed preference available at the time of cardiac arrest, and the

TABLE 22.1. Resuscitation (CPR) of competent patients—physician's assessment in relation to patient's preference

Physician's assessments	Patient favors CPR*	No preference	Patient opposes CPR*
CPR would benefit patient	Try CPR	Try CPR	Do not try CPR; review decision†
Benefit of CPR unclear	Try CPR	Try CPR	Do not try CPR
CPR would not benefit patient	Try CPR; review decision†	Do not try CPR	Do not try CPR

*Based on an adequate understanding of the relevant information.
†Such a conflict calls for careful reexamination by both patient and physician. If neither the physician's assessment nor the patient's preference changes, then the competent patient's decision should be honored.
From: *Deciding to Forego Life-Sustaining Treatment.* President's Commission for the Study of Ethical Problems in Medical and Biomedical and Behavioral Research. Washington, D.C., U.S. Government Printing Office, 1983, p 244.

best decision would be to proceed with CPR. The President's Commission correctly stresses the necessity for hospitals and institutions that care for seriously ill patients to be explicit in their institutional policies. This need is even more important for teaching hospitals with several levels of providers including medical students, student nurses, interns, and attending physicians, all of whom have some degree of responsibility for patient care. The problem is complicated further by the regular rotation of teams in the inpatient service, with each team having a different and perhaps conflicting set of opinions about each case.

The tendency of U.S. institutions to have such strict guidelines for decisions about resuscitation is in marked contrast to the more informal system in other countries, such as Great Britain. Bayliss (2) has written that he feels there are at least five reasons for the U.S. position:

1. To improve efficiency of management and remove doubt
2. Because American patients are better informed on medical matters
3. Because of the tremendous financial implications of the decision for the individual and the family
4. Because of laws which may differ from state to state
5. Because of the expectations of some of the public and some of the medical profession.

Whatever the reasons, there are guidelines for decisions regarding resuscitation available in each teaching institution and these should be closely followed by both teachers and learners. Miles and his colleagues have reviewed the do-not-resuscitate procedures in teaching hospitals and developed a series of guidelines that consider house staff and attending physician responsibility and family and patient wishes (3).

A very important aspect of this case is conflict between individuals who are caring for the patient. Such conflict is usually resolved in a hierarchical system by recognizing that the individual with the highest legal authority has the final decision. In teaching situations, this hierarchy is less clear. Certainly, the attending physician has legal responsibility for decisions, but the attempt to be a good teacher—to help residents and students reason to solutions in a logical manner—will make an authoritarian position regarding such decisions less than optimal. What seems clear in this case is that the attending and inpatient team did not adequately follow the guidelines for discussing the decision not to resuscitate. If there was formal discussion between the patient, family, and physicians, it was not transmitted to the house staff or to the medical record. Thus the decision to proceed with CPR was justified, because it was based on the best, properly gathered, and recorded information available at the time. The long-term solution entails greater responsibility for precision in the record, a willingness to communicate with patients about their choices, and clarity about role priorities when physicians also serve as teachers.

A final point in this case has to do with the conduct of the house officers toward the other members of the health team. Although highly charged situations occur frequently in the course of hospital care, the breach of professional etiquette, which causes those in positions of power (physicians) to embarrass or reprimand those who are not, is unacceptable. Bosk (4) has emphasized that this type of behavior violates the normative conduct of a physician. Respect for the feelings of colleagues is as important as other generally accepted professional norms such as confidentiality. The *International Code of Medical Ethics,* adopted in London in 1949, states that "a doctor ought to behave to his colleagues as he would have them behave to him" (5). Clearly, respect for others and appropriate private expression of concern, rather than public displays of anger, help the quality of patient care through collaborative and supportive work of the health team.

CASE 3: "PUBLISH OR . . ."

A junior faculty member is involved with a survey research project that is being directed by a more senior member of the department. The project is expected to yield useful information about the relationship of year of graduation of medical residents to the level of confidence of graduates in carrying out various medical procedures and in caring for hospitalized critically ill patients. The junior faculty member is involved in refining the research question, designing the questionnaire to be used, and organizing the data for analysis.

As the survey information begins to be collected,it is obvious that there are going to be significant differences between 5-year cohorts of graduates, and that men and women graduates will have less significant but important differences. The senior faculty member then directs his junior colleague to prepare two manuscripts for submission to different journals. One will report survey data that emphasizes only the 5-year cohorts, and another will use the same survey but report only the differences between men and women physicians.

The junior faculty member feels that the data really are sufficient only for a single publication and expresses this to his superior. He is instructed to prepare the two manuscripts with the rejoinder that "This is the way it is done here. You need the publications, and this form of research reporting is quite common in academic medicine."

Discussion

There has been a great deal of scrutiny in recent years of the ethical conduct of research reporting. There are a variety of pressures, particularly on junior faculty members, to produce large numbers of published articles in order to be promoted or to obtain prestigious postgraduate fellowships or faculty appointments. One controversy has focused on the overt falsification of data by members of a research team, and the responsibility of all members of the team for the validity of the research data (6). Review committees at medical institutions have now been formed to review potential violations of ethical conduct in research and to recommend appropriate action. This important committee is partly a response of the country's medical schools to the potential loss of confidence by both major funding institutions and the public in the conduct of scientific research. There are now clearly prescribed guidelines that outline the procedures to be followed if there is a suspicion of falsification of data, plagiarism, or "otherwise misrepresentation of the results of activities associated with research" (7).

In this case, the junior faculty member is not being asked to falsify data. He is being asked to engage in a lamentably common practice in the reporting of data, which is that of lumping data in one manuscript and splitting data in another in order to report the data to two separate academic journals. This has been referred to as duplicate submission and publication.

It is common in medical scholarship to "squeeze the most out of a research project," but that practice does not necessarily improve the quality of reporting. This has been referred to as writing the "Least Publishable Unit" (8). It is the practice of reporting a new research design, a preliminary report, a publication of the completed study, and a small follow-up study. By this means an author can manage to milk four publications from a single research project. This should be distinguished from the legitimate process of reporting a new research instrument or data collection method in a methodologically oriented paper, and the application of that method to a research protocol, and reporting the results in a subsequent paper. Generally, the second paper in this process will refer to the first one as a background paper for those who wish to know more of the research design and methods. This process requires reference to the previous work and a clear indication to reviewers and editors that there has been other work by the same authors in the same general area.

However, it is a different thing altogether to send two papers, analyzing the same data in different ways, to different journals without an acknowledgment of the existence of each by the other. Each journal thinks that it is reviewing new data and makes decisions about publication based on that assumption. Most journals ask for assurance that the submitted manuscript is not being considered for publication by any other journal. In the case under discussion, although an identical manuscript is not under consideration by another journal (thus satisfying the letter of the law), the process does appear to violate academic rules of conduct. The rationale given by the senior faculty appeals to the most common sort of experience. If taken seriously, it would reduce ethical deliberations to custom ("This is the way it's done here"), raw self-interest ("You need the publications"), and morality by consensus ("Quite common in academic medicine").

None of these statements is a persuasive moral argument. They have more the character of excuses than reasons. Thus, the junior faculty member is showing a higher level of ethical conduct, and acting on good moral grounds by refusing to report the research in two separate manuscripts. However, the risk to himself and his career may be significant if he should act against the wishes of his supervisor. It is an uncomfortable and difficult ethical decision. If resolved badly, the young faculty member may rationalize such conduct to himself, perhaps permitting him to rationalize similar future situations to his supervisor. This process diminishes in a small way the work of science. There is a responsibility all of us have to our own values, to the principles that govern our behavior, and to the greater norms that involve us as professionals.

The burden of ethical conduct in the realm of research and scholarship lies with the senior members of a program or academic department. Through their counsel and tutoring of younger faculty, they determine the tone of the research and the rigor with which it is carried out. Much of the discussion about fraudulent research of late has focused on the lack of supervision by senior faculty in charge of large laboratories with large numbers of postdoctoral students and junior faculty.

Case 4: Politics and Religion

A young woman faculty member is reviewing cases from the morning with a senior resident in the ambulatory health center. The resident presents the case of a young woman who came to the practice for advice on contraception. The patient is a married working mother of two young children who are in day care. She has used oral contraceptives in the past, but the side effects have caused her considerable discomfort. She and her husband have talked about using barrier methods, such as a diaphragm, but are reluctant to use them because they feel it might inhibit spontaneity in their sex life. After a great deal of consideration they have decided to ask that an IUD be inserted. The patient is familiar with some of the potential problems associated with IUDs but wants to have one inserted anyway.

The resident presents the case fairly straightforwardly until the point at which he states that he feels that IUDs are not good methods of contraception. He gives a detailed review of studies that show higher than expected rates of hospitalization of women with IUDs and also higher rates of pelvic inflammatory disease. When the faculty member questions whether the resident feels comfortable with the patient's choice even if it is not the doctor's, the resident admits that he does not like IUDs and does not personally insert them. When asked whether this is based on any other reasons, the resident confesses that it is against his religious beliefs to recommend IUDs and to insert them himself. He says that their mechanism of action is effectively to produce an abortion and that he is against abortion.

The faculty member points out that he is wrong in his approach and that he is doing an inadequate job as a primary physician not to offer the patient an IUD as an option. She strongly feels that women have the right to choose their method of contraception and, moreover, that male physicians often use their biases to influence contraceptive choices of their patients. Both the resident and the faculty member sense that the problem is a conflict between the resident's religious convictions and the faculty member's political beliefs. No consensus is reached.

Discussion

There are two conflicts in values here: one between the resident and the patient who desires the IUD, and one between the attending physician and the resident. The conflicts interlock because the second problem arises as a response to the first. Both have emotional undercurrents that need probing to identify the values involved.

The resident has personal religious values that make it impossible for him, in his mind, to consider an IUD as an alternative contraceptive choice. A number of aspects of this position require examination.

First, the resident surely has the right to hold personal values and to apply them to his own life. The ethical problem becomes evident when he attempts to apply his personal values to his patients. Most physicians agree that patients have the right to participate in decisions regarding matters that affect them and their families, assuming that there are a number of alternatives that have equal or roughly equal risks. Some physicians feel that even if there are higher risks to a particular option, patients have the right to choose that option if they fully understand the risks and possible benefits, and are knowledgeable about the alternatives.

Informed consent is an area of great discussion and debate in recent years. This case is not one in which that is the central issue. However, if this patient had been less well informed about the options for contraceptive choice, and had come to this resident for information about choices, there might have been the possibility that the resident would have skewed his explanations away from an IUD, not only by completely ignoring it, but also by misrepresenting the risks and benefits of IUDs to the patient.

Physicians are often asked by patients what they would do in a given situation. Although that is ordinarily an appropriate question, personal opinion can be expressed as such and still enable the patient to choose differently from the physician. Personal opinion couched in "medical objectivity" is not only dishonest but also deprives the patient of the real objectivity necessary to make informed decisions. We are unable to determine whether this resident used medical risks associated with IUDs to mask his own strong feelings about the moral value of their mechanism of action. The teacher, however, might create a hypothetical situation for this resident, where the patient was truly undecided, and see how the resident would explain options. Physician attitude strongly affects patient choice in any situation. Confusion between medical fact and personal values therefore needs to be avoided.

The resident's chief problem seems to be the conflict between his personal religious beliefs and the extent of his professional obligations to the patient. It is not clear whether he actively discourages use of IUDs because of his personal values, or whether he is able to discuss objectively all options and then accept his patient's choice. Most physicians would argue that they should not be required to act in a way that conflicts with their religious beliefs. However, the male resident has a professional obligation to assure that the patient will receive the treatment of her choice. He can do this by arranging for a colleague or consultant to insert an IUD after informing the patient why he is referring her to another physician.

In clarifying the intentions of the resident, the attending physician might ask him what he feels the patient should be able to do in exercising choice, and what his professional obligations are to the patient. She might reinforce the resident's decision to accept the patient's choice for an IUD. Instead, she does not examine the resident's actions or their intent but chooses to bring her own values and assumptions into the discussion. The teaching encounter becomes a conflict of what is perceived as religious or political values when, in fact, not enough is known about what the resident said or how he resolved the conflict that was present in this case. A common problem in many teaching situations is the lack of facts, which makes a discussion of principles difficult if not impossible. What is clear in this situation is that the strong feelings of both teacher and resident flavor and, in fact, impede the learning that could have come from a discussion of this case. The teacher's use of the word "wrong" decreases the possibility of any discussion. We are unclear what has actually happened with this patient—what the outcome of the office visit was. What we are clear about is the lack of exploration by the teacher of the resident's intentions and assumptions and options for management and resolution of the dilemma.

A less directive, and more questioning, facilitative teaching style is appropriate for discussing ethical questions. The implication of a self-evidently right or wrong answer in these situations will discourage discussion of the issues, and decrease opportunities for residents to grow in their understanding of themselves as moral agents. Just as we require the resident in this case to avoid confusing personal religious values with his professional obligation to the patient, we need to remind the teacher to do the same. Teachers have the opportunity to help residents become more ethically sensible and better decision-makers but can do this only if they are themselves attuned to ethically significant issues and appropriate decision alternatives.

References

1. President's Commission for the Study of Ethical Problems in Medicine and Biomedical and Behavioral Research. *Deciding to Forego Life-Sustaining Treatment.* Washington, D.C.: U.S. Government Printing Office, 1983.
2. Bayliss RIS. Thou shalt not strive officiously. *Br Med J 285:*1373–1375, 1982.
3. Miles SH, Cranford R, Schultz AL. The do-not-resuscitate order in a teaching hospital. *Ann Intern Med 96:*660–664, 1982.
4. Bosk C. *Forgive and Remember: Managing Medical Failure.* Chicago: University of Chicago Press, 1979.
5. *International Code of Medical Ethics.* London: Third General Assembly of the World Medical Association, 1949.
6. Altman L, Melcher L. Fraud in science. *Br Med J 286:*2003–2006, 1983.
7. Office of the Chancellor. *Policy on Ethics in Research.* The University of North Carolina at Chapel Hill, 1984.
8. Broad WJ. The publishing game: getting more for less. *Science 211:*1137–1139, March 13, 1981.

Suggested Readings on Medical Ethics

The works listed below are but a fraction of those that might be cited. These were chosen for their demonstrated usefulness, the variety among them, or because they are standard works commonly quoted.

Journals

Hastings Center Report, 360 Broadway, Hastings-on-Hudson, New York, 10706.

Journal of Medical Ethics. Professional and Scientific Publications, 1172 Commonwealth Avenue, Boston, MA 02139.

Social Science and Medicine. Pergamon Press, Maxwell House, Fairview Park, Elmsford, New York 10523.

Monographs and Anthologies

Beauchamp T, Childress J. *Principles of Biomedical Ethics,* ed 2. New York: Oxford University Press, 1983.

A widely used book that approaches the area through a careful explication of four basic principles. Material from 29 cases are interwoven into the text. The best of the rationalist approaches to ethics.

Brody H. *Ethical Decisions in Medicine*, ed 2. Boston: Little, Brown, 1981.

Written as a self-instructional book, this work emphasizes problem-solving, is replete with cases, and is carefully cross-referenced.

Dyck AJ. *On Human Care: An Introduction to Ethics.* Nashville: Tennessee, Abingdon Press, 1977.

A general introduction to ethics, with a focus on medical issues; interesting chapters on relativism and the ideal moral judge.

Harron F, Burnside J, Beauchamp T. *Health and Human Values: A Guide to Making Your Own Decisions.* New Haven: Yale University Press, 1983.

A clear, readable, and well-conceived book written for laypersons and summarizing some important perspectives in individual and policy issues.

Hauerwas S. *Truthfulness and Tragedy.* South Bend, Indiana: University of Notre Dame Press, 1977.

A collection of essays that takes Christian motifs as basic tools to analyze such problems as personhood, rationality in ethics, caring for the retarded, and the limits of medicine.

May WF. *The Physician's Covenant: Images of the Healer in Medical Ethics.* Philadelphia: Westminster Press, 1983.

An exploration of the physician's role and responsibility in five images; parent, fighter, technician, teacher, and convenantor, from the perspective of a Protestant theologian.

Ramsey P. *The Patient as Person.* New Haven: Yale University Press, 1970.

A somewhat dated book in its detail (death criteria, organ donation) but forcefully argued; still one of the best sustained statements for human sanctity, stated mostly in nontheological categories.

Reiser SJ, Dyck HJ, Curran WJ (Eds). *Ethics in Medicine: Historical Perspectives and Contemporary Concerns.* Cambridge, Massachusetts: MIT Press, 1977.

The most comprehensive of the anthologies available, containing historical material as well as contemporary essays. Excellent documentary resource.

Smith H. *Ethics and the New Medicine.* Nashville, Tennessee: Abingdon Press, 1970.

Another older book but carefully argued exposition of some key moral-medical issues. It achieves the difficult feat of being a responsible theological treatise and speaking to a cultural situation that is avowedly secular but de facto Judeo-Christian.

Smith, HL, Churchill, LR. *Professional Ethics and Primary Care Medicine.* Durham, NC: Duke University Press. 1986.

An examination of the moral implications of primary care, dealing with both individual doctor-patient interactions and the social responsibilities of physicians.

Veatch RM. *Case Studies in Medical Ethics.* Cambridge, Massachusetts, Harvard University Press, 1977.

A collection of 112 case studies, organized topically; generally helpful, sometimes overwritten.

Reference Volumes

Reich WT (Ed). *Encyclopedia of Bioethics.* New York: Free Press-Macmillan, 1978.

A four-volume work, generally quite accessible to the layperson and with an extensive appendix of professional codes.

Walters L (Ed). *Bibliography of Bioethics.* Detroit, Michigan: Gale Research Co. Issued annually since 1975.

Index

Abortion, 341–342, 353, 384
Absolutism, 342
Abstracts, 299–300
Academic freedom, 31
Academic medical centers, 3–9, *see also* Hospitals
  career goal "fit" and, 23–24
  departmental organization of, 7–9
  disorganization of, 6–7
  environmental constraints and, 65–66
  organizational health and, 55–58
  organizational problems of, 63–67
  political dynamics of, 86
  reward system and, 6
Academic physicians, *see also* Clinical instructors
  antipathy toward management responsibilities, 13–14
  autonomy and, 5–6, 8, 56, 60, 362
  burnout and, 14
  career stages and, 25–27
  development activities for, 112–113
  disclosing vs. withholding patient diagnosis, 344, 346–351, 356
  evaluation of learners and, 132–133, 135, 141–144
  goals of, 253
  institutional "fit" and, 28
  integration into working groups, 56–58
  and patient relationship, 35
  percentage of time for committees, 37
  percentage of time for research, 182
  in primary care, 181
  priorities of, 6
  professional competence, 129
  professional development, *see* Professional development
  professional system and, 156–157
  roles and responsibilities of, 4–5, 54–55, 346–348, 368, 373–374, 376
  self-examination of career plans, 23–25
  stress and, 11–16
  stress management techniques for, 16–20
  tenure and, 25, 33–34, 173
Academic purism, 54
Academic standards, 6
Accountability, 54, 56–57, 61, 64
  program evaluation and, 149, 151, 168
Accreditation, 65, 152
Achievement standards, 143
Active voice, 290
Activity analysis, 27–28
Ad hoc committees
  benefits and liabilities of, 47
  composition of, 48–50
  determining tasks of, 48
  management of, 52
  relationship with initiator, 50–51
Administrative responsibilities, 54–55
  career conflict and, 13–14
Admissions, 132, 135
Affective reactions, 137, 345, 366
Agendas, 40, 43–45
Alliteration, 306
Alphanumeric visuals, 315
AMA/NET, 256
Ambiguous professional role, 12–13, 16–17

Ambulatory care education, 98–100
American Medical Association Code, 347
American Society of Information Science, 257
Anarchy index, 56
Area health education centers, 4
Arena behavior, 40–41
Assumptions, ethics and, 350–351, 356, 367–368
Attending physicians, 98–99, 377–381
Attitudes, clinical instruction and, 103, 109
Attitudes Toward Learner Responsibility Inventory (ATLRI), 111–115
Attributable risk, 227–228
Audience
  oral presentations and, 307–310
  professional presentations and, 282, 294–297
  visual media and, 322–324, 326
Audit, 179–181
Authorship, 301–302
Automation, 252–253, *see also* Computers
Autonomy, 5–6, 8, 56, 60, 153, 362
  of patients, 347

Bakke case, 135
Bar charts, 321, 326, 329–330
Bargaining, 8
BASIC (programming language), 260–261
Bias, 190, 199, 214–215
  peer review and, 303
  teaching and, 363
Bibliographic references, 298–299, 303
Bibliographic Retrieval Services (BRS), 255
Bibliographic searching, 185, 241–242, 247, 252, 255–257
Biosis, 256
Blueprints, 318
Board certification, 74, 125, 127, 129, 132, 148, 168
Book chapters, 296
Boolean logic, 321
Bottom line, 55
Boundaries, educational programs and, 157
Breast stimulation, 185–195, 205–210, 212
Budgets, 66, 221–222
Bureaucracy, 13
Burnout, 14

Cancer, 214, 222, 227, 340, 344, 359
Cardiac arrest, 377
Cardiopulmonary resuscitation (CPR), 377–381
Career conflict, 13–14, 17
Career continuum, 130, 142
Career planning
  activity analysis and, 27–28
  assessment of needs, 23–25
  objectives and, 23–27
  performance appraisal and, 31–34
Case-control study, 211–212, 214–215
  measures of effect and, 227
Cathartic function, 346
Causal relationship, 162
Centralization, 59, 61
Certification, 74, 125, 127, 129, 132, 148, 168
Chairperson, 42–47
  ad hoc committees and, 51–52
  departmental, 58–59, 64–65
Character ethics, 340
Charts, 310, 315–316, 318, 320–321
Chemical Abstracts, 256
Child abuse, 370
CIPP Model, 160–161, 167
Citations, 298–299, 303
Clerkships, 77, 80, 85
  learning values and, 359
  patient encounter data and, 127
Clinical audit, 179–181
Clinical competence, 144
Clinical instruction, 101–104, 165, *see also* Curriculum; Medical education; Teaching
  ambulatory care and, 100
  case studies for, 117–122
  closure and, 110
  definition of, 98
  exercises for, 111–122
  goals and, 140, 143
  guidelines for evaluation of, 111–113, 117

learners' developmental stage and, 102–106
organization of, 106–111
patient encounter data and, 126–127
purpose of, 100–101
quality of, 106
set expectations for, 117
vertical integration in, 81
Clinical instructors, 104–106, *see also* Academic physicians
evaluation of learners and, 110
and learner interaction, 107–109
problem-solving skills and, 108–109, 117
student control over learning and, 111–115
Clinical investigation report, 283
Clinical research, *see* Research
Clinical trial, 189–201, 211–213, 215
Competence tests, 128–129
Computerized bibliographic databases, 185, 241–242, 247, 252, 255–257, 321
Computers, 168, 241–242, *see also* Microcomputers
data management and, 252–254, 260–264, 267
graphics and, 274, 276
for reference files, 258
visual materials and, 312, 329–332
volume of data and, 259
Confidentiality, 347, 355, 374, 377
of research data, 221
research subjects and, 241
Conflict management skills, 56
Confounders, 213
Consultants, 235–237
Context integration, 81
Contraception, 367–368, 383–385
Contracts, 243–244
Control group, 194, 200, 211–212
cohort study and, 213
Coordination of departments, 55–56, 58
Corporate funding, 247
Correlational techniques, 162–163
Counseling, 99
Crisis management, 20
Critical thinking, 363–364
Cross-sectional study, 211–212, 215–216
Curriculum, *see also* Medical education
characteristics of, 80–81
definition of, 77
design, 78–79
dictated by specialty boards, 148
goals and, 77–79, 82–84, 88–89
hidden, 123
planning, 77–80, 85–86
research skills and, 250
Cybernetic development model, 319, 323, 328

Data analysis, 165–166
Data collection, 232–233
curriculum planning and, 78, 83, 86
data entry and, 201, 260
educational feedback and, 128
evaluation of learners and, 125, 135
falsification and, 382–383
forms for, 197–200
from patients, 105
instruments, 191–193, 201–202
interpretation of evaluation measures, 140–141, 143
key variables and, 165
Data collection protocol, 196
Data management, 251–254
input steps, 259–260, 264–265, 267
project development and, 254–259
reports, 272–274
summarizing data, 267–270
Data management systems (computerized), 260–264, 267
Data processing facilities, 241–242
Database management system (DBMS), 260
Databases, 241–242, 247, 255–257
dBase II, 261
Decentralization, 60–61, 64
Decision making, 38–39, 46
ethics and, 343–352, 355–356, 365–366
instruction in the process of, 108–109
leadership and, 57
learner evaluation and, 74
learning and, 103
participative, 60
professional competence and, 129

Decision making (*cont.*)
program evaluation and, 148–149, 164, 166
resuscitation and, 377–381
Decision tree, 320
Delegation of authority, 60, 64
Delegation of work, 20
Departmental management, 7–9, 22, 54
coordination and, 55–56, 58, 63–64
intradepartmental conflicts and, 65
Dependent variables, 224, 227
Depression, 224
Design professionals, 328
Developmental psychology, 101–103
Diagnosis
ambulatory care and, 100
disclosing vs. withholding of, 369
of learning problems, 113
Diagnostic tests, 229
Diagonal slice approach, 49–50
Dialog, 255
Didactic instruction, 80
Digoxin toxicity, 180
Discrimination, 375
Disease, 215–216, *see also* Health
change in prevalence and pattern of, 176
prevention, 85, 99
risks and, 213–214, 227
Doctors, *see* Academic physicians; Attending physicians
Domino effect, 353
Drug studies, 244, 247

Editing, 289
Education, *see* Clinical instruction; Curriculum; Medical education
Emergencies, 378
Environmental constraints, 65–66, 156–158, 160
Errors, 267
Ethics, 103, 119–120, 337–341
case studies
cardiopulmonary resuscitation and, 377–381
conflicting responsibilities, 373–377
politics and religion, 383–386
publications and, 381–383
clinical instruction and, 109
clinical trial and, 212
decision making process and, 343–352, 355–356
definition of, 340
failure of, 352–355
intentions and, 348–349
obstacles to, 342–343
options and, 349–350, 368–370
the situation and, 354–355
teaching and, 364–371, 385–386
tragedy and, 355–357
value of, 343
weighing consequences and, 347–348, 369–370
Etiquette, professional, 381
Euthanasia, 353
Evaluation, *see* Learners, evaluation of; Performance appraisal; Program evaluation
Excerpta Medica, 255–257
Existentialism, 341
Experimental techniques, 162–163

Faculty, *see* Academic physicians
Fallacy of False Quantification, 140
Family abuse, 370
Family medicine
curriculum development and, 75, 82, 84–91
evaluation of residents in, 129–130
program evaluation and, 149, 159
Federal grants, 245–246
Feedback, 20
learning and, 104–105, 110, 113, 117, 122, 127–128
Fellowships, postdoctoral, 245
Field experiments, 162–163
Films, 322–323
Financial resources, 65
Flow charts, 331
Follow-up study, 211, 213–214
Footnotes, 298–299, 303
Foundations, 245–246

Galley proof, 304
Generalists, 178

Goal achievement, 159
Goals, 63
academic physicians and, 253
clinical instruction and, 140, 143
curriculum and, 77–79, 82–84, 88–89
learners and, 106
medical education and, 127–128, 160
research, 206–207, 220
time management and, 20
Golden Rule, 351–352
Government regulations, 157
Grammar, 290, 298, 306
Grants, 236–238, 243–249
Graphics, 315–316
computers and, 329
data management software and, 274, 276
Graphs, 316, 320–321
Group behavior, 40–41
GTE Telenet Medical Information Network, 256

Hawthorne effect, 208
Health, *see also* Disease
problems of modern society, 177
promotion, 99
Health maintenance organizations, 4
Hidden agenda, 40, 44–45
Hierarchical committee approach, 48–49
Hippocratic ethics, 347
Histograms, 316, 329
Holistic research approach, 177
Horizontal integration, 81
Hospital consortia, 4
Hospitals, *see also* Academic medical centers
guidelines for resuscitation, 380
program evaluation and, 156
research and, 175
university owned, 3
Hypothesis, 81–82, 187, 200, 314
operational, 209–211

Illness, *see* Disease
Image-related graphics, 316
Independent learning, 104
Independent study, 77
Independent variables, 224–225, 227
Index Medicus, 255, 299
Indoctrination, 362–363
Information management systems, 168
Information retrieval, 181
computerized bibliographic databases, 241–242, 252, 255–257
libraries, 181, 186, 241–242, 247, 255, 331
Informed consent, 384–385
Inpatient clinical instruction, 99
Inputs, 160
Institutional research grants, 245
Instruction, *see* Clinical instruction
Insurance, 157
Inter-library loans, 242
International Classification of Diseases, 257
International Classification of Health Problems in Primary Care, 257
International Code of Medical Ethics, 381
Interval level measurement, 225, 227
Iterative committee reports, 52
IUD, 383–385

Jargon, 291
Journals, 295–297, 383

Kantianism, 337

Labor (child birth), 185–195, 205–210, 212
Law of the Instrument Fallacy, 140
Leadership, 57–58, 64–65
organizational growth and, 59–61
Learners, 102–104
assumptions and, 367–368
clinical instruction and, 98–101
case studies, 117–122
decisions about competence of, 128–129
development of, 101–104
documenting experience of, 126–127
evaluation of, 101, 104, 110, 122–123, 159–160

Learners (*cont.*)
    consequences of, 135
    context of, 130–134
    criteria and standards for, 134–135, 141, 143
    effects of, 150
    management of, 141–144
    planning and, 142, 144
    professional system and, 157
    purpose of, 125–130, 136, 142
    satisfaction of faculty expectations and, 134
    technology of, 136–141
    values and, 144–145
  and faculty relationship, 107–109, 133–134, 361–362
  failure and, 143
  goals of, 106
  learning and, 363–364
  moral resolution and, 366–367
  perspective on clinical instruction, 111, 113, 116–117
  problem-solving and, 108–109, 117
  as research assistants, 242
  responsibility for learning and, 103–105, 111–115, 122–123
  role of, 99
  set expectations for, 107–108
Learning Vector, 102–104, 106, 123
Lectures, 80, 98
Letter of understanding, 236
Letter to the editor, 283, 286–287
Levels of measurement, 225–227, 230–231
Libraries, 181, 186, 241–242, 247, 255, 331
Licensure, 74, 125, 129, 130, 136
Literature search, 185–187, 208–209, 255–257
Logbook, 224
Longitudinal study, 211, 213–214

Management information systems, 66
Manuscripts, *see* Written presentations
Maternity leave, 374–375
Matrix system, 57
Measurement, 224–232
  learner evaluation and, 136, 137–141, 143
Measures of effect, 227–228
Media, *see* Visual materials
Medical career continuum, 130–132, 142
Medical centers, *see* Academic medical centers
Medical codes of ethics, 347
Medical competence, 132–133
Medical education, 4, 63, 73–74, *see also* Clinical instruction; Curriculum; Learners; Teaching
  diversity and, 80–81
  educational management and, 268
  ethics and, 340, 358–361, 364–372
  evaluation of programs, *see* Program evaluation
  faculty priorities and, 6
  goals of, 127–128, 160
  program components, 154–155, 158, 159–161
  purpose of, 77
  quality of, 129
  vertical and horizontal integration in, 81
Medical ethics, *see* Ethics
Medical illustrators, 328, 331
Medical libraries, 181, 186, 241–242, 247, 255, 331
Medical objectivity, 385
Medical records, 214–215
Medical schools
  admissions, 132, 135
  complexity of, 22
  family medicine and, 85
  priorities and, 6
  research and, 241–242, 245, 250
  variety of, 3–4
Medical students, *see* Learners; Medical education; Residents
MEDLINE, 185, 241, 255–256
MeSH (Medical Subject Headings), 255–257
Methodology, 251, 283–284, 288
  faulty, 182
Microcomputers, 241, 252, 256, 329, *see also* Computers
  data management systems and, 260, 262, 267
  visuals and, 312
Minutes of meetings, 46–47
Modifiers, 290

Moral
absolutism, 342
levels of activity, 345–346
problems, 339–341, 343–345
quagmire, 352–353
reasons, 352, 370–371
relativism, 342–343
tragedy, 356–357
Morale, 220–221

National Board of Medical Examiners, 129
Needs assessment, 78, 83, 86, 160
Negotiation, 8, 12, 17
career planning and, 32–33, 35
committees and, 48, 51
No Code, 377, 379
Nominal level measurement, 225, 227, 230–231, 251
Null hypothesis, 187

Odds ratio, 227
Online databases, 185, 241–242, 247, 255–257, 321
Open negotiation, 8
Oral presentation
clarity and directness in, 290–291
development of, 287–289
manuscripts and notes, 306–307, 309–310
peer review of, 293
planning a, 281–283, 305
rehearsal, 307–308
structure of, 283–287
visual materials for, 321–329
writing for the ear, 306
Ordinal level measurement, 225, 230–231
Organizations
climate of, 78
growth stages of, 58–62
health of, 55–58, 61–62
Orientation, 33
Outcome measures, 149–150, 158, 161
clinical research and, 195
Outlines, 297–298
Outpatient clinical instruction, 99
Output measures, 158
Overhead transparencies, 310, 322–324, 326, 328
Overload, 14–15, 17–18, 20
Oxytocin, 185–187, 206

Paragraph, structure of, 289
Participative decision making, 60
Pascal (programming language), 260–261
Passive voice, 290
Patient case report, 283–285
Patient consent form, 191, 194–196
Patient encounter data, 126–127
Patient management problems, 140
Patients, *see also* Primary care
autonomy, 347
care, 4, 13, 63, 98
coordination problems and, 56
decisions concerning, 377–381
faculty priorities and, 6
learning management skills for, 159
demographic characteristics and, 156
disclosing vs. withholding diagnosis from, 344, 346–351, 356, 369
physician's influence on, 385
psychosocial management of, 99–100, 103
research and, 173, 232
rights, 119–120
PDMS (Pascal Data Management System), 263–275
Pearson correlation, 230–231
Peer review, 239–240, 246
of manuscripts, 293, 300, 303
Performance appraisal, 31–34, 56
of research staff, 220
techniques, 128
unclear criteria for, 64
Personnel supervision, 220–221
Pharmaceutical firms, 244, 247
Physicians, *see* Academic physicians; Attending physicians; Clinical instructors
Pictorial graphics, 316
Pie charts, 321, 329
Pilot study, 184, 188, 209–210
purpose of, 254
reliability and, 229
research process and, 251
resources for, 240, 243, 245

Pitocin, 185, 189, 194, 199, 205
Placebo group, 211
Plagiarism, 382
Planning, 19, 66
  agendas for committees, 43–45
  curriculum, 77–80, 85–86
  rational planning model, 78, 82–83
Podium presence, 308–311
Policy making, 7
Political process, 6–7, 86, 383–386
Postdoctoral fellowships, 245
Posters, 322, 326–328
Posture, 308–309
Pragmatic planning model, 79
Pre-training assessment, 128
Preceptorships, 77, 85, 99
  in ambulatory care unit, 100
President's Commission for the Study of Ethical Problems in Medicine and Biomedical and Behavioral Research, 379–380
Prevalence study, 211–212, 215–216
Preventive health, 212
Primary care, 63, 121, 127, *see also* Patients
  research and, 176–181, 185
  setting, 100, 105
Principles, 352, 354, 356, 370
Private foundations, 245–246
Private practice, 12
Problem-solving, 45
  defined by method of measurement, 140
  ethics and, 368
  instruction in, 108–109, 117
  moral, 339–340, 344–345
Professional development, 1–2
  clinical instruction and, 102–104, 106
  definition of, 98
  skills of, 7–9
Professional Development Contract, 29–31, 33–35
Professional distance, 345
Professional etiquette, 381
Professional journals, 296
Prognosis, study of, 179
Program evaluation
  assessing program effects and, 150
  criteria and standards for, 158–159
  defining program dimensions, 154–158
  formative, 149
  how results will be used, 151–152
  impetus for, 151
  internal versus external, 152–153
  methodology for, 162–166
  outcome measures and, 149–150
  program components and, 159–161
  purpose of, 148–152, 167
  reporting results and, 166
  resources and constraints, 152, 157
Program Evaluation Review Techniques (PERTs), 331
Programming languages, 260–261
Project grants, 243–244
Project reference files, 257–259
Proposals, 244, 246
Protection of the Rights of Human Subjects, 191, 239
Protocol design, 190, 192–193
Psychinfo, 256
Psychology, developmental, 101–103
Psychosocial issues, patient management and, 99–100, 103
Public Health Service agencies, 246
Public speaking, *see* Oral presentation
Publication process, 302–304, *see also* Written publications
  credit and, 220
  duplicate submission and, 382–383

Qualitative evaluation, 163–166
Quality control, 202
Quandary ethics, 340, 365
Quantitative evaluation, 162–165
Quasi-experimental techniques, 162
Quinlan, Karen, 353–354

Random allocation, 212
Rational planning model, 78, 82–83
Rationalism, 345
Reductionist model, 176–177
Referred journals, 296
References, 298–299, 303
Relative risk, 227–228
Relativism, 342–343
Reliability measures, 228–229
Religion, 383–386

Report writing, 166, 222–224, *see also* Written presentations
Request for Proposals, 244
Research, 4, 13, 173–174
  budget, 221–22
  coordination problems and, 56
  data collection, *see* Data collection
  data management, *see* Data management
  definition of, 175
  design, 216, 259
  desire to know and, 205–206
  ethics and, 381–383
  faculty priorities and, 6
  funding for, 181, 184, 243–249
    peer review and, 239
    using consultants and, 236–237
  goals, 206–207, 220
  historical review of, 175–177
  hospital staff attitudes and, 199–200, 202
  implementation of, 200–201
  literature search and, 185–187, 208–209, 255–257
  objectives, 187, 206–207, 210
  organization and management, 201, 217–224
  pragmatism concerning, 55
  in primary care, 176–181
  procedures, 184
  process, 251–252
  promotion and, 182–183, 222
  quality control and, 202
  resources for, *see* Resources
  sources for work in progress, 186
  sources of ideas for, 189, 205–206
  staffing and, 202
  strategy, 211–216
  what "not" to do, 203
Research game format, 81
Residents, *see also* Learners; Medical education
  backgrounds of, 129
  clinical instruction and, 99
  clinical instruction case studies and, 117–122
  conflict between 377, 381
  developmental learning stages and, 106
  exposure to a variety of cases, 270
  patient encounter data and, 127
  training grants and, 245
Resources
  for education, 78–80, 83, 86
  program evaluation and, 152, 157, 160
  for research
  colleagues, 234–240
  funding and, 243–249
  organizational support for, 240–243
  strategies for promoting, 249–250
Resuscitation, 377–381
Review articles, 283, 285–286
Review process, 302–303
Role
  ambiguity, 12–13, 16–17
  creation, 8, 12
  modeling, 98, 105, 119, 122, 144, 359
Rules, 370, 377
  ethics and, 352–354

Sample size, 191, 196, 212, 214, 259
Scales, 225, 227
Schematics, 318
Science Citation Index, 255–256
Scientific journals, 295–296
Scientific method, 81–82
SCISEARCH, 256
Seed grants, 243
Seminars, 77
Senior house officers, 99
Sentences, 290, 301
  oral presentations and, 306
Simulation exercise, 75, 81–84, 86–97
  curriculum planning and, 80
Situation ethics, 354–355
Slides, 310, 322–324, 328, 330–331
Smithsonian Science Information Exchange, 247
Social issues, 157
Social justice, 347
Social Scisearch, 256
Socialization, 103–104
Software, *see* Computers; Data management systems
Specialization, 175
Specialty boards, 74, 125, 127, 129, 132
  need for accountability and, 168
  residency curriculum and, 148
Speeches, *see* Oral presentation
Stage fright, 308

Standardized achievement test scores, 125–126
Standards, 64
academic, 6
accreditation, 65
learner achievement, 143
for program evaluation, 158–159
Statistics, 165, 230–231, 259
using consultants for, 236
Stress, 8
definition of, 11–12
diagnosis of, 15–16
management of, 16–18
sources of, 12–15
Students, *see* Clerkships; Learners; Medical education; Preceptorships; Residents
Study design, 189–190, 196
Study population, 207
Style, 290, 298
speaking and, 309
Suffering, relief of, 341
Symbolic graphics, 316
Systems theory
curriculum planning and, 78
education programs and, 154–157, 160

t statistic, 230
Tables, 319–320
Task force, *see* Ad hoc committees
Teaching, 4–5, 13, *see also* Clinical instruction; Medical education
coordination problems and, 56
ethics and, 364–371, 385–386
moral values in, 358–361
objectivity and, 362–363
respect for students and, 361–362
student learning and, 363–364
Technical skills, 103
Technology, 157, 168, *see also* Computers; Visual materials
Tenure, 25, 33–34
research and, 173
Tertiary care settings, 100
Textbooks, 296
Time management, 14, 18–20
Timetables, 219–220, 331
Training grants, 243–245
Transparencies, 310, 322–324, 326, 328
Truth, ethics and, 344, 346–350

Uniform Requirements for Manuscripts Submitted to Biomedical Journals, 298–299
Universities, *see* Medical schools
Utilitarianism, 337

Validity, 229–230
Values, 81, 103–104, 385
clinical instruction and, 109
ethics and, 340–341, 343–345
evaluation of students and, 144–145
medical education and, 358–361, 364–365, 368
program evaluation and, 159
Values clarification, 365
Vancouver style, 299
Variables, 165, 209, 224–226
computerized data file and, 264
Vertical integration, 81, 85
Videotape, 322–323
Visual materials, 310, 312–314
computers and, 329–332
effective use of, 328–329
for oral presentations, 321–329
types of, 315–318
for written communication, 318–321
Vocabulary, 291–292, 298

Word processor, 297
Work Perception Profile, 16
Workaholics, 12
Workplan, 218–219, 220
Written presentations
clarity and directness in, 290–291
development of, 287–289
ethics and, 381–383
peer review of, 293, 300, 303
planning, 281–283, 297–298
preparation for publication, 294–297
reports, 166, 222–224
revising, 301
structure of, 283–287
visual materials for, 318–321